A Journey with
Fred Hoyle
The search for cosmic life

A Journey with
Fred Hoyle
The search for cosmic life

Chandra Wickramasinghe

Cardiff University, UK

edited by Kamala Wickramasinghe

 **World Scientific**

NEW JERSEY · LONDON · SINGAPORE · BEIJING · SHANGHAI · HONG KONG · TAIPEI · CHENNAI

Published by

World Scientific Publishing Co. Pte. Ltd.

5 Toh Tuck Link, Singapore 596224

USA office: 27 Warren Street, Suite 401-402, Hackensack, NJ 07601

UK office: 57 Shelton Street, Covent Garden, London WC2H 9HE

British Library Cataloguing-in-Publication Data
A catalogue record for this book is available from the British Library.

A JOURNEY WITH FRED HOYLE
The Search for Cosmic Life

ISBN 981-238-911-3
ISBN 981-238-912-1 (pbk)

Typeset by Stallion Press
E-mail: sales@stallionpress.com

Printed in Singapore by World Scientific Printers (S) Pte Ltd

To Priya, whose steadfast support was a source of supreme strength

Foreword

This is an autobiographical account of a remarkable collaboration between two of the best known astronomers of our time — Chandra Wickramasinghe and the late Sir Fred Hoyle. Chandra's early life in Sri Lanka forms the backdrop for a remarkable scientific contribution of the 20th century. Investigations of the composition of cosmic dust begun in the 1960s led Wickramasinghe and Hoyle to the unlikely conclusion that the Universe is teeming with microbial life, and that such life can be transported from one cosmic location to another. Seen from this point of view, we are part of a connected chain of being that extends from the Earth to the far reaches of the Universe. Science fiction writers have for decades talked about our genes coming from stardust. Diligent and painstaking research over 30 years has turned a once heretical idea into mainstream science. It is a case of science fiction turning into science fact.

Ironically, Sir Fred developed this idea in his science fiction novel *The Black Cloud*. What a pity he did not live to see its possible realization.

Sir Arthur C. Clarke
Fellow, King's College, London

Contents

Prologue

Harmony:
The stars shine
I gaze at them.

Among a myriad stars
I stand alone

And wonder
How much life
And love
There was tonight.

Written by the author in 1956

After Fred Hoyle had published his second volume of autobiography entitled: "Home is where the wind blows" he told me on one of his last visits to Cardiff:

> *"I have made only a passing reference to our long collaboration, because it seemed disjoint from the thesis I was developing there. Ours is an even bigger story that is certainly worth telling. Perhaps you would like to do that some day?"*

The journey described in this book leads to an inescapable truth about the origins of life — humans, animals, plants and indeed all life on Earth share a cosmic ancestry. Our genetic heritage is derived from the wider universe; our genes neatly packaged within bacteria came from space.

This radical position was by no means rashly conceived. The long series of steps, over four decades, in reaching it were taken with utmost caution and always with a measure of trepidation. Every single unorthodox step was taken only after more conservative alternatives were carefully evaluated. A suite of ideas concerning the cosmic nature of life that was considered outrageously heretical twenty-five years ago, is now sliding imperceptibly into the domain of orthodox science.

Yet the word "panspermia", the concept that life — actual and potential — pervades our universe has its origins in classical Greece and dates back to the pre-Socratic philosopher Anaxoragas (c 500–428 BC) who posited that living seeds or "spermata" have been ever-present in the Universe. In India, in Vedic traditions that go back still further (c 1200 BC), the notion of life being an ever-present cosmic attribute is well-chronicled. These ideas had little impact in the development of Western philosophy, however. The philosophy of Aristotle (c 384–322 BC), based on empiricism, gave the Earth a position distinct from the rest of the Universe, which was thought of as an intangible abstract concept not amenable to reasonable study. Aristotelean ideas that dominated Western thought for centuries included the doctrine of the spontaneous generation of Earth-based life. Fireflies, for instance, were said to emerge from a mixture of warm earth and morning dew, a fanciful explanation which was nonetheless held as irrefutable fact.

The idea of spontaneous generation received its first serious challenge in the 1860's with the work of Louis Pasteur (1822–1895), which suggested that even the very simplest known organisms could not arise independently of a parent organism. This led to the view that each generation of every plant or animal present today must be preceded by an earlier generation of that plant or animal. The same causal link, including the effects of evolution, must persist all the way through the record of fossils and microfossils on the Earth, to the moment when the first microbial life appears. The fact that we now know this moment to have occurred during an epoch of intense cometary bombardment some 3.8–4 billion years ago, possibly means that the life-from-life connection could be extended to a time before the Earth itself existed. This logical outcome of Pasteur's life-from-life paradigm was recognised by several distinguished scientists in the latter part of the nineteenth century, notably John Tyndall, Lord Kelvin and Hermann von Helmholtz. Helmholtz' succinct summary of his position speaks eloquently for them all:

> *"It appears to me to be fully correct scientific procedure, if all our attempts fail to cause the production of organisms from non-living matter, to raise the question whether life has ever arisen, whether it is not just as old as matter itself, and whether seeds have not been carried from one planet to another and have developed everywhere where they have fallen on fertile soil..."*

> (H. von Helmholtz in *Handbuch de Theoretische Physik*,
> Vol. 1, Braunschweig, 1874)

Despite this early advocacy of panspermia, it was Svante Arrhenius (1859–1927) who first elaborated upon the idea in a quantitative way, and adduced evidence for survival of spores, whilst also proposing an explicit mechanism for interstellar transport (Svante Arrhenius, *Worlds in the Making*, Harpers, London 1908).

Arrhenius' ideas quickly fell into disfavour partly because it was felt that microorganisms could not be expected to withstand space conditions, and partly because the theory was regarded as intrinsically untestable. The former objection has turned out to be largely false — bacteria are incredibly space-hardy, but the latter point concerning lack of testability remains an issue in Arrhenius' rendering of panspermia. The claim that four billion years ago life on Earth evolved from a primordial seed or spore from space is not testable. But if it could be demonstrated that living seeds still reach the Earth from space, the essence of his theory would be vindicated. Arrhenius asserted that the process he described would be exceedingly difficult, if not impossible, to detect. Science does not look kindly on untestable hypotheses, so Arrhenius' version of panspermia was destined to fall on fallow ground.

Our own approach to the theory of cosmic life, seventy years later, led us at every stage to propositions that were imminently testable or verifiable. Very many of our predictions have indeed been verified, and a multitude of other tests are now in progress.

Chapter 1

Origins: Prelude to the Journey

"There was nothing in the idea of evolution; rock pigeons were what rock pigeons always had been"

Wilberforce, 1860

During the debate on evolution which took place at a meeting for the British Association for the Advancement of Science in Oxford, June 1860, Bishop Wilberforce turns to Thomas Huxley: "Sir, is it on your grandfather's or your grandmother's side that you claim descent from a monkey?" Famously Huxley replied that he would sooner have an ape for an ancestor than accept the dogma of the church. This much mythologised altercation took place a year after the publication of Darwin's *Origin of Species*. Up until this time, the Judeo-Christian account of creation had dominated Western thought and Darwin's impacting new theory posed a significant threat to the Anglican Church, establishing the autonomy of Science. Darwin's theory of evolution grew in strength throughout the early 20th century with expressions of it being evident in literature, philosophy and social policy. Some sectors of the Church were to assimilate the secular aspects of his theory into their doctrines, but others remained staunchly opposed and an undercurrent of dissention remains even in the present day.

All thinking people were required to have an opinion on the subject of creation and even as late as the 1920's and 1930's students seeking admission to Cambridge University had to satisfy the examiners in a paper called "Paley's Evidences". My first sight of such papers came from my father's documents — he was a Senior Scholar in Mathematics at Trinity College Cambridge from 1930 to 1933. William Paley (1743–1805) was a theologian and philosopher whose influence prevailed over the intellectual world for over two centuries. In his treatise *Natural Theology*, Paley introduced the renowned image of the "watch (that) must have a maker" to expound a teleological argument for the existence of God. The tenor of his work was almost scientific and quite distinct from the doctrines of a miraculous creation. He looked at the intricate workings of the natural world and conjectured that such perfection must imply the work of a Creator: "The marks of design are too strong to be got over. Design must have a designer. The designer must have been a person. That person is GOD".

Darwin was familiar with the writings of Paley, "the logic of ... Natural Theology gave me as much delight as did Euclid", and his own works echo Paley's style and methodology. Both men sought to explain evolution in terms of mechanistic, natural processes, albeit to different ends. Where Paley would always revert to metaphysics to explain the first cause, Darwin remained elusive. Indeed in this respect the title of his *Origin of Species* is somewhat misleading, as he traces the evolutionary processes that led to the fact of Man, but seems to slide away from looking into the causes of the first stirrings of life. However, in a landmark letter to Hooker in 1871, he writes:

> "*It is often said that all the conditions for the first production of organisms are now prevalent which could ever have been present. But if (and oh! What a big if!) we could conceive in some warm little pond, with all sorts of ammonia and phosphoric salts, light, heat, electricity, etc. present, that a protein compound was chemically formed ready to undergo still more complex changes. At the present such matter would be instantly devoured or absorbed, which would not have been the case before living creatures were formed.*"

It is interesting that Darwin, who never relished the prospect of being viewed as anti-establishment, chose the intimate medium of a letter to put forward his views on the origins of life. And the idea expressed here marks the beginnings of the primordial soup theory of life's origins that came to govern scientific thinking throughout much of the 20th century.

By 1939, the year of my birth, the primordial soup theory had gained so much credence that it was the only scientific way of looking at the problem of the origin of life. Indeed science proudly considered it to be a problem solved. The scientists Oparin and Haldane had set the ground rules for discussions of this theory that life on Earth originated on Earth. Recognising that the production of organic molecules in the present-day atmosphere of the Earth was unlikely, Oparin and Haldane argued for a hydrogen-rich primordial atmosphere in which the organic molecules required for the origins of life would be formed. The process first required the break-up of inorganic gas molecules such as water, methane and ammonia into reactive fragments or radicals through the action of solar ultraviolet light and electric discharges. Next, the radicals recombine through a cascade of chemical reactions, and in this process a trickle of organic molecules is formed. Such molecules, which are the chemical building blocks of life, then rain down into the primitive oceans forming a dilute primordial soup. It is from such a soup, through a multitude of chemical reactions over millions of years, that life is supposed to have begun.

This model of life's origins had rapidly acquired popularity and kudos in the post-Darwinian era. Haldane, a distinguished geneticist, physiologist and philosopher was reputed also to possess great charisma which he employed successfully to popularise his theory which eventually came to be written into biology textbooks as irrefutable fact. The reasons for its success were multifarious — it reassured man to consider himself to be autonomous with terrestrial origins, and in an age where religious beliefs were dwindling, Science was stepping in to offer man a "logical" explanation of his origins. The only problem with the primordial soup theory was that it had no foundation in fact. In the 1920's and 1930's, there was no empirical basis whatsoever for its justification, and a judgement had to be made solely on aesthetic or philosophical grounds. For the Soviet scientist Oparin, an overall consistency of this theory with Marxist materialistic ideologies appears to have been an important consideration. Haldane too had espoused Marxism in the 1930's and was for several years editor of the *Daily Worker*. In Haldane's chequered career he later became disillusioned with communism, as indeed with British imperialist policy, and in 1957 emigrated to India where he spent the rest of his days.

Retrospectively, I suspect that Haldane's communist leanings may have contributed to the alacrity with which Fred Hoyle embarked on our challenge of the Oparin–Haldane model in the 1970's. Fred Hoyle, though he came from working class parentage — his father was a wool merchant — was always a staunch supporter of British Conservative politics. His distrust of communism and Marxist philosophy may have added to his suspicion of theories — even scientific theories — that had sprung from such a system of thought. But his political leanings quickly became incidental in our opposition to the received theory of the origins of life — our bold refutation emerged as a natural outcome of our fervent interests in astronomy.

Whilst scientists in the modern age like to believe that their activities are always free of prejudice, such a position cannot be further from the truth. At the deepest level, science, particularly when it comes to fundamental questions such as the origins of life, is inextricably linked to cultural traditions. That includes political as well as religious prejudice. Although subconsciously ignored or sublimated they remain as invisible constraints.

Some aspects of my personal life are relevant to the thesis of this book only in so far as they connect to the remarkable story of my journey with a man who was amongst the most original and imaginative of scientists in the 20th century. My collaboration with Fred Hoyle from 1960 onwards led me, over four decades, to question one of the most cherished paradigms of science. We did not meet until the 21st year of my life. But the period of my life up to this time that was spent in my native land of Sri Lanka was a preparation for the unique adventure that was to follow.

My early years spent in Sri Lanka followed a more or less predictable course *vis-à-vis* my circumstances. Sri Lanka (Ceylon as it was then called) was an outpost of the British Empire, in many ways overshadowed in importance by the neighbouring subcontinent of India. It is a fertile island of supreme versatility accommodating mountains, rain forests and beaches within its relatively small confines. The interest it held for its successive colonisers was, however, mostly commercial, in its precious stones, spices, coffee and later tea. The Dutch, the Portugese and the British prized this colony mostly for these commodities as well as for the strategic location of its natural harbours on sea routes to the Far East. Sri Lanka has a history stretching back over two millennia with sprawling sites of ancient ruins testifying to epochs of past splendour. However, four centuries of colonial domination had to a great extent left Ceylon demoralised in a state of national lethargy, and even its struggle for independence from Britain, which was achieved in 1948, was a pale shadow of the emotions expressed in the independence struggles of India.

In my earliest recollections Ceylon was an impoverished feudal society with a sharply visible division between rich and poor. The rich privileged class had access to good schools, whilst the poorer underclass was only minimally literate with limited access to education. The country was also sharply divided between the amenities available in a few major cities (e.g. Colombo, Kandy and Jaffna) and a multitude of villages of which the country was comprised. My own home was in the capital Colombo, and my school Royal College Colombo, established in 1838, was modelled on the traditions of the English Public School system. My teachers, specially in mathematics and physics, conveyed to me their own passion for these subjects, and were a source of inspiration in my formative years. I was lucky too in that my father was a talented mathematician who obtained the highest honours in this subject both in Ceylon and in the Mathematical Tripos in Cambridge, where he became a B star wrangler in 1933. Not only did I have this added source of stimulus at home but I was also surrounded by my father's collection of mathematical and astronomical books which included classics such as Eddington's *Mathematical Theory of Relativity* and Brown's *Lunar Theory*, not to mention an extensive popular list.

A benefit of living in Ceylon in the 1950's was that the environment was still pristine and unpolluted. There were no bright street lights in the suburb where we lived and hardly any pollution from cars and buses, so that the pageant of the night sky was magnificently brilliant. We lived close to a beach by which a rail road ran connecting Colombo with smaller cities in the south. I would often walk along this beach in the evening, sometimes along the railroad sleepers, and watch the sun set over the Indian Ocean. I vividly recall my childhood experience of spectacular sunsets such as I have never since seen. Within minutes the sunset disappears into a wide black canvas overhead studded with millions of stars. Looking up at the myriads of stars that populate the Milky Way, contemplations about man's place in the universe were inevitable.

It is rare to see such a spectacle nowadays in our modern cities with their deplorable output of light pollution. Our inability to enjoy our natural heritage of the night sky leaves us far poorer, and also less able to make a connection between ourselves and the wider cosmos — a connection that was deeply felt by our ancestors.

Sri Lanka is steeped in Buddhist traditions and its influence is inescapable. The island is strewn with ancient temples and 2000 years of Buddhism literally permeates the land. Buddhist descriptions of cosmology that date way back to the early Christian era are distinctly post-Copernican. In a Buddhist text *Visuddimagga* (written in Sri Lanka in the 1st century AD) it is stated that:

> "... as far the these suns and moons revolve shining and shedding their light in space, so far extends the thousand-fold universe. In it are thousands of suns, thousands of moons ... thousands of Jambudipas, thousands of Aparagoyanas ..."

the latter being translated as meaning extraterrestrial abodes of life. The billions of galaxies of modern astronomy could be identified in statements found in other contemporary Buddhist texts which referred to the entire Universe as "this world of a million, million world systems". Such passages made a significant impact on me in my young years, and I noticed a striking similarity between these ideas and the ones expressed by James Jeans in his *Mysterious Universe*.

My resolve to study astronomy was strengthened by an astronomical event that was fortuitously connected with my homeland.

A total eclipse of the Sun, visible from Sri Lanka, was to take place on June 5, 1955. I was 16 at the time and my serious interest in science was just beginning to develop. Sri Lanka, which had hitherto been a scientific backwater, was suddenly transformed into a hive of professional scientific activity. This particular eclipse was to have the longest period of totality since AD 699 and several important scientific experiments were being planned. Scientists from Britain, USA, France, Germany and Japan all converged here and the local newspapers where full of news about these momentous scientific events. One experiment that was planned was a test of Einstein's Theory of General Relativity which predicts a bending of light by a small predictable amount (1.75 arc sec) as the light of a star passes close to a massive object like the sun. The project was designed to validate an experiment of a similar kind carried out by a team led by Eddington during the solar eclipse of 1919.

As a keen amateur photographer, I had set up my own experiment with a simple camera fixed at the end of a home-built telescope to capture the event. We were of course warned about the dangers of looking directly at the Sun, so at the appointed hour, we had our darkened glasses and basins of water in place to watch the progress of the eclipse. I watched with bated breath as the Moon slid ominously over the Sun's disc, casting an instant gloom over the landscape as in an impending thunderstorm. Then total darkness descended suddenly, lasting for an interminable seven minutes. There was a noticeable chill in the air and a denatured atmosphere. Lotuses began to fold their petals inwards, animals cowered and crows cawed wildly. The fabled spectacle of the solar corona with its outstretching flaming tongues was visible intermittently through transient clearings in a thin veil of drifting cloud. Then it was all over, the noon day sun mystically reappeared. I felt more than ever before the indomitable power of the cosmos.

As often happens with observations in astronomy, most of the observing teams in 1955 were disappointed because clouds intervened, but a few stations were able to make successful observations that led to new science. To a teenage admirer of science these events provided a thrilling experience. Science was happening at my own doorstep, and the events of 1955 played no small part in my determination to pursue astronomy as a career.

Nowadays most students enter astronomy through undergraduate courses in physics, or physics and astronomy. When I asked around for advice on how one became an astronomer the answer I was given by informed persons, not least my father, was through mathematics. So I entered the University of Ceylon in 1957 as an entrance scholar in Mathematics and had set my sights on my father's old University, Cambridge, should things work out for me as I had hoped.

The transition from school to University was an easy one as I attended University from home. The University of Ceylon, Colombo was only 20 minutes away by bicycle or ten minutes by car. I took my three years of University studies in my stride, following most of the courses in applied mathematics, because I felt this was the most important tool for exploring the Universe. I was lucky to have some excellent teachers who inspired me. A person who influenced me greatly in those early years was the professor of mathematics C.J. Eliezer who was himself a distinguished Cambridge product, a former Fellow of Christ's College and a pupil of the illustrious physicist Paul Dirac, whose major success was to reconcile special relativity and quantum mechanics. Through Eliezer's lectures I obtained exciting insights into the theory of electromagnetism, a subject in which I was later to specialise. I did not realise at the time that Eliezer and Fred Hoyle were Cambridge contemporaries and that they both had associations with Dirac. Because of the connection between these three people — Dirac, Eliezer and Hoyle — it turned out by a really curious coincidence that Hoyle was to be the external examiner in mathematics for the University of Ceylon in the same year that I sat my final degree examination. It amused me later on to think that Fred would have read my examination scripts long before he ever set eyes on me.

In the summer of 1960, I graduated with First Class Honours in Mathematics and was awarded a Commonwealth Scholarship by the British Government to pursue postgraduate studies at Cambridge University. I applied for a place to do a PhD in Theoretical Astronomy at Trinity College, my father's old College, and I was delighted to be accepted, and more so to be told that I would be supervised by Professor Fred Hoyle of St John's College who was Plumian Professor of Astronomy and Experimental Philosophy at the University of Cambridge. Whilst at the University of Ceylon I had already read two classic books by Hoyle: *Nature of the Universe* and *The Frontiers of Astronomy*, both of which had made an indelible impression on me. So when I received a handwritten letter from Fred Hoyle at my home in Colombo, recommending a list of books to read prior to coming to Cambridge in October 1960, I was naturally overjoyed.

Chapter 2

Cambridge and a First Meeting

In September 1960, I found myself preparing to leave home for the first time. In those days the normal way to travel from Sri Lanka to England was by ship. Although air travel was rapidly coming into fashion it remained a considerably more expensive option reserved mostly for business travellers and the rich. On a warm September evening I sailed away from the port of Colombo aboard the P&O Liner *SS Orcades*, wistfully watching a palm-fringed coastline recede slowly into the distance. A two week voyage took us through the Suez Canal, via Naples, Gibralter and Marseilles to Southampton. Nowadays such a voyage would be regarded as a luxury cruise. But my enjoyment of this new experience was hindered by sickness due to rough seas. I spent a lot of the journey in my cabin feeling sorry for myself whilst reading and re-reading Fred's *Nature of the Universe*, a book I still regard as one of the greatest classics of popular science.

We docked in Southampton on a cold grey autumn morning and I was met on board by a lady who was a representative of the Commonwealth Scholarship Commission who had sponsored my visit to the UK. This was the first year of awards in the Commonwealth Scholarship Scheme and the authorities were very much at the stage of finding their feet as to what should be done. I recall being herded along with students from other Commonwealth countries to an initiation week in the capital, London. Many of the aid programs sponsored by affluent Western countries to assist the development of the Third World are conceived with trade-benefits in view, and many set up to conscience cleanse for past colonial exploitation. I was more than a trifle surprised at the Commission's ignorance of the needs of

11

their clientele from the Commonwealth. They presumed that all the beneficiaries of their scholarships came from African villages or from rural areas of India or Sri Lanka. Most of the advice proffered as to how to make the transition from ex-colony to the hub of empire, was hardly relevant to myself or, I suspect, the vast majority of the Commonwealth Scholars who came from more or less Westernised backgrounds in their home countries.

The literature of England enthralled me throughout my childhood and adolescence. And the history of the British Isles, its culture, landscape and even details of place names, now coming into context, contained a dreamlike familiarity for me. What I found difficult to enjoy, however, was the climate — inclement weather or persistent grey skies, that ice-breaker of conversation among the English. The long dark evenings of winter still, after 40 years, make me yearn for a more equable tropical climate!

My first impression of London, despite the magnificence and grandeur of its history and its buildings, was one of alienation. My own personal sense of detachment from the people and the environment did, however, seem to mirror a more general malaise. The density of people, the frenetic pace of life, the catacombs of the tube and the pollution all added to my sense of estrangement. Such highly subjective early impressions altered only much later as I came to discover the theatre, music and art galleries that make London unique.

Cambridge, on the other hand, had an instant appeal. I arrived at Cambridge station on a brilliant autumn morning, and following the instructions I have received in London, took a taxi to the Porter's Lodge of Trinity College for my first meeting with Dr. Robson, Tutor for Graduate Studies at 11 am. I recall vividly the sense of awe that swept over me as I walked across the Great Court of Trinity surrounded by the ghosts of legendary figures in both science and literature. This was the College of Isaac Newton, Bertrand Russell and William Wordsworth, and a brochure I had in my hand directed my attention to Newton's rooms above the porter's lodge, where his classic experiments on prisms and light were done.

Dr. Robson's welcome was eminently professional. He made me feel relaxed with a glass of sherry and offered a personal touch to our meeting by talking about my father's distinguished record as a student in 1930–1933. This, I later discovered when I myself became a Fellow and Tutor, was a trick that had to be learnt! After my first lunch in the College Hall and a few other meetings with Junior Bursars and the like, I arrived at my lodgings (a room) in a Trinity College house for graduate students in Burrells Field. Among a pile of inconsequential letters that awaited me was a handwritten letter by R. A. Lyttleton (Ray Lyttleton) of St. John's asking me to see him in his rooms to discuss matters relating to my supervision. That surprised me somewhat because I had already received a letter from Fred Hoyle in Sri Lanka saying that he would be supervising me.

When I met Lyttleton at the appointed hour I learnt that Fred Hoyle was in the USA for most of the Michealmas term and that he, Lyttleton, would supervise me to start with. My encounter with Lyttleton was a little short of formidable as I now remember it. After inquiring whether I had any specific problems in mind that I wished to work on, he informed me that astronomy was an extremely difficult subject to research. Most of the easy problems had already been solved, and what remained unsolved were insurmountably difficult! Only much later did I come to realise that these remarks were probably true to an extent if one confined one's attention to areas of classical astronomy that excluded applications of physics, astrophysics as it is now called. At this first meeting he asked me to take a look at a rather abstract problem in the theory of stellar structure concerning "solutions of Emden's equations for polytropes". He also very generously gave me a copy of his book on comets (which I still possess), in which he described the dust bag theory of comets, following from work he had done in the 1940's. The book was in fact an elaboration of a paper he co-authored with Fred Hoyle on the accretion of interstellar dust by the sun. Lyttleton, it should be noted, was the person who turned-on Fred's interest in Astronomy at a time when he was beginning to find Nuclear Physics a fallow field of research in the 1940's.

Apart from Fred Hoyle and Hermann Bondi, Lyttleton had hardly any other collaborators or graduate students as far as I know. Shin Yabushita, who arrived in Cambridge a year after I did, became his student, but it appears that Yabushita went his own way receiving little or no guidance from Ray Lyttleton. It was clear at the outset that I was not going to get much guidance from Ray, and I do not think I saw him again during my first few years in Cambridge. I had enough reading to get on with so I decided to bide my time until Fred Hoyle returned from the States. I also attended a few advanced lecture courses that were being offered to students of the Mathematical Tripos. These included Paul Dirac's course on Quantum Mechanics and Mestel's course on Cosmic Electrodynamis. All this was in preparation for when Fred Hoyle would return from the USA.

The next significant event I remember from my early Cambridge days was an invitation to tea by Jayant Narlikar. This was my first meeting with another of Fred's students who started research in the academic year 1960/61. Jayant, who later became a good friend of mine, was a star pupil in the Mathematical Tripos that summer (he was a star wrangler and a Tyson Medallist in the Tripos). When I met him in October, Jayant had already started serious work on a research problem in cosmology. Fred Hoyle had clearly spotted his exceptional mathematical talent and set him on a course to assist him in a long and bitter cosmological battle that lay in the years ahead.

Jayant and I shared a hut that was the temporary home of Theoretical Astronomy which was an outpost of the Department of Applied Mathematics and Theoretical Physics. My conversations with him made it clear that the 1960's were turning out to be a watershed era for cosmology. A right royal battle between the Steady State Theory of Hoyle, Bondi and Gold and the rival Big Bang Theory was set to begin. Cambridge Radio astronomers led by Martin Ryle had been claiming in the late 1950's that the Steady State Theory could be disproved by their study of the distribution of galaxies in the universe emitting radio waves (radio source counts). Hoyle claimed that these arguments were far from secure, and indeed were most likely to have been contrived. It turned out that Ryle's early analysis of the matter was based on limited surveys and consequently poor statistics.

Throughout successive surveys of radio sources made at the Mullard Radio Observatory in Cambridge, Ryle and his team pressed home the point that there was enough evidence to disprove Steady State Cosmology. In a simplistic interpretation, the universe appeared, on the basis of source counts, to be more compact at the earliest epochs, favouring the concept of a Big Bang origin. Jayant Narlikar came on the scene amid gathering storm clouds, and the atmosphere in Cambridge was highly charged when I first arrived.

Fred Hoyle had set Jayant the task of re-examining Ryle's radio source count claims before he had gone to the States earlier in the Summer. Without embarking on an extended diversion on cosmological history I should say that this particular storm was eventually weathered. Radio source counts did not say anything very definite. But other conflicts were to follow. Together with Fred, Jayant had embarked on a valiant course of defending steady state cosmologies, an endeavour that was to occupy most of his professional career.

The insatiable appetite for denigrating Steady State Cosmology, often with the flimsiest of evidence, that I witnessed in the 1960's still puzzles me. I cannot help thinking that the reasons have a deeply cultural basis. Without going into the technical details on either side of the argument, my own cultural predilection was for a steady state universe of some kind. Such a cosmology is consistent with the philosophical world view that pervades the Indian subcontinent, and in particular it is in harmony with Buddhist traditions that are prevalent in Sri Lanka. Another aspect of the Cambridge debacle that astonished me was the component of jealousy that entered a scientific controversy. In my naivety I had believed that science was pursued in a cold detached manner independent of personalities or social constraints. This was far from the truth. It was clear from the situation I witnessed that the two contesters in this argument were worlds apart in their personalities and ideologies. Fred was a forthright and candid Yorkshireman; Ryle a stiff upper lip product of the public school system. Never would their differences be reconciled.

When my turn finally came to meet Fred Hoyle, problems of enormous gravity would have been on his mind. As I walked from my lodgings on a frosty afternoon in December, across the backs of the river Cam, over a desolate playing field of St. Johns, I wondered what my long-awaited meeting would be like and what it might eventually lead to. Would I encounter a man weighed down by the strain of a huge controversy, preoccupied and terse; or a sparkling communicator, the author of the some of the most stimulating books I had read? But my anxieties were quickly dispelled when I arrived at the door of 1 Clarkson Close.

Barbara Hoyle greeted me with an unprecedented warmth of affection and generosity. Fred, I was told, was just finishing an interview with a journalist and I was ushered in to their dining room to take tea with Barbara and her mother, Mrs Clarke. The ladies were so welcoming that in minutes I felt completely at ease, and within half an hour, that I had known them for a lifetime.

When the journalist had left I was taken into a spacious, thickly carpeted open plan living room that served as both library and study. Large patio windows extended over the entire length of the room on the far side, looking out on a slightly undulating lawn. There was Fred seated on his easy chair by the window, with a writing pad on his knee, fountain pen in hand scribbling a calculation with intense concentration. It was hard to decide whether he was pleased or not to be disturbed, but as Barbara introduced me as his new Ceylonese student, he switched into a more relaxed mode, stood up, shook my hand and uttered his standard greeting that all his students came to gleefully imitate, "Oh Hellow!"

I cannot remember exactly how our conversation went, but I do recall a remark that embarrassed me such as "I hear you write poetry?" He must have seen such a declaration on the application forms that were sent to the University. On confessing that I had published a slim volume of poems and had some poems included in an *Anthology of Commonwealth Poetry* that was published by Heinemanns that year, he was noticeably impressed. We had established a connection at a deep level — an abiding passion for creative writing and a love of the English language.

We talked about the Lake poets, about Malowe and Ezra Pound. We talked about politics, about cricket, about Sri Lanka and my old college professor there, who had been a Cambridge contemporary of Fred's. We even talked about the weather. In fact we talked about everything except science. I took all this to mean that he had not yet given much thought to what I might do as a research project, probably because of his preoccupation with other more incumbent matters at the time. He did however direct me to his own monograph on Solar Physics and a book by Cowling on Magnetohydrodynamics, but without any explicit application to think about. He also suggested that I attended a Part II course on Electromagnetic theory and a Part III course on Stellar Structure both of which were to be given by Fred in later terms during the academic year. I left Clarkson Close that evening feeling greatly relieved and excited. And I had finally met one of my childhood heroes in science.

During the next two months I must have seen Fred on half a dozen occasions, always at home. The explicit problem to which I was first directed by Fred, was not in any of his main areas of concern at the time. It was the problem of the origin of the Sun's polar magnetic field and of its alternations of polarity (north pole becomes south and vice versa) through a 11 year solar cycle. Fred's idea was that the polar magnetic field was built up by small-scale magnetic loops carried outwards by streams of charged particles. My suspicion was that Fred had already worked through this model in his head, even to the minutest detail and required me only to confirm his hunches with a few straightforward calculations in electromagnetic theory. This I did with relative ease. It seemed strangely appropriate that my first research project should concern the sun's polar field given that it was a solar eclipse that had provoked my desire to become an astronomer. Within only a few months of our first meeting I found myself becoming co-author of my first scientific paper with Fred which was published in the Monthly Notices of the Royal Astronomical Society (F. Hoyle and N.C. Wickramasinghe, *Mon. not. R. Astron. Soc.* **123**, 51, 1962).

Chapter 3

A Hike in the Lake District

Seeing our first paper in print gave me a thrill, even though my own part in the project had turned out to be relatively minor. But although the problem of the polar field was interesting enough in itself, I could not muster enough enthusiasm to think deeply about it. I began to wonder how this problem with its rather limited scope could pan out into a substantial programme of research occupying the full three years of my Commonwealth scholarship.

I spent most of the Lent and Summer terms of 1961 profitably engaged in reading widely around my subject and attending a variety of lecture courses. I enjoyed the new freedom of being able to learn without the overhanging threat of examinations. This was also a time of readjustment to a new country and a different way of life for me.

And the way of life in the Western world at large also seemed to be undergoing rapid change. There was the expansion of global air travel and the advent of the concept of multinational industries. The stage was set for the modern world of instant communication, although the internet itself was still a couple of decades in the future. More importantly in relation to the subject of this book, we were on the threshold of the Space Age that was destined to transform astronomy for ever.

A less tangible transformation of social attitudes had also begun. The election of John F. Kennedy as President of the United States was seen as a new beginning. A quiet confidence and optimism began to pervade Western society. The lean post war years of the 1950's had given way to an unprecedented economic boom that had an effect on all our lives and a feel good factor was evident in people's psychology. Problems that were looming large on a distant horizon — the Cold War, the Arms Race, the growing dissension in South Africa, conflicts in the Middle East and the slow beginnings of global terrorism — made no impression on the day-to-day lives of people in the West. Along with slogans for combating communism came clamours for freedom, liberalism and permissiveness. *Lady Chatterley's Lover* found its acrimonious entry into the literary canon, and the cause of feminism was taken a step further with the election of the World's first woman Prime Minister in Sri Lanka. For me the new found sexual freedom that was given abundant expression on the streets of Cambridge came as a shock to my sequestered upbringing in an ex-colony with distinctly Victorian values.

The 1960's could also be seen as the halcyon days for Science. Most areas of research advanced steadily and with the same confidence that characterised society as a whole. The dawn of the new decade was marked by several notable technical triumphs. The laser, a light source of unprecedented intensity, made its debut in 1960. The quark theory proposed that particles like protons and neutrons, hitherto thought to be the fundamental units of matter, were in turn made up of even more basic units called quarks. Man walks on the Moon. The exploration of space was well under way.

All these developments seemed to flow naturally and with ease, but it should not be forgotten that the groundwork — perhaps the hardest groundwork — in many areas was already laid in the 1950's. It was in the mid-1950's that radioastronomy began, and amongst its rich harvest are discoveries that would turn out to be germane to the questions that will be addressed later in this book. For instance, the study of how hydrogen gas is distributed in the galaxy, and the discovery of molecules connected with life, all stemmed from developments in radioastronomy. The mid-1950's also saw the successful completion of the work by Geoff Burbidge, Margaret Burbidge, Willy Fowler and Fred Hoyle that led to an understanding of how the chemical elements, including the elements of life, are formed from hydrogen in the deep interiors of stars. At about the same time, there was the monumental discovery by James Watson and Francis

Crick of the famous double-helix structure of our genetic material — DNA. Shortly afterwards Fredrick Sanger analysed the nature of a protein (insulin to be specific), showing its detailed sequence of constituent amino acids; and Harold Urey and Stanley Miller completed their classic experiments in which they showed how the most basic chemical building blocks of life might be synthesised from inorganic matter.

It might seem ironical that the burgeoning of liberal traditions in society had little effect in promoting a spirit of free and unfettered inquiry on the scientific scene. Although science in general, and astronomy in particular, continued to flourish, there was a decline in the development of new ideas. Newly invented experimental techniques generated a vast body of factual data. But there was precious little being done in the way of a critical re-appraisal of old hypotheses. The belief grew that the really important problems were either close to being solved, or that they were really so hard that we should not even begin to worry about them. In practical terms it was a case of turning out more of the same type of factual data, attempting to consolidate existing theories, seeking to tie up the last loose ends. An openness of mind required for a vibrant scientific culture was noticeably lacking.

Such then was the general backdrop against which my specific experience of research continued through the early part of 1961. As mentioned earlier Fred Hoyle was engrossed at the time with trying to resolve the radio source count conflict with Martin Ryle — a battle that was becoming ever more bitter. Fred, however, never slackened his determination to defend steady state cosmology, in the conviction that the alternatives were far less credible, and the data adduced in their favour either contrived or indeterminate. His collaborations with Jayant Narlikar on many aspects of Steady State Cosmology were gathering momentum at this time.

Although Fred saw cosmology as his prime battle field in the 1960's, his interests were by no means restricted to cosmology. His knowledge and experience across the entire field of astronomy and astrophysics was impressively encyclopaedic. By 1961, there were few areas of this subject that had not been embellished by his imagination and genius in some way. In the year I joined him in 1960, Fred took on a record number of research students — four in all. Besides Jayant Narlikar there were John Faulkner and Ken Griffiths who were set working on problems of stellar evolution, and Sverre Aarseth who was working on the gravitational N-body problem.

By working on a wide range of problems through his several students Fred's attention may have been diverted from the bitterness of the cosmological conflict. Another welcome diversion he had was an alternative career as a science-fiction writer. In 1959, a year before my arrival in Cambridge, he had already published his novel *The Black Cloud* which in a sense was a forerunner to his much later collaborations with me on life in the cosmos. I acquired a copy of *The Black Cloud* in the spring of 1961 and remember reading it avidly, particularly because I was able to connect the political nuances and intrigues described in the book with what I had seen enveloping Fred's life in recent months. The hard scientific content of the novel also impressed me enormously. His arguments for complex molecules acting collectively as intelligent entities intrigued me even before I began to work on organic molecules in space. Fred's science fiction novels were not merely plausible, they were entirely in the realm of what was possible according to what was definitely known (or *almost* definitely known) about the real world.

Fred often gave an impression of cold detachment and indifference when he came to dealing with people. But with his collaborators and students he always displayed a profound insight as to their temperaments, special aptitudes and capabilities. In the spring and summer of 1961, I was still groping to find my next problem to tackle, and thoughts about the nature of interstellar matter began to enter my head. I saw Fred a couple of times and recall expressing more than a passing interest in the scientific content of *The Black Cloud*. He informed me then that he had found it difficult to publish his calculations that showed the hydrogen molecule and other molecules to be abundant in interstellar space, and that his novel was a way of expressing these ideas at a time when science would not countenance them.

I discovered that he was at this time also engaged in another science fiction project with TV producer John Eliot. Together they were writing the script for *A for Andromeda* which became a highly successful BBC TV serial when transmitted in 1961 and 1962. Later on it was to become a cult classic of sci-fi fanatics. In the script, a newly built radio-telescope picks up intelligent signals from the constellation of Andromeda which are fatefully interpreted as instructions for building a gigantic super computer. Once built, the computer begins to relay the information it receives from Andromeda and the security of humankind falls under dangerous threat. By presuming the presence of intelligent life in a far flung corner of the Universe, albeit through the guise of science fiction, Fred was moving unmistakably towards some form of astrobiology as early as 1961.

When my interest in *The Black Cloud* matters became evident Fred directed me to a review article on "Interstellar Matter" by Jesse Greenstein who was at the time working at the Mt. Wilson and Palomar Observatories. Greenstein had discussed the composition of interstellar dust, considering the arguments for two separate classes of dust grain — dielectric icy particles and metallic iron particles. At that time, however, there was no clear idea as to how either of these grain types could populate interstellar clouds. My reading introduced me to many questions that seemed to be in urgent need of answering.

In my initial discussions with Fred at Clarkson Close it was clear that he was not happy with any of the theories that were around at the time. It was only later that I realised the precise reason for his disquiet. He was at work at the time with Willy Fowler on a fragmentation model for star-formation and needed to have certain opacity and volatitlity properties for the dust. The properties they required did not conform with any of the existing grain models.

My 40-year long journey with Fred Hoyle was now about to properly commence. I arrive at the door of 1 Clarkson Close on a balmy evening in the late summer of 1961 and, as usual, I am welcomed by Barbara Hoyle. Before taking me to see Fred I was, as usual, brought up-to-date the family's doings. Barbara seemed more socially disarming than usual when she announced that she had decided that I should accompany Fred on a walking holiday in the Lake District. Here I could experience first hand the romance of the Wordsworth country I had read so much about, and, as she put it "you men could discuss your work".

Walking in the Lake District was Fred's way of escaping from the tribulations of academic politics that had lately plagued him. In the mountains of the Lake District he found repose and time to think. Here he would experience for a while what he saw to be the real challenges of life, those that a mountaineer had to face, struggles with the elements and the terrain. In this way the irrelevant squabbles in the cloisters of university would fade momentarily into insignificance.

Needless to say I was overjoyed at the prospect of joining Fred on one of his walking escapades. Poised beside him in his two-seater Sprite, we set off on our five-hour long journey. As we passed through mountainous landscapes, Fred would describe how this magnificent scenery had been gracefully sculptured by ice floes as ice ages came and went over long periods of geological time. He was thinking aloud as he often did when he felt relaxed in appropriate company, and listening to his passing thoughts was always an invigorating experience.

We arrived at the Old Dungeon Ghyll Hotel, then a modest guest house in Little Langdale (near Ambleside). It has since been transformed into a sumptuous hotel, but in those days it was a serious hikers retreat, run by the mountaineer Sid Cross and his wife. It was early evening and there was just enough time to unpack before we sat down to a sumptuous dinner cooked by Mrs Cross. The evening meal at the Old Dungeon Ghyll was the focal point of the day and was more than ample compensation for its somewhat spartan accommodation.

After browsing through the day's newspapers, the conversation in front of a blazing log fire turned briefly to the scientific matters we had discussed at Clarkson Close. What would be an acceptable alternative to ice and iron for the composition of cosmic dust? Mesmerised by the flames rising above the burning and sputtering logs, I remember posing a question without too much thought. Could the dust in space be carbon like the soot that lofted up from the fire? Fred's eyes appeared to light up, with excitement or disbelief I could not say. Carbon in the cosmos was after all Fred's baby. It was his prediction of the 7.65 MeV resonance in the nucleus of carbon-12 that opened up new vistas of astronomical thought. Fred, together with Willy Fowler and Geoff and Margaret Burbidge told the world how carbon was synthesised in stars and expelled into interstellar space by supernovae.

But carbon in interstellar dust did not make an immediate impression on Fred, no more than iron did. We all had become used to thinking, following the trend set by Dutch astronomers of the 1940's, that the bulk of interstellar dust had to be formed *in situ* in the gigantic gas clouds of space. And of course oxygen is more abundant there than carbon by a factor of 1.6, so most of the carbon would tend to be tied up in the strongly bound molecule CO. I would guess that these would have been the thoughts that raced through Fred's head at the suggestion of a possible carbon composition of dust.

The following morning after breakfast I joined Fred on his habitual trek over the hills, heading I was told to Bowfell. He had decided that the more ambitious climb to Scafel Pike was not for first time hikers like myself. Although I was armed with a pair of mountaineering boots and an appropriately warm anorak and headgear, I felt hopelessly ill equipped for the task ahead. Fred had a battered-looking rucksack on his back in which he carried sandwiches and bars of chocolate supplied by Mrs Cross. I remember Fred saying that a single bar of chocolate would have enough calories to make good the energy we would expend all the way to Bowfell Pike and back. We set out under an azure blue sky almost cloudless save for a stray cumulous cloud or two drifting lazily overhead. The walk began as a more or less gentle amble along the valley which suited me fine, but before long the proper climb began and I was finding it increasingly difficult to keep up with a seasoned hiker like Fred. I believe that Fred noticed my predicament and accordingly modified the route to take us only so far as a lesser peak than Bowfell on this first day.

The gentler walk was welcome, if only to enjoy my first experience of the exhilarating scenery of the Lake District. The landscapes that I had dimly apprehended, thousands of miles away, via the works of the romantic poets, now revealed themselves in all their sheer physical glory. Walking here in the company of Fred Hoyle had a particular poignancy. I recalled Wordsworth's Preface to his *Lyrical Ballads* in which he compares the aspirations of the poet and the scientist:

> "*The knowledge of both the Poet and the Man of Science is pleasure; but the knowledge of the one cleaves to us as a necessary part of our existence, our natural and unalienable inheritance; the other is a personal and individual acquisition, slow to come to us, and by no habitual or direct sympathy connecting us with our fellow beings .... Poetry is the first and last of all knowledge — it is immortal as the heart of man ...*".

Quite early in my association with Fred I could see how the poet and scientist could come together in a single individual.

Some three hours later, we stopped for a spot of lunch. As every mountaineer will tell you the Lakeland weather is extremely fickle, changing suddenly without any warning. So it seemed to be today. The sun had slipped behind an ominously dark cloud and a shower of rain might have been imminent. We found a flat crag to sit upon and as Fred unpacked our sandwiches from his rucksack he glanced thoughtfully at the grey sky. When I casually enquired, "Is it going to rain?" I could not have predicted that his answer would turn out to be a defining moment in my scientific career.

"Not necessarily", said Fred, "These clouds could be saturated with water vapour, but for rain to fall, condensation nuclei are required. These could be charged molecular fragments (ions) or fine dust, but such condensation nuclei need to be present before rain could form". With a moments further reflection he added, "Some have argued that meteor dust could supply nuclei for rain".

From my long experience of Fred I have since realised that a thought such as this rarely remained in isolation in Fred's head: he would begin to make connections with the widest range of problems. With a little prompting from me the connection with dust in interstellar clouds soon came to the fore. If it is difficult to form water droplets in the densities that prevail in the terrestrial atmosphere, how could ice particles condense in the exceedingly tenuous clouds of interstellar space, where hydrogen densities are in the range 10–100 atoms per cubic centimetre? Was the nucleation problem really solved for the case of ice grains in interstellar space? These were questions we pursued in the remaining few evenings by the fireside in the lounge of the Old Dungeon Ghyll Hotel. Fred's writing pad and pen were used for extensive scribbling and we decided that the problem could not really have been solved by the Dutch astronomers. It seemed incumbent on us, therefore, to find a denser place than the interstellar medium to resolve the nucleation problem of interstellar grains.

Chapter 4

Betwixt the Stars

Little did I know as I travelled back with Fred to Cambridge that our walks in the hills of Cumbria in the autumn of 1961 marked the beginnings of a line of research that would, decades later, lead to a new theory of the origins of life. The "Black Cloud" of Fred's novel seemed destined to spring into life.

The black clouds of deep space had, however, to be set in the context of what else there was in the Universe. They show up conspicuously as complex shapes and structures against the more or less uniform distribution of stars in the Milky Way. The Milky Way itself, our Galaxy, is a hundred thousand light year-wide collection of a few hundred billion stars, each one more or less similar to the sun. And our galaxy is one of many billions of similar galaxies that populate the observable Universe.

Black clouds — interstellar clouds — are by no means restricted to our own galaxy. External galaxies often show conspicuous dust lanes, a striking example of which is seen in the case of the Sombrero Hat galaxy NGC4594, which is a galaxy very similar to our own viewed edge-on. The dark lane across the middle represents an overlapping complex of interstellar clouds that collectively obscures the light of background stars. The interstellar clouds in NGC4594, as in our own galaxy, are the birthplace of new stars and planets. So they must clearly have a very special importance in the scheme of things. Depending upon where the clouds lie in relation to stars and how dense they are, they take on a wide range of temperatures and properties.

Interstellar clouds are on the average about 10 light years across, and the typical separation between neighbouring clouds is about 300 light years. It should be noted, however, that there is a fairly wide spread in the sizes of these clouds and in their separations one from another. Some are more compact and uniform in their disposition, whilst others are extended and irregular. The more extended clouds appear as giant complexes, showing a great deal of fine-scale structure as cloudlets and filaments. These so-called "giant molecular clouds" are often associated with the formation of new stars.

An interstellar cloud may contain anywhere from ten to many millions of atoms per cubic centimetre. Even the higher values in this density range are considerably lower than the densities that can be attained in laboratory vacuum systems. So it should be remembered that all our intuitive ideas of how gases behave under normal conditions could prove wide off the mark when it comes to understanding what happens under the rarefied conditions of space.

Hydrogen makes up the overwhelming bulk of material in interstellar clouds and occurs in one of three forms: neutral atomic hydrogen (intact atoms with no electrons lost), ionized hydrogen (atoms stripped of their outer electrons), and molecular hydrogen (atoms paired in molecular form as H_2). Molecular hydrogen was first detected using ultraviolet spectroscopy in the late 1960's after my own researches into interstellar matter had begun, although its existence was predicted by Fred Hoyle in the 1950's and indeed exploited in the science fiction novel *The Black Cloud*. Hydrogen molecules are to be found mostly in the denser interstellar clouds that are able to screen off the ultraviolet starlight that could destroy them. A large fraction of all the hydrogen in the galaxy is found to be in molecular form, H_2, and the total mass of the hydrogen is billions of times the mass of the sun.

But what else exists in interstellar clouds besides hydrogen? Information derived from several sources, including solar and stellar spectroscopy and the direct examination of meteorites (rocks of extraterrestrial origin) all have a bearing on the overall composition of interstellar material. Next to hydrogen in order of abundances comes the element helium, which accounts for close to a quarter of the total mass of interstellar matter, although from our point of view this element is inert and uninteresting. Then comes the group of chemical elements carbon, nitrogen and oxygen that together make up several percent of the mass of all the interstellar matter. It is these elements that are of course crucial for life. Indeed, life depends for its function on the unique range of properties of the carbon atom including its high levels of chemical reactivity and its ability to combine into many millions of interesting carbon-based compounds. Next in line are the elements magnesium, silicon, iron and aluminium, which again account for a percent or so of the total interstellar mass. Then a group including calcium, sodium, potassium, phosphorus is followed by a host of other less abundant atomic species. All these chemical elements are synthesised from hydrogen in the deep interiors of stars in the manner worked out by Fred Hoyle and his colleagues in the 1950's. The synthesised elements are injected into interstellar space through a variety of processes, including mass flows from the surfaces of stars. In the case of the most massive stars, the end product of their evolution is a supernova (an exploding star), and it is through supernova explosions that life-forming chemical elements are injected into the interstellar clouds.

The discovery of interstellar molecules (assemblages of atoms) by methods of radioastronomy and millimetre wave astronomy got properly under way a full decade after my journey with Fred had begun. Next to molecular hydrogen the second most abundant and widespread molecule in space turns out to be carbon monoxide. A significant fraction of all the interstellar carbon in our own galaxy, as well as in external galaxies, seems to be tied up in the form of this molecule. Next in line of importance comes the all pervasive molecule formaldehyde, H_2CO, which is present in gaseous form, both in clouds of high density as well as in clouds of relatively low density. In the denser interstellar clouds, particularly in clouds associated with newborn stars, vast amounts of water in gaseous form are found. Water is an important molecule for life, and its close association with newly formed stars and planetary systems would have a vital relevance to our story.

The spatial distribution of interstellar molecules in the galaxy shows wide variations depending on physical conditions such as ambient temperature and density as well as the proximity of clouds to hot stars. As a rule, denser, cooler clouds contain the larger and more complex molecules, whereas lower density clouds and those nearer to hot stars have simpler molecular structures. A region that is particularly rich in organic molecules (molecules that could be connected with life) is the complex of dust clouds in the constellation of Sagittarius, located near the centre of the galaxy. It is in this region that the first tentative detection of an interstellar amino acid, glycine (a component of proteins) was reported, as was the molecule of vinegar and a sugar glycolaldehyde. All these molecules are detected by tuning radio receivers to precisely the frequencies at which the molecules absorb or emit radiation. An inherent difficulty of this technique lies in the need to determine beforehand the correct radio frequencies that characterise particular molecules. Once these are known, the technical problem of finding the molecules is relatively simple. But the actual number of detections could represent a mere tip of an iceberg: a vast number of more complex molecules would inevitably remain undiscovered.

Interstellar organic molecules have been detected by other techniques besides radio astronomy, notably using the methods of infrared astronomy, of which I shall have more to say in future chapters. An important class of organic molecule that is found to be present in quantity are the polyaromatic hydrocarbons or PAH's. These molecules (just as CO) are well-known by-products of the combustion of fossil fuels as occurs, for instance, in automobile engines. They are largely responsible for the suffocating smog that pollutes our larger cities. We shall argue later that even in interstellar space such molecules are most likely to have a direct biological connotation, possibly representing the break-up or degradation products of biological material.

The existence of complex organic molecules in interstellar clouds poses one of the most challenging problems for modern astronomy. The conventional viewpoint is that complex molecules form out of atoms and simple molecules through reactions that take place in the gaseous phase. But because the gas densities in space are so exceedingly low (far lower than that present in a laboratory vacuum, as we have said) the reactions occurring between interstellar gas molecules would be too slow to produce any appreciable quantities of the most complex organic molecules. Their presence would have to be explained in some other way.

Comparatively high densities of organic molecules tend to be associated with regions of the galaxy where new stars (and presumably comets and planets) are forming at a rapid rate. The Orion nebula is a spectacular example of such a region where young stars are evident in large numbers, some even with discs of newly formed planetary material around them. Large quantities of organic molecules are associated with the denser parts of the Orion cloud complex, and it would be tempting to link the formation of such molecules with the formation of stars, planetary systems, and perhaps life itself. But more of this would emerge at a later stage of our journey.

In addition to atoms and molecules, interstellar clouds contain an all-pervasive and enigmatic dust component to which I have already referred. Astronomy has struggled to understand the precise nature of this cosmic dust and to discover the circumstances under which such particles are formed. In the autumn of 1961 when I started reading on these matters in earnest I discovered that it was almost an article of faith amongst astronomers that interstellar dust grains were comprised of dirty ice material — frozen water with perhaps a sprinkling of other ices — ammonia-ice, methane-ice and a trace quantity of metals. Furthermore, the firmly held belief was that these particles had to condense, more or less continuously from the gaseous atoms and molecules that were present in the interstellar clouds.

The basic ground rules in this area of astronomy had been laid down in two classic papers and a PhD thesis written in the mid-1940's by the Dutch astronomer H.C. van de Hulst. My researches had led me to a large body of work in an area of physics known as "homogeneous nucleation theory" that was immediately applicable to the interstellar problem. It was easy enough to check the claims of van de Hulst and his colleagues against straightforward predictions of the theory. I soon discovered that van de Hulst's ideas were flawed in several important respects. I was quickly able to verify Fred Hoyle's conjecture about the nucleation problem — the insuperable difficulty of forming dust in the exceedingly tenuous gas clouds of interstellar space. The rain cloud analogy I was introduced to in the Lake District applied here, only more dramatically. When I communicated my findings to Fred it was evident that he was pleased. His immediate response was that I should see van de Hulst in Leiden and confront him with all the technical details and the difficulties as we perceived them.

So in November 1961, I made the journey by boat from Harwich to Hook in Holland and thence to Leiden. I was charmed by Leiden which was not unlike Cambridge in many ways. It is an historic University town, somewhat younger than Cambridge — the University being founded in the 16th century — but with a very similar ambience. There were as many bicycles in Leiden as in Cambridge, but unlike Cambridge's single water way the River Cam, Leiden was criss-crossed with boat-lined canals. The birthplace of Rembrandt, the city is saturated with art galleries and museums.

But the purpose of my visit was to tackle the great man H.C. van de Hulst. This too turned out to memorable, albeit in a somewhat negative way. As affable as he was it was clear at the outset that he was not inclined to engage in any sensible scientific dialogue with me. I sensed that he was possibly affronted by my challenge of his views. Only much later did I come to appreciate that Professors in Europe are traditionally placed on a high pedestal, their authority rarely being questioned by younger members of staff let alone students. Be that as it may, van de Hulst in 1961 seemed unable or disinclined to offer a defence of his ideas of grain nucleation that he had pioneered a decade earlier. Perhaps he had lost interest in the problem, having moved on to new areas of research. He promptly directed me to his colleague, Professor J. Mayo Greenberg, at the Rensselaer Polytechnic in Troy New York, who had now taken up the cudgels of defending the old ice grain theory against its critics.

My encounters with Mayo Greenberg were still months away. To conclude the present chapter I shall summarise the first major objection to the ice-grain model that surfaced as soon as I began to think seriously about it. True enough, water molecules, if they already exist in interstellar space, could condense on pre-existing particles of dust. But if there is no supply of such dust particles to serve ice grain nuclei, then no condensation could occur. Thus the requirement for any ice-condensation process in space is that condensation nuclei must be injected into the interstellar clouds at a steady rate. In view of the very low gas densities that are present, these nuclei could not be generated at an adequate rate from within the interstellar medium itself. The problem is akin to that of the seeding of water-vapour clouds in the terrestrial atmosphere. One could have highly supersaturated clouds in the atmosphere, but no rain will fall unless condensation nuclei are somehow supplied.

Chapter 5

The Route to Carbon Dust

Faced with the almost insurmountable difficulty of perceiving how ice grains form in interstellar space we took a different approach to the problem of dust formation in 1962. What if the dust was not made of water ice but of carbon, similar to the particles of soot that were rising into the chimney from the log fire we watched at the Dungeon Ghyll Hotel? In this case the formation of carbon dust could occur at much higher temperatures, perhaps, for instance, in the outer atmospheres or envelopes of some cool stars? Such dust grains would also have the advantage of being able to survive in interstellar regions of much higher temperatures.

But what do the astronomical observations tell us about the properties of interstellar dust? What are their precise optical characteristics? How do they behave in relation to the scattering and absorption of starlight? Dust shows up as conspicuous patches of obscuration against the background of distant stars. But several more precise quantitative statements about the nature of the dust were already possible to make in the 1960's. The earliest quantitative investigations were mainly restricted to the way the dust dims and reddens starlight, just as a street light is dimmed in a fog. The first attempts to obtain measurements of interstellar dimming — or extinction, as it is called — of starlight were made in the 1930's. It was found that at a single wavelength (colour) close to 4500 Å the dimming of starlight amounted to a reduction of intensity by a factor of about 2 for every 3000 light years of passage through interstellar space. From this one piece of information alone it was easy to infer that interstellar dimming could only be reasonably attributed to solid particles that have dimensions comparable to the wavelength of light. Much smaller or much larger particles would have to be present with implausibly large densities if they were to produce the observed amount of dimming. And so the dust grains in space were just about as efficient as they could be in blocking the light of distant stars.

With the advent of new techniques in observational astronomy it became possible to measure accurately how interstellar dimming varies with the wavelength of light. This is done by comparing the spectra of two stars which are intrinsically similar, one of which is more dimmed by interstellar dust than the other. This is analogous to comparing two cosmic street lamps, one nearby and another dimly seen through a fog. Such comparisons provide information on the wavelength dependence of extinction caused by interstellar dust. The relationship between extinction and the wavelength of light — what astronomers call the extinction curve — provides an important item of information which has a bearing on the properties of interstellar dust grains. In 1961, this extinction curve was known only over the wavelength interval from about 9000 Å in the near infrared to about 3300 Å in the near ultraviolet. Over much of this wavelength range it was known that the opacity of interstellar dust was proportional to the inverse of the wavelength — in other words, when the wavelength is doubled, the opacity was approximately halved. And most remarkably it turns out that precisely the same type of relationship, the same extinction curve, was found to hold over wide areas of the sky. The requirement for this is that the dust with almost identical sizes and properties exists throughout large volumes of galactic space. In 1961, the best interstellar extinction data over the range of wavelengths 3300–9000 Å was obtained by Kashi Nandy at the Royal Observatory Edinburgh, and this is shown by the points plotted in the upper panel of Fig. 1.

When a set of data points as in Fig. 1 are given, the process of interpreting it involves the construction of a "model" or "hypothesis". In this case a "model" would constitute an informed guess as to what type of interstellar dust could give rise exactly to the relationship between extinction and wavelength as expressed in Fig. 1.

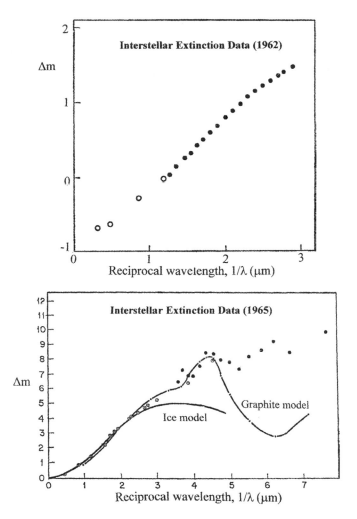

Fig. 1 The manner in which starlight is dimmed by interstellar dust. The points are observations of interstellar extinction: upper frame — as it was known in 1961, lower frame — as known in 1965, with extensions into the ultraviolet. The curves are calculations for particles made of ice and graphite.

A model that one might consider then leads to a calculation of an extinction versus wavelength curve that may or may not match up with the data. If it does match the data, the model can be regarded as being consistent with the observations and would be a valid representation of the data. The same model could also lead to other predictions — data that are not yet obtained, and if the predictions are satisfied by later observations, the case for the model is strengthened. If, on the other hand, a prediction is found to be not true, the model will be falsified. In this way the scientist hopes that he would ultimately gain a knowledge of the true nature of the world.

The limited wavelength range of the extinction data available in 1961 permitted an interpretation in terms of a wide range of dust models, including ice and iron. For each grain model, however, a narrow definition of sizes was required. To fit the astronomical extinction data to any particular theoretical model one needed to calculate the scattering and absorption cross-sections of particles of various radii using Electromagnetic Theory. Such calculations that had already been made for iron grains and ice particles established these models as being possible candidates for interstellar dust, with ice particles having the edge over iron in certain crucial respects. Particularly so because it could be argued that there was insufficient iron in the galaxy to make up the required mass in the form of the interstellar dust.

In 1961–1962, in the days before PCs, the calculation of the optical properties of any new grain model was a major computational undertaking. The absorption and scattering cross-sections of a uniform spherical particle comprised of materials with known bulk optical constants (refractive index n and absorptive index k) was amenable to calculation using formulae derived from Maxwell's equations of electromagnetism. The formulae were mathematically derived by Gustav Mie in 1908, and my task was to program these so-called Mie formulae for use on a high speed electronic computer. This is what I set out to do in the winter of 1961 for a preliminary exploration of our carbon dust models. The optical constants of bulk carbon (graphite) were available at the time only for a few wavelengths, but extrapolations were possible to include the whole wavelength range for which observational data was available. I devised a computer programme in the then popular language FORTRAN to carry out the calculation, and ran this on the EDSAC2 computer of Cambridge University.

Computers in those days were large clumsy devices. Transistors and printed circuits had not yet been invented and the use of vast numbers of thermionic valves meant that computers were not only large and sprawling, they had also to be cooled efficiently with stacks of air-conditioning fans. This, together with the constant clattering of card readers and output devices, made the environment of the Computer Centre a very noisy place indeed. EDSAC2 occupied several large rooms in the Computer Centre and had less computing power than that of the average desktop computer that is currently available in 2004. Although my problem was by no means a very difficult one, each computing run on EDSAC2 had to be booked in advance, so developing a new program, particularly debugging it was a tedious affair. After numerous trips to the computer centre in December 1961 and January 1962, I made a remarkable discovery. I found that as long as the diameter of carbon particles were less than about a tenth of a micrometer (0.1 μm), the predicted extinction curve was almost indistinguishable from the interstellar extinction observations.

Fred was particularly pleased with this result because it possibly freed the model of any parameter dependence. An identical result was obtained as long as the diameter was less than 0.1 μm. For ice grains a very precise distribution of sizes was required in order to match the same observations. In view of the invariance of the observed interstellar extinction curve from one direction to another this constraint on the sizes of ice grains was hard to understand and appeared to be a deficiency of the model. The interstellar medium is extremely inhomogeneous with regard to density and one would expect ice particles (if they grew in space) to have wide fluctuations in size from one region to the next. This being not the case we were able to argue that carbon grains had a distinct edge over ice grains in explaining the observed behaviour of interstellar extinction, at any rate over the waveband over which observations were available in 1961. It was possible to calculate how much in the form of carbon grains we required, and the answer turned out to be one to two percent of the total mass of the interstellar clouds. That is to say an equivalent of one to two percent of the mass of interstellar hydrogen had to be in the form of carbon according to our model. This was consistent with the availability of carbon in interstellar space.

Carbon in the form of graphite is a highly stable, highly refractory material with a sublimation temperature in excess of 2500 K. Searching for places where carbon grains might form, Fred suggested we might turn to cool giant stars where the surface temperatures were generally below 3000 K. But the major class of oxygen rich giant stars, known as the Mira variables, were not appropriate. With an oxygen excess over carbon in the Mira stars, most of the atmospheric carbon will be in the form of the strongly bound molecule CO and there will be none left for graphite formation. There was however a class of red giant stars known as the carbon stars, the so-called R and N variable stars. The N-stars have surface temperatures that varied cyclically between 1800 K and 2500 K over a period of about a year, and they are known to have more atmospheric carbon than oxygen. Thus although oxygen would again from CO, there would be an excess of C that is able to condense into solid particles when the temperature fell below some critical value. Again a computer programme had to be deployed to determine the physical state of the excess carbon as the temperature varied between 1800 K and 2500 K. Using standard theories of nucleation in a homogeneous saturated gas, we showed that carbon particles were able to nucleate and grow in the stellar atmosphere as soon as the temperature fell towards 2000 K. We further showed that graphite particles of radii of a few hundred Angstroms would grow from initial nuclei and be expelled into interstellar space, the expulsion being caused by the pressure of light from the parent star. There is observational evidence which points strongly to the existence of dust around carbon stars. The variable and highly luminous carbon star R Corona Borealis is a spectacular example. Here we see direct evidence of a star erratically puffing out clouds of carbon soot into the interstellar medium. More recently astronomers have also detected thermal infrared radiation from carbon stars which is consistent with the presence of heated graphite particles.

Throughout much of 1962, Fred was busy with Willy Fowler's visit to Cambridge and their mammoth project on nucleosynthesis.

This did not, however, deter him from an active engagement in our own work on carbon grains as the story began to unfold. It did not take more than five months after returning from the Lake District for us to produce a paper entitled "On graphite particles as interstellar grains" for the Monthly Notices of the Royal Astronomical Society. This was submitted at the end of May 1962 and our paper appeared later that year (F. Hoyle and N.C. Wickramasinghe, *Mon. not. R. Astron. Soc.* **124**, 417, 1962). This publication represented our first step in the direction of cosmic biology although we did not recognise it as such at the time. I soon followed up our first paper with a series of others in Monthly Notices describing detailed calculations of the properties of carbon particles, including the possibility of ice mantles condensing around them in interstellar clouds. I also made a prediction of the properties of interstellar grains based on new measurements of the optical constants of graphite, which were brought to my attention by the Belgian astronomer C. Guillaume. The prediction was that if the extinction curve was extended into the ultraviolet, a strong absorption feature centred at 2200 Å would be seen (N.C. Wickramasinghe and C. Guillaume, *Nature* **207**, 366, 1965).

Throughout most of 1963, my interaction with Fred was confined mostly to reporting progress on new developments of the carbon grain story. This was my opportunity to establish myself as an independent researcher which I did with much enthusiasm. Early in 1963, I began writing my PhD thesis and thinking seriously about what I might do when my Commonwealth Scholarship came to an end in September 1963. Our work on the carbon grains and on grain formation was opening up new vistas of research in this field, so I did not relish the prospect of returning to a University position in Ceylon, where I may not have been able to pursue my research.

On Fred's advice I decided to apply for College Fellowships in Cambridge. The competition for such fellowships is notoriously stiff. Competitors include young researchers from all academic disciplines, so the best in one field had to be compared with the best in others, a very difficult task for electors, as I was to later discover. So I was delighted when Jesus College elected me to their Research Fellowship in 1963. This opened a new chapter in my life as a research scientist and a Cambridge don. The Fellowship included a modest stipend, free rooms and meals in College. I moved into my rooms in a new building in North Court to finish my PhD thesis whilst also taking on a modest amount of undergraduate teaching for the College.

Compared to my somewhat isolated life as a graduate student at Trinity, I began to enjoy the fellowship of my colleagues at Jesus who came from a wide range of academic disciplines. The President of Jesus College at the time was a mathematician Alan Pars, who had taught Fred Hoyle as well as my father in the 1930's. Alan Pars very generously took me under his wing and made me feel comfortable in my new surroundings. Jesus College had no great traditions in astronomy at the time, but had in its history an eminent alumni: John Flamsteed (1646–1719), the original Astronomer Royal of England. Flamsteed's monumental publication of the first extensive star catalogue set the highest standards for observational astronomers who were to follow him. His work was also a great boon to navigation, providing a basis for the accurate determination of position at sea. In the congenial setting of my rooms at Jesus I was able to pursue my researches on various aspects of interstellar grains. It seemed clear to me that a major paradigm change was round the corner — a shift from volatile ice grains to refractory grains, grains that had to be based largely on the element carbon. It was also a shift from ideas of grain formation in diffuse interstellar clouds to condensation of dust in much denser stellar environments. The transition, however, was not as smooth and painless as I had anticipated it to be.

The publication of our papers on interstellar graphite led to a head-on clash with proponents of the ice grain theory in the USA, particularly J. Mayo Greenberg and his band of collaborators. New techniques in observational astronomy were now making it possible to study the behaviour of interstellar dust outside the visible wavelength range — at longer wavelengths beyond red (the infrared) and shorter wavelengths beyond blue (the ultraviolet). Observations in the infrared waveband were soon to be used for discriminating between grain models. R.E Danielson, N.J. Woolf and J.E. Gaustad were the first to search for a characteristic absorption feature of water ice at the infrared wavelength of $3.1 \, \mu m$ in the spectra of highly dimmed stars. By 1965, the lack of a 3.1 μm water ice band in the spectra of several stars led to the conclusion that ice particles if they exist at all can make at most a very minor contribution to the the interstellar dust. Although 3.1 μm bands were later detected in some astronomical sources, these most probably arose from matter local to the sources themselves, from dense circumstellar clouds not from dust in the general interstellar medium.

Spectroscopic data in the ultraviolet region of the electromagnetic spectrum at wavelengths shorter than 3000 Å were first obtained by T.P. Stecher, R.C. Bless and A.D. Code using equipment carried on rockets and satellites. The most conspicuous feature in the ultraviolet interstellar extinction curve observed by T.P. Stecher was a broad hump centred on the wavelength 2175 Å. This was exactly as we calculated for spherical graphite particles of radius 0.02 μm, as shown in the comparison of the lower panel of Fig. 1. The ice grain model could not produce such an ultraviolet feature and must therefore be deemed to be inconsistent with observations.

The new infrared and ultraviolet data provided enough reason for a conference on interstellar grains to be convened. A conference was promptly organised by J. Mayo Greenberg and held at Rensselaer Polytechnic Institute in Troy New York and held from 24 to 26 August 1965.

This was my first visit to the United States and memories connected with it still linger in my thoughts. Landing at La Guardia Airport in New York I spent a day walking around in the shadow of skyscrapers in Time Square. But after the initial bedazzlement the stark reality of American society begins to impinge. Concepts of aesthetics and grandeur differ markedly from those of the old world.

The next day I arrived at Troy for my first scientific conference. I felt a heavy burden of responsibility to defend the graphite grain theory that Fred and I had proposed. I had tried to persuade Fred to join me on this trip, but it turned out to be impossible for him. Greenberg, the local host of the conference, and the person van de Hulst had referred to, was our sworn enemy. He was determined to defend the ice grain model using every device that was available to him. His most gallant attempts at producing 2200 Å extinction features using "trimodal size distributions of infinite cylinders made of ice", were more ridiculed than lauded at this meeting. Other papers presented here included one by Bertram Donn and Ted Stecher from NASA's Goddard Space Flight Centre on a graphite model, which was not dissimilar to ours. Perhaps the most remarkable paper of the entire meeting was one by Fred M. Johnson in which he argued that many unidentified diffuse absorption bands in stellar spectra at visual wavelengths can be caused by a derivative of chlorophyll (a porphyrin, the green colouring substance in plants) — a paper that was several decades ahead of its time.

The period between 1965 and 1967 was particularly difficult for Fred. His negotiations with Cambridge University and the Government for starting an Institute of Theoretical Astronomy had suffered many time-consuming setbacks. There was also an important development in cosmology which was quickly seized upon by his adversaries. In 1965, Arno Penzias and Robert Wilson accidentally discovered a diffuse background of microwave radiation emanating uniformly from all directions in the sky with a temperature of 2.7 K. Excluding all other possible causes of this radiation they interpreted this signal as the relic radiation from a Big Bang Universe. This so-called cosmic microwave radiation was described as the last nail in the coffin of the Steady-State Universe. With the enormous weight of the propaganda that was mounted it appeared to me that Fred began to accept defeat for a while. But that did not last too long.

Chapter 6

A Theory Takes Shape

After two years of working as a postdoctoral researcher on interstellar grains the time had come to take stock of what had been achieved. When I first started my studies in 1960, interstellar dust was considered to be a barren field for research. To most astronomers the presence of dust was a nuisance, its only effect being to hinder the observations of distant stars. All they needed were simple rules to correct measured intensities of starlight to compensate for the presence of dust, beyond that interest in dust was minimal. In the few years of my research this situation seemed to be changing. Interstellar grains were certainly coming into vogue with new observational opportunities and techniques paving the way to new ideas, and to ambitious programmes of work. It cannot be denied the entry of Fred Hoyle and myself into this field had a part to play in this transformation of attitude.

The group of international astronomers who attended the conference in August 1965 were aware of my work on graphite grains and new avenues became open. Many were sympathetic to the idea of carbon grains, evincing more than a passing interest to challenge the ailing ice grain paradigm. Bertram Donn of NASA's Goddard Space Flight Center was particularly keen in pursuing further the nucleation aspects of the carbon dust theory, to understand how a gas of carbon atoms could form condensation nuclei and thence grains. Some years earlier, he and John R. Platt had argued the case for large unsaturated organic molecules nucleating in interstellar space. He now felt that a similar process could more easily operate in carbon stars.

In the so-called "Platt particles", the extinction of starlight (absorption and re-emission) occurs due to electrons jumping between discrete energy levels. Although I had pointed out to Donn that it was difficult to explain the details of the observed interstellar extinction by this process, the formation of carbonaceous particles certainly merited careful study. The work of Platt and Donn in 1960 gave what was perhaps the first hint of large organic molecules occurring in space. It was therefore not surprising that Donn thought it worthwhile to explore further aspects of our ideas on the theory of carbon grains. He accordingly arranged with the Department of Physics and Astronomy at the University of Maryland for me to be appointed a Visiting Professor in the Summer and Fall of 1966. In this way I would be able to interact with his Astrochemistry research group at the Goddard Space Flight Center in Greenbelt which was only a few miles from College Park along the freeway.

But long before my meeting with Donn, Fred had arranged, through his friend Willy Fowler, that I spent the fall semester of 1965 at the Kellogg Radiation Laboratory at Caltech in Pasadena. To accommodate all these arrangements I obtained leave of absence from Jesus College for the entire academic year 1965/66 to work part of the time in the United States and the rest in Sri Lanka. The Sri Lankan stint was arranged in order to take up an appointment as a Visiting Professor of Mathematics in a newly-formed Vidyodaya University in Kelaniya. Here I thought I might be able to assess a possible long-term option of returning eventually to Sri Lanka. The research task that I set myself to do during the Sri Lankan spell was to complete a technical monograph on "Interstellar Grains" that was commissioned by Chapman and Hall (London) in their International Astrophysics Series. This, I felt, was an important step that would put our new ideas on grains firmly on the scientific map.

The first phase of my plan for 1965/66 did not turn out to be as successful as I had hoped it would be. Indeed my semester at the Kellogg Radiation Laboratory in Caltech took off to a bad start, when on my first evening I decided to take a stroll along the wooded streets just outside the campus precincts. Within minutes a police car pulled up beside me and a slow-witted, thick-set cop jumped out to interrogate me as to my business of walking on the street. Everyone, he explained, went by car and walking here was simply against the law. My crime, for which I was let off with a warning, was compounded by the fact that I did not carry any money or form of identity. This incident, marking an infringement of my civil liberties, was an unfortunate introduction to life in the United States.

I had expected to be greatly stimulated by the scientific atmosphere of Caltech, one of the most distinguished centres of scientific research in the world. But this was not to be. During my first weeks at Kellogg all my attempts to interest astronomers with problems connected with grains did not meet with too much success, and this was a disappointment to some extent. Most astronomers here were preoccupied with much bigger problems of cosmology and nucleosynthesis. The recent discovery of the cosmic microwave background gave an added impetus for such a narrow focus. This lack of enthusiasm for problems connected with interstellar grains might sound strange from the perspective of 35 years on; Caltech was later to take a leading role in the development of infrared astronomy, which of course had a direct bearing on the nature of interstellar grains. Infrared astronomy reached maturity around 1968, so my visit in 1965 was perhaps only a few years too early. I left Caltech at the end of November 1965, a little sooner than I had anticipated, and headed home to Colombo, via Hawaii and Tokyo.

Landing at Katunayake Airport with the prospect of a full five-month stretch in my homeland inspired me with joy. Compared with the maze of freeways around Los Angeles that I had just left, the journey from the airport to my home in Colombo appeared somewhat medieval. Pedestrians, cyclists, bullock carts, cars and buses jostled for space along a narrow winding road. Disorderly humanity wound its own path along crowded routes and a twenty-five mile journey took well over an hour and a half.

In a span of five years the city of Colombo as well as the immediate environs of my family home had changed. The capital had become more congested and polluted, and one sad consequence for me was that the magnificent spectacle of the night sky that I had so often enjoyed from my front garden, had now become a rarity. Street lights just outside our house and a more general haze of light pollution had spoilt for ever our capacity to enjoy the beauty of the cosmos.

The experience of being installed as a Professor in a new Sri Lankan University turned out to be more daunting than I imagined. I found myself being instantly drawn into a great deal of arduous administration. This was perhaps to be expected in a new university institution, but I did not find this part of my work particularly agreeable or rewarding. A poorly stocked University Library made any chance of continuing front-line research in my highly specialised field somewhat remote. This would of course have been different had the internet come 40 years earlier! Fortunately I had brought with me all the material that I needed for writing my book on Interstellar Grains. This is what I was able to do in the time that was left for me between my other commitments. And for the rest I simply had to bide my time until I returned to Cambridge, back to my Jesus Fellowship and eventually to a staff position at Fred Hoyle's new Institute of Theoretical Astronomy. The Institute itself became a statutory reality in July 1966, with its buildings in Madingley Road to be completed the following year.

By far the most momentous event in my life, one that was to have a profound influence on the course of my journey with Fred, happened during my sojourn in Sri Lanka. This was my meeting with Priya Pereira, a law student at the University of Ceylon, and our subsequent marriage in April 1966. A beautiful, intelligent, talented girl aged 20 enters my life and this story. As she still continues to complain in jest, she was wrenched from her family and a highly promising legal career (her father was a leading lawyer) to become my wife. As our work on the grains turned into directions that became ever more controversial, Priya's steadfast support and encouragement, and most of all her combative temperament was a crucial factor in our progress. It could be said with all honesty that Priya helped me endure the slings and arrows of outrageous fortune, in the instinctive belief that the ideas so carefully thought out by Fred Hoyle and myself must turn out to be right! Little did Priya realise in 1966 that she was stepping in with me into a lion's den of acrimony and controversy in the years that lay ahead.

But the worst of all that was still a full decade into the future. In April 1966, I went on the second sea voyage of my life, this time in the delightful company of Priya, from Colombo to Southampton, bidding farewell to family and friends, and the country of our birth. We arrived together in Cambridge to set up our first house in a flat in Jesus Lane that was provided for us by Jesus College. I had just two months to resettle in Cambridge and introduce Priya to all my friends, particularly the Hoyles.

Then we were on our travels again. This time by ship *SS France* to New York, and then on to College Park, Maryland where we would be based for three months. Although Priya and I were agreed that we would not be living full-time in the States, we made the most of our stay here, spending our weekends touring the national parks of Virginia, Baltimore and Washington in an old Chevrolet that we had bought for $50.

My office on campus was located in the Department of Physics and Astronomy headed at the time by the Dutch radio astronomer Gart Westerhout. My appointment as a Visiting Professor was in the University of Maryland, but since the funding for my post came from NASA, most of my collaborative links were with Bert Donn's Astrochemistry group at the Goddard Space Flight Centre in Greenbelt. It took me a while to get accustomed to the new environment and particularly to the work ethic that prevailed. I had got used to working Cambridge-style on an individualistic basis, even in my collaborations with Hoyle. Here at Goddard I was drawn into more restrictive communal research atmosphere. People liked to talk a great deal and to work, or appear to work in big teams. There was always an impression of great industry, with large numbers of people working at a frenetic pace on a single problem. But the output was not always commensurate with the manpower or effort that was expended.

As agreed with Bert Donn my task in the three months at Maryland was to investigate further the problem of the formation of graphite grains. The earlier work on this subject that was already published by Fred and myself in Monthly Notices came to be carefully re-examined and elaborated upon by a team of four authors led by Bertram Donn. The work as I remember it involved a tiresome succession of meetings and conferences occupying a great deal of time. And even at the end of my stay a final manuscript for submission was not agreed upon. Many letters and drafts had to be exchanged across the Atlantic, and the ultimate result was the publication two years later of a paper in the *Astrophysical Journal* (B. Donn *et al.*, *Astrophys. J.* **153**, 451, 1968).

American scientists in 1966 were clearly ahead of the British in the practice of counting research papers and citations in a race to justify their existence, and to secure continued public funding for their projects. The levels of stress generated by this process were visibly detrimental to the health and well being of the scientists concerned and also to the progress of science. Nowadays almost every scientist has come to take this practice for granted, with universities and research groups vying to get the maximum number of papers published in the shortest time and in the so-called high impact journals.

Our papers on interstellar grains published throughout the period 1962–1968 and beyond were all in high impact journals, and impact they certainly had in helping to change astronomical fashions from volatile ice grains to refractory grain models. A connection between grains and organic material slowly entered my thinking when I began to look at the detailed thermodynamic equilibrium compositions of the atmospheres of cool stars. In carbon stars, for instance, the equilibrium calculations I performed whilst I was in the States showed clearly the presence of large numbers of hydrocarbon molecules which would inevitably be expelled into interstellar space along with the grains. I had suggested a stellar origin for some of the organic molecules detected in the interstellar clouds, but an explicit link between grains, biochemicals and life had to remain in gestation for some years to come.

Chapter 7

The Institute of Astronomy:
The Vintage Years

In the summer of 1967, the Institute of Astronomy in Cambridge came into physical existence in a well-equipped open plan building in the midst of a meadow off Madingley Road. It seemed to be strategically placed between two friendly institutions — the Cambridge University Observatories on the one side and the Geophysics Department on the other. Fred's cosmological adversaries at the Mullard Radio Astronomy Observatory were located only a couple of miles away on Madingley Road, but as far as interaction was concerned they may as well have been as far away as the Moon. Martin Ryle and his team were continuing in their single-minded pursuit of disproving Steady State Cosmology. From their studies of the counts radio sources to various intensity levels, they claimed that radio emitting galaxies appeared to be closer together as one goes further back in the Universe, showing that the Universe could not be in a steady state.

The real crisis for Steady-State Cosmology was, however, not radio source counts but the new discovery of a cosmic microwave background with a temperature of 2.7 degrees above absolute zero (2.7 K). To offer any credible defence of Steady-State Cosmology it was imperative to explain this background by a process that was unconnected with an early hot phase of a Big Bang Universe.

After my return from the United States, Fred and I had many discussions exploring possible non-cosmological interpretations of the cosmic microwave background. The question as to whether grains could play a role naturally arose, but I had difficulty in arriving at any tenable model. We even briefly entertained the possibility that an isotropic cloud of dust enveloping the entire solar system could somehow succeed in degrading sunlight energy to a mere 3 degrees! Nothing was working. Fred alerted me to several strange coincidences that may be relevant to this issue. The density of energy of starlight in the galaxy was very close to the density of energy in the cosmic microwave background. Further he pointed out that the energy released in the conversion of hydrogen to helium averaged over a cosmological volume (a large part of the entire observable universe) had a density that was also similar to the energy density of the microwave background. The implication of all this is that a perfectly absorbing and emitting object (a black body) placed either in interstellar space or intergalactic space would take up a temperature close to that of the microwave background, 2.7 K. Could all this be dismissed as a miraculous coincidence? I thought not. But the challenge was to find approximations to black body absorbers and radiators in space. If some mechanism can be found for thermalising the energy of starlight, there would be a hope for explaining the microwave background in terms of ongoing astrophysical processes within a steady state universe. But the situation was by no means easy. If one thought of solid interstellar grains of radii of about a tenth of a micrometer, they would be exceedingly efficient absorbers of visual starlight, but very poor radiators at long wavelengths. The reason being that such a small particle acts as a very inefficient antenna for re-emitting the absorbed starlight at very long wavelengths, that is to say wavelengths that are much greater than the size of the grain itself. Thus a normal interstellar dust grain would heat up due to an internal greenhouse effect, and take up a temperature of 15–30 K, much higher than the blackbody temperature which is nearer to 3 K.

If, however, there are oscillators (radiators) within the grain that can vibrate at microwave frequencies 10–100 Megahertz (MHz) the situation could be different. Grains could then yield high outputs of radiation over a microwave waveband. This is precisely what we discovered in 1967 for the case of impurity atoms that are very weakly bound in solids. Fred Hoyle and I argued in a paper in *Nature* (**214**, 969–971, 1967) that the cosmic microwave background could perhaps include a contribution arising from these types of grains. Jayant Narlikar and I followed this up with two attempts to work out detailed spectra and to estimate isotropy of a background (how uniform the radiation is in all directions in the sky) arising from such a process (*Nature* **216**, 43–44, 1967; *Nature* **217**, 1235–1236, 1968).

Yet another attempt at a non-cosmological explanation of the microwave background came from a collaboration with Vincent Reddish, during one of the many visits to the Royal Observatory in Edinburgh that I made in those days. We discovered from published data relating to solid and liquid hydrogen available at the time that the condensation of solid hydrogen onto grains could occur in interstellar clouds at a temperature close to 3 K — very close indeed to the temperature of the microwave background (N.C. Wickramasinghe and V.C. Reddish, *Nature* **217**, 1235–1236, 1968). For Fred, though, this was an opportunity to revive his enthusiasm for steady state cosmology. A very elegant logical argument could be made. If the most abundant chemical element in the Universe, hydrogen, could freeze at 3 K several important consequences would follow. Freezing of hydrogen in dense clouds reduces gas pressure within the clouds, and hence leads to the collapse of clouds. The formation of stars and galaxies that follows, tends to increase the energy density of starlight, because they put out more radiation into space. This in turn cuts off the process as soon as the local temperature rose above the freezing point of hydrogen. A feedback loop of this type acting like a thermostat would maintain the background temperature at precisely 3 K. This impressive argument was published in 1968 (F. Hoyle, N.C. Wickramasinghe and V.C. Reddish, *Nature* **218**, 1124–1126, 1968). Unfortunately for us it turned out that the thermochemical data used by us in 1967 was inaccurate and came to be revised. The new data, which was brought to our attention by our old friend J. Mayo Greenberg, showed that the temperature of freezing of hydrogen in interstellar clouds was nearer 2.3 K than 2.7 K. This of course dealt a death blow to the solid hydrogen explanation of the cosmic microwave background. We were back to square one.

But there were more urgent matters that demanded our attention. The Institute was now in full swing and Fred had entrusted me with the job of getting a core library together, a task which took up several weeks of my time. And of course there were other interesting astrophysical problems to turn to. A state of the art IBM Computer up and running in a building adjacent to the main Institute beckoned us to tackle computational problems that were hitherto beyond our reach. I continued my attempts at modelling various aspects of interstellar grains, including their infrared emission and absorption properties, whilst also beginning to think about another interesting computational project. This was the problem of the condensation of the planets.

The problem itself was not new and goes back to the old nebular hypothesis first suggested by Immanuel Kant and formulated in a scientific framework in 1796 by the French mathematician Pierre Simon de Laplace. It is generally believed now that the material of the sun and planets once comprised a single diffuse cloud of interstellar gas and dust. We see similar clouds widely distributed between the stars of our Galaxy, and as we have pointed out earlier it is clear that stars condense from such clouds. The manner in which such a cloud of gas and dust can separate to form the sun and the planets is less clear, however. An early theory suggests that the first step was the formation of a double star system consisting of the sun and a much smaller companion star going round one another; then the smaller star is supposed to have exploded spreading its material in a disc out of which the planets condensed. Another theory states that a single star, the sun, contracted from the cloud leaving behind a debris of gas and dust from which the planets subsequently condensed. Yet another possibility is that the planetary material left the contracting sun during an early stage of its evolution.

Fred and I argued in favour of this latter possibility, believing that the others are more or less untenable. The vital clue lies in the very simple fact — that the sun rotates about its axis once in about 26 days. If the sun contracted from a cloud of gas and dust, similar in size and speed to the clouds we see between stars, then it is possible to predict quite unequivocally that the sun must have rotated several hundred times faster than it now does. In fact it should rotate on its axis once in a fraction of a day. This is on account of a physical law which asserts that the rotational momentum of a mass of material, which is not acted on by external forces, cannot change. The consequence must be that the compressed smaller object must spin faster in order to contain all the rotational momentum of the original dispersed cloud. The situation is analogous to ballerina who spins faster as she draws in her arms. What has happened to this rotational momentum of the sun? If one adds up all the rotational momentum of the planets it turns out that here indeed is all the missing rotational momentum of the sun. It appears that, in some way, the contracting sun had given up most of its rotational momentum to the much smaller mass of the planetary material.

There is one very simple manner in which such a transference of rotational momentum can take place. As the cloud of gas and dust which formed the sun contracted, it started off by taking in all the rotational momentum. The cloud would thus rotate faster and faster as the contraction proceeded. A critical stage is reached when the cloud's gravity cannot hold up to the strong rotary forces that begin to develop, particularly near the flattened equatorial regions. The result will be that a disc of material which later formed planets is squired out from the equator.

At this stage we can estimate that the young Sun would have been about 10 times larger than at present, and that its surface temperature was about 3000 K, about half its present value. The planetary matter ejected at this time is slightly ionised (that is, separated into nuclei and electrons) and thereby anchored to the parent sun by magnetic fields in a manner discussed by Hannes Alfven. It is this anchoring (acting like spokes in a bicycle wheel) that actually caused the transference of rotational momentum from the sun to the planetary material, resulting in the slowing down of the rotation of the sun. The speed of ejection of the material must have been sufficiently great for the planetary disc to have been able to reach out to the distances of the outermost planets, Uranus and Neptune. It is at such distances that large quantities of the lightest gases, hydrogen and helium, must have escaped into space. Indeed it appears that as much as 6/7 of the total mass of the planetary disc must have been lost to the solar system in this way.

My brief now was to deploy the Institute's computer to work out the chemical history of the planetary material from the stage the disc leaves the Sun's equator in the form of a gas to the time of planet formation. The actual calculation took up about 25 hours of computing time on the Institute's computer, although nowadays on my laptop at home the same calculation would take only a few minutes!

When the planetary material leaves the sun its temperature is about 3000 K. It consists of gaseous atoms and ions in uncombined forms such as Hydrogen, Helium, Oxygen, Carbon, Nitrogen, Silicon, Magnesium, Iron and Sulphur. In the course of its journey outwards, the temperature, density and composition of this material changes in a manner which can be calculated from known physical and chemical theories. Various types of molecular species begin to form as the gas recedes from the Sun and consequently cools off. It is possible to calculate the distances from the Sun at which various types of solid particles and liquid droplets begin to condense out of the gaseous mixture. Our calculations indicated that the first such condensations must take place almost precisely at the distances of the innermost planets — within the orbits of Mercury, Venus, Earth and Mars. The temperature of the gas at this stage is about 1500 K, and we found that fine particles of iron, magnesium oxide and various silicates begin to form — precisely the constituents of the terrestrial planets. At these high temperatures iron particles tend to be very sticky and collisions between such particles would lead to the formation of quite large sized objects measuring 1–10 m across. Whilst a fine smoke of particles would tend to be swept along with the gas, objects of 1–10 m across would fall out of the expanding disc at the distances where they are formed. These metric sized blobs of iron and magnesium silicates then aggregated further and eventually led to the formation of the inner planets, including the Earth.

The most notable success of this theory was that we could give a plausible explanation for why the Earth and inner planets are made up mainly of iron and magnesium silicate in various mineral forms, and why they are found at their present distances from the sun. It seems that these features were determined almost entirely by the density and temperature at the surface of the sun when the planetary material left it, and by the thermochemistry of the various gases involved. Water and carbon dioxide could have been trapped only in rather small quantities at the Earth's distance — probably in the form of hydrated silicates and carbonates. This again is in agreement with respect to the quantities of such substances we find on the Earth.

It is only when the planetary disc expanded out to the present orbital distances of the outer planets Uranus and Neptune that carbon dioxide-ice and water-ice could condense, first into solid particles, then into 1–10 m sized bolides and next into hundreds of billions of cometary-sized icy objects. Collisions between these icy bodies occurring over hundreds of millions of years led to three further effects:

1. Some icy comets were deflected into orbits that made them collide with the Earth and other terrestrial planets, leading in the case of the Earth to the acquisition of volatiles such as water and carbon dioxide. The Earth's oceans and atmosphere came in this way.
2. Some cometary objects were deflected outwards to form a shell of comets around the sun.
3. Approximately half of the icy objects ended up in planetary accumulations we now recognise in the planets Uranus and Neptune.

The attractive feature of this theory is that it is able to account for the observed chemical compositions and distances of the various planets in the solar system in a sensible way.

During the time of the accumulation of the outer planets, the entire solar system remains immersed in the remnants of the molecular cloud from which it formed. In a contact sweeping through the cloud, the shell of comets mops up large quantities of organic molecules and dust of the kind we described in an earlier chapter. It is in this way that comets came to incorporate pristine interstellar material.

Our new theory of planet formation was published in *Nature* in February 1968 (*Nature* **217**, 415–418, 1968). Its impact on ideas relating to interstellar grains and to life in space eventually turned out to be more far-reaching than we imagined at the time. As a source of interstellar grains the planet forming processes were just as important, if not more important, than sources from cool giant stars. When the inner planets were accumulating not all the iron and siliceous dust would have coalesced into larger bodies that remained gravitationally bound to the solar system. A significant fraction would have been carried outwards and expelled into interstellar space through the action of radiation pressure by an early superluminous sun. This would provide one source of refractory metallic and siliceous dust in interstellar space. We discuss later how nascent planetary systems possessing comets like our own solar system could also provide a source of organic interstellar dust, perhaps even life.

Fred was working at this time with Willy Fowler on nucleosynthesis in exploding supermassive stars in the centres of galaxies. The question naturally arose whether dust could condense directly from the heavy element enriched gas that flowed out from such exploding supermassive stars. To tackle this question I had once again to deploy my chemical equilibrium program. In a calculation very similar to that described earlier for the case of the planetary disc I was able to show that, with solar abundances at the source, magnesium oxide, iron and silica particles must again form when the temperature of the expanding supernova shell cooled below about 1200 K. We thought now only in terms of spherical dust particles for all these three compositions — including iron. It would later turn out that iron was much more likely to form as long whiskers, a feature that would have a relevance to the problem of the microwave background that was discussed earlier in this chapter.

Chapter 8

Winds of Change

Just as I was beginning to feel settled in our life in the UK, racial tensions began to grow on both sides of the Atlantic. On April 11th 1968, Civil Rights leader Martin Luther King was gunned down in Memphis and a wave of race riots spread across major US cities. Within an amazingly short space of time ripples of racial disquiet reached Britain. On April 21st, Enoch Powell made his historic "rivers of blood" speech. Powell, a distinguished classical scholar, made his point most eloquently: "As I look ahead I am filled with foreboding. Like the Romans I see the River Tiber foaming with much blood". He went on to say that Britain must be, "mad, literally mad as a nation to admit 50,000 dependents of immigrants into the country every year". The present situation he concluded is like a nation "busily engaged in heaping up its own funeral pyre". Conservative Party leader Ted Heath was quick to denounce Powell's speech and expel him from the shadow cabinet, but for those of us who were attempting to adopt Britain as our home, these developments were a source of anxiety and insecurity.

My work was a solace. Fred and I now continued to explore non-carbon based contributions to interstellar dust. Our papers on condensation sequences of solids in the expanding solar nebula and in supermassive stars (*Nature* **217**, 415–418; *Nature* **218**, 1126–1127, 1968) detail our views at the time. We were considering then the possibility of particles comprised of iron and siliceous material coexisting alongside with graphite in interstellar space. However, because the elements silicon and iron were down in abundance from carbon and oxygen by a factor of more than 10, we did not think there was an easy way of detecting such grains from the overall extinction behaviour of interstellar dust.

Whilst this work was in progress, Fred was involved in a government project to build a telescope in collaboration with the Australians. He played a key role in the choice of site for a southern hemisphere observatory, and also in the negotiations for securing UK funding. In fact it was the formidable Margaret Thatcher, then UK Education Secretary, who had to be convinced that Britain should make a financial commitment of this kind. After trying various approaches, Fred informed her that television viewing figures for recently screened programmes connected with astronomy ran into several millions, far more than for any other factual programme. That, according to Fred, instantly won her support.

The summer of 1969 had a special significance for Priya and myself. Fred's wife Barbara visited Sri Lanka with her friend Viv Howes, and we had the pleasure of looking after them. Whilst they were in the capital they stayed in my parents' house in the pleasant residential suburb of Colombo 4, but they spent many days touring the island. We accompanied them on most of their trips and together with our visitors we saw things and places we had never seen before. The 21 years of our lives that we spent in Sri Lanka were evidently not enough to have explored all the remarkable sites of this richest of isles. Our trips took us to the ancient ruins of Anuradhapura and Polonnaruwa, to the hill country capital Kandy, to tea plantations and to the beaches on the West coast of the island. Whilst in the hill-country station of Ella, the place that made the greatest impression on all of us was "Land's End". In the first morning mists at the guest-house where we were staying, we experienced that peculiar sense of the infinite, of our dwelling here on Earth giving way to the sky and the eternal cosmos. Gigantic stone statues of the Buddha set among the arid planes of central Sri Lanka reinforced this feeling. Anita (infinity) lies at the heart of the Buddhist philosophy — a world without end or limit, into which the individual may at last be sublimated.

The year 1969 witnessed two great triumphs of our technological civilization — in air travel and space travel. A long 65 years after the Wright brothers took flight in their primitive pedal aircraft in 1904, the first supersonic aircraft took to the air. And in the realm of space travel the first human being set foot on the moon. In the small hours of July 21, flickering pictures of Neil Armstrong taking his first steps on the lunar soil beamed back to our television screens and to a world agog with excitement and wonder. The first moon walk was undoubtedly a great technological achievement and a commendable success for NASA. Indeed, it remains Science's most popular gesture to date, creating the appealing impression of Man as the supreme ruler of the Universe. Its relevance to the advancement of astronomical knowledge was, however, a matter of dispute.

However, the events of 1969 paved the way to a new era of astronomical observations conducted from space. A new generation of telescopes and instruments carried aboard satellites came into operation. For example, NASA's Orbiting Astronomical Observatory 2 (OAO2) began to yield a wealth of new data on ultraviolet spectra of stars that had a direct bearing on the composition of interstellar dust. The existence of a conspicuous ultraviolet absorption feature centred at about 2175 Å was confirmed, and a continued rise of interstellar extinction further into the ultraviolet was discovered. At this time Fred and I were still pretty much tied to the idea of graphite grains causing the mid-ultraviolet extinction hump, with other grain materials being responsible both for the visual extinction of starlight, and the further rise of extinction into the far ultraviolet. A striking result that emerged from the OAO2 data was the invariable wavelength placement of the centre of the 2175 Å absorption hump from star to star. This, according to our models, demanded the presence of spherical graphite particles with radii almost exactly fixed at 0.02 micrometres. Even as early as 1969, Fred and I were beginning to feel a little uneasy about the artificiality of these assumptions. But a properly formulated alternative to graphite lay a few years into the future.

The period 1967–1970 saw also the emergence of the new discipline of infrared astronomy. The first deployment of liquid nitrogen-cooled detectors operating at 2.2 micrometres at the Mt. Wilson and

Palomar Observatory yielded a catalogue of nearly 20,000 infrared sources, mainly cool stars. Further refinements of infrared detection techniques and the use of ground-based telescopes at high, dry mountain sites, provided a wealth of new information about the cosmos. A new window in the electromagnetic spectrum was open to the observation of astronomical sources. The first discovery of infrared astronomy directly relevant to our story came from the work of John E. Gaustad and his colleagues confirming that the infrared spectra of highly reddened stars showed no evidence of water ice (*Astrophys. J.* **158**, 151, 1969). Then followed a spate of other discoveries all showing a spectral feature of dust over the 8–12 μm mid-infrared waveband. The feature was observed in emission in a wide variety of astronomical objects including oxygen rich cool stars and in the Trapezium region of the Orion Nebula. (See for example, Fig. 2.) The results were published in the form of several papers in the same issue of the *Astrophysical Journal Letters* in 1969. The authors E.P. Ney, F.C. Gillett, J.E. Gaustad, W.A. Stein, N.J. Woolf, R.F. Knacke, D.A. Allen and R.C. Gilman worked at various telescopes operated by the Universities of California and Minnesota. One of the papers in this volume interpreted the new results as evidence that the dust was made of a mixture of silicates — combinations of magnesium, silicon and oxygen as they occur in the rocks of the Earth. Their conclusion, however, was premature.

The papers in *Astrophysical Journal Letters* were soon followed up by a report by Nick Woolf and his colleagues describing the discovery of the same 10 μm feature now seen in absorption against an extended infrared source located at the centre of the galaxy. This was of course a clear indication that the feature in question was not confined to localised sources of dust around stars, but was a property of interstellar dust along a 10 kiloparsec path length to the centre of the galaxy.

I believe that Fred was given a few week's advance notice of the appearance of these papers by his friend Ed Ney, who was at the time the chairman of the astronomy department of the University of Minnesota. He asked me whether I thought silicate grains could fit the astronomical data that we had been modelling now for some years.

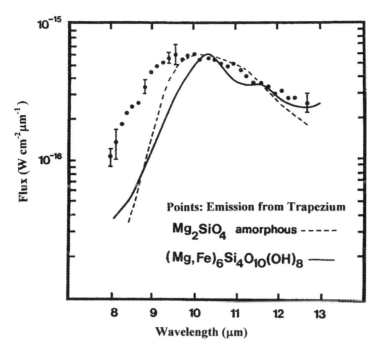

Fig. 2   The infrared emission by heated dust in the Trapezium nebula observed in 1969 (points). Curves show the behaviour of amorphous and hydrated silicates. The mismatch evident over the 8–9 μm micrometre waveband led us to consider alternative organic compositions of the dust.

What struck me immediately on studying the *Astrophysical Journal Letters* papers was how poorly the newly observed astronomical feature at 10 μm actually fitted the behaviour of any known silicate or mixtures of silicates. Since mineral silicates also have absorptions spanning the 8–13 μm waveband, there was of course a crude match to be seen. The "best" silicate fits to the Trapezium data are shown in Fig. 2 as curves for amorphous and hydrated silicates. But matches of this quality were not by any means restricted to silicates. Other chemical systems, including some that involved carbon, could be candidates. Whilst one could not dispute that some quantity of silicate dust might exist in space, how would this compare with contributions from the more cosmically abundant element carbon?

Our search for a carbon-based interpretation of the 10 μm interstellar feature began as early as 1969. An examination of spectroscopic literature showed that soot containing hydrogen admixtures had spectra that might be satisfactory from this point of view. Fred and I next calculated the behaviour of two component mixtures comprised of silicates and impure graphite grains, and showed that they could, at least approximately, fit the entire range of astronomical observations that were then available. A description of this work was published in the form of an article in *Nature* (*Nature* **223**, 459–462, 1970).

We were, however, aware of some glaring shortcomings. The fit of our synthetic model spectra to the new mid-infrared data in the $10\,\mu$m waveband, although better than for silicates, still left much to be desired. With the quality of the observational data that was now available, particularly in the extinction curve of starlight over the visual and ultraviolet bands, a more perfect agreement with models was required. There was also the question we referred to earlier, of justifying the artificial requirement of perfectly spherical graphite particles all of one size. And finally, how could we account for the almost invariant shape of the visual extinction curve? With a view to resolving at least some of these questions we began to consider three component, rather than two component grain models — particles comprised of a mixture of graphite, silicate and iron grains. Better results emerged, still not perfect. (Wickramasinghe and Nandy, *Mon. not. R. Astron. Soc.* **153**, 205–227, 1971). Increasing the number of constituents in our model with relative proportions that had to be fixed arbitrarily was also a source of concern.

For the next couple of years Fred's direct involvement with this programme of work had to be curtailed by the many strenuous commitments he had undertaken. Whilst the old cosmological wrangles still continued, Fred also became involved in matters relating to national astronomical policy. The volume of committee work that this involved robbed him of time for research and diminished his enormous creativity. But most soul destroying of all for him were the negotiations that were under way with Cambridge Universtiy and the Government for renewing support for his Institute of Theoretical Astronomy. Fred had quite definite views as to how it should continue into the next quinquenium, others amongst his adversaries held different views. Despite the enormous success of the Institute in a brief span of four years we are told that its very continuity was being threatened as early as the summer of 1971. Fred relates this part of the story in great detail in his autobiography "Home is where the wind blows". I shall not reiterate the details, but merely say that things turned out to be so intolerable for him that he submitted his resignation to the University in January 1972.

The news came as a shock when it finally reached us in the early Spring. Cambridge at this time was looking at its seasonal best. The colleges were shedding the austerity and grandeur of their winter identity and the backs were now softening to a new ministry of snow-drops, crocuses and daffodils. Stone courtyards, grey too long, glimmered again in sunlight, revealing their glorious red and purple depths. It seemed a cruel fact that ill winds could blow through so exquisite a setting. For the next few weeks I felt a little numbed at the thought that great changes were in the air. Dining with my wife at a Ladies Night feast at Jesus College in April 1972, I wondered if this would be my last such experience! Such feasts at Cambridge Colleges (there were several each year) were unique social occasions, lavish to a certain extent and mostly supported by endowments that could not be used for another purpose. Feasts aside, I had benefited from the many privileges that my decade-long Fellowship had afforded. But most of all I had enjoyed the privilege of working so closely with Fred Hoyle.

Chapter 9

The Cardiff Era

I spent most of the summer of 1972 in Sri Lanka with my wife and family. The era of mass communication had not yet dawned — faxes, emails and the internet were nearly two decades away and international dialling was expensive. So this chit of an isle in the Indian Ocean seemed blissfully but strangely remote.

From an early age the sea held an irresistible attraction for me. This is true also for very many people. The theory that life sprang from oceans, accounting for humanity's fascination with the sea, I was later to refute. Fred would always favour the exigencies of rugged landscapes as a means to reflect and meditate, but I would always choose the beaches and the ocean where communion is instantaneous. Wave after wave rolled forward from the distant ocean, interrupted only by the occasional rumbling of a train passing behind me. And vexations would momentarily recede, drowned by the great swell of the sea. Every evening the sun would set around 6.30 pm. The period of sunset that follows is brief because we are so close to the equator. The sunset colours depend very much on the disposition of the clouds. Sometimes the sunsets are spectacular, at other times a disappointment, mirroring the vicissitudes of life itself.

My thoughts this summer were co-mingled with nostalgia for my days at Cambridge that I feared were drawing to a close. It seemed also that my collaboration with Fred may have come to its logical end and I kept wondering what shape my career might take. Priya was pregnant with our second child, so thoughts about the future were even more resonant.

I returned to Cambridge that October and the reality of the situation dawned on me even more powerfully. As I walked into the Institute I felt a sense of sadness sweep over me, like returning to a house of bereavement — its life and soul had departed leaving what seemed an empty shell. Fred had left the Institute in mid August, never to return, and shortly afterwards sold their house in Clarkson Road. Fred and Barbara had entered the next phase of their lives and were now in the process of resettling at Cockley Moor, near Pernrith in Cumbria in their beloved Lake District.

My own appointment at the Institute was assured only until the end of the academic year. What was to happen after that would be decided upon by the next Director. Without Fred the Institute lost its attraction for me and I could not see myself carrying on there, even if I were given the chance. The option of looking for openings in a Sri Lankan University was one I had considered but did not entertain seriously for too long. I was still fired with enthusiasm to carry forward the many promising lines of research I had begun, and the scientific culture of Sri Lanka could not permit the funding of such ventures. The country had more pressing economic problems to deal with.

I then began to look for university jobs in the UK. One of the first that came to my notice was the Chair of Applied Mathematics and Mathematical Physics at University College, Cardiff. Fred had not yet left the UK astronomical scene, so I was able to consult him on this matter. He strongly advised me to apply. I did so and my application was successful. At the interview for selection the Principal of University College Cardiff, Dr. Bill Bevan, made it clear to me that he was very keen to start Astronomy in Cardiff, and furthermore that he would like me to involve Fred Hoyle in this venture. I could not have expected a more attractive offer, so I accepted the position and agreed to assume duties in the summer of 1973.

My remaining few months in Cambridge were occupied with completing my book *Light Scattering Particles for Small Particles with Applications in Astronomy* that I had started to write earlier in the year. A major part of the book was a set of tables and graphs intended to save astronomers the tedious job of computing optical cross-sections each time they were needed in a problem. The book appeared in 1973 (*Light Scattering Particles for Small Particles with Applications in Astronomy*, John Wiley, 1973). For several years this book was much used by the growing number of workers on interstellar dust who did not have immediate access to the relevant programmes and high speed electronic computers. This situation changed dramatically in the 1980's with the advent of affordable personal computers, and later the internet. This type of information is now readily accessible on the internet, although I believe my book is still being used in some circles and occasionally even referenced. The one person for whom my "big red book" was immensely useful was Fred Hoyle. Fred could never be persuaded to use a PC let alone even consider using the internet. A hand held programmable HP calculator and a fax machine were as far as he could be pushed to acquire in the early 1980's.

Early in the summer of 1973 we sold our house in Barton Cambridgeshire and moved to Cardiff with our two young children. Our new home in Lisvane, a suburb of Cardiff, overlooks gently undulating mountains — a refreshing change from the flatness of the Fens around Cambridge. At the University I felt at first a little overwhelmed by the responsibilities entrusted to me. The task was by no means easy. There was jealousy and hostility of the existing staff of the department to contend with, particularly so as I had set out to steer the department in a direction that appeared quite alien to them. However, with the unstinting support of Principal Bill Bevan and senior members of Senate I was able to institute great innovations. Within a year of my appointment I had changed the name of the Department from "The Department of Applied Mathematics and Mathematical Physics" to "The Department of Applied Mathematics and Astronomy" and we were soon on the way to appointing 4 new lecturers in Astronomy. I had a dilemma of deciding upon the research fields in which to choose the new lecturers. Had I opted to have all appointments in the area of interstellar dust we would now probably have had the most powerful research group in the world in my own specialist field. Following the example of my mentor Fred Hoyle and the experience of the Institute in Cambridge

I opted instead to develop as diverse a group as I possibly could. The first appointee was in the area of plasma physics, the second in the chemical evolution of galaxies, the third in star formation theories and the fourth in relativistic astrophysics. A year later I obtained support for a Chair of Observational Astronomy, so a highly diverse and balanced group was begun. The choice as I saw it in 1973 was like the difference between sowing a packet of mixed flower seeds and one of a single species of flowering plant. The mixed seeds if they took would lead to a glorious splash of colour. This is what has now happened in Cardiff, leading to the evolution of an astronomical research centre that is one of the best in the UK. Such heterogeneity is much rarer in the present day as subjects have become so overly specialised that people from different fields barely speak the same language as each other, and often cannot recognise that they are trying to address the same problems.

By the late Spring of 1974 I had set all the major changes in train and was headed again to North America for a short respite — this time to Canada and to the University of Western Ontario in London, Ontario. Priya and I instantly took to Canada which we can adapt better than in its neighbour America. The country was more akin to Europe in terms of its culture and value systems, thus we found it more recognisable. It was during my three-month stint in Canada that I made a breakthrough that was to become a defining moment in this story. I mentioned earlier that we had not been satisfied with the quality of the fits that had been obtained with silicate models for the 8–13 μm astronomical spectra. So I was still searching for a better solution.

We had already found that hydrogen impurities in soot could offer a partially improved solution, but the absorption strengths in such bands were exceedingly weak. In any case it was clear that if an alternative to the currently fashionable silicate explanation was to be credible, the fit to the data had better be close to perfect. As these thoughts were running through my head, I was beginning to feel the 8–13 μm band mismatch of the silicate model must somehow conceal a much bigger story. The carbon with hydrogen model described in our 1969 paper was in a sense skirting around the possibility of an organic grain model. What if the carbonaceous component of the dust was not simply graphite but made of organic materials, organic polymers in fact? The ice grain model of van de Hulst was argued as a plausible composition because in principle at least oxygen could combine with hydrogen to form the stable molecule of water. In the same way carbon atoms in interstellar space might be combined with hydrogen and oxygen to form an extraordinarily vast variety of organic chemicals. In terms of the basic chemical elements at least there would be more than enough mass to explain the properties of interstellar dust.

At this time the molecule formaldehyde H_2CO had been discovered to exist ubiquitously in interstellar clouds. It was present in dense molecular clouds as well as in the less dense interstellar medium. What if such molecules started to condense and polymerise on the surfaces of pre-existing graphite or silicate grains expelled from stars? Looking though the books in the library of the University of Western Ontario dealing with properties of formaldehyde, I soon discovered that it could readily polymerise on the surfaces of silicates. It also turned out by a stroke of luck that the Chemistry Department here had one of the leading experts on formaldehyde polymers and I had the benefit of many discussion with him. A simple calculation showed that under interstellar cloud conditions substantial mantles of formaldehyde polymers would indeed grow. Next I began to look at the optical and infrared properties of many types of formaldehyde polymers. It turned out that such polymers were dielectric (that is to say, non-absorbing) in the visual waveband as the observations demanded. And most strikingly polyformaldehyde (polyoxymethylene) had absorption bands over the 8–12 and 16–22 μm wavebands, with the former absorption fitting the astronomical data better than any known silicate, as shown in Fig. 3. This was a breakthrough moment and within a couple of weeks my paper entitled "Formaldehyde Polymers in Interstellar Grains" was submitted to *Nature*. It was published (*Nature* **252**, 462, 1974) and made a splash in the science news columns of the broadsheets. This was the first ever suggestion of the widespread occurrence of organic polymers in the galaxy,

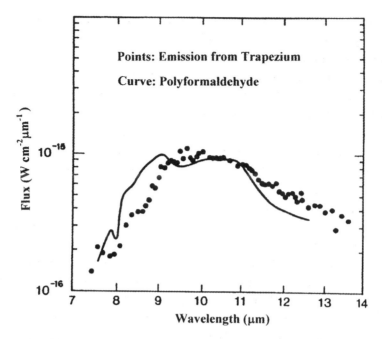

Fig. 3   The infrared emission from the Trapezium nebula (points) compared with the behaviour of polyformaldehyde dust (curve) heated to 300 K. This was the first argument used in favour of organic polymeric dust in 1974.

and it was also the first paper to come from the newly reconstituted Department of Applied Mathematics and Astronomy in Cardiff.

Within days of the publication of my paper, Professor V. Vanysek of Charles University, Prague came to visit me with a proposition to be considered. Polymerised formaldehyde and also other organic polymers, could form a major component of comet dust as well. Comets were thought of at this time to be dirty snowballs, mostly comprised of inorganic ices with siliceous and metallic dust occurring as minor impurities. When a comet approaches the inner parts of the solar system, the molecular species in the developing coma were thought to be break-up products of ices which were regarded as "parent molecules." With polymerised formaldehyde being a component of the comets, coma radicals such as OH, CN, C_2 could be interpreted as the break-up products of such a polymer. When we began to explore this idea together it appeared to us to seem more and more plausible. Since formaldehyde polymers remain stable up to temperatures 500 K, changes in the 10 μm spectrum of the new Comet Kohoutek (1973f) as it came within 0.5 AU of the sun were shown to be consistent with this model. The idea of an organic comet was thus born, and its justification was described in a joint paper published in *Astrophysics and Space Science* (V. Vanysek and N.C. Wickramasinghe, *Astrophys. Space Sci.* **33**, L19–28, 1975). This publication establishes our priority for the idea that both interstellar and cometary dust could be comprised of organic polymers.

Up to this point in my story there was no major discord with mainstream astronomical ideas, and indeed there were even some grudging plaudits being accorded to us for making such an innovative suggestion. I soon acquired a research student, Alan Cooke, who started doing more detailed modelling of formaldehyde grains, and a grant from SERC to do experimental work on the optical properties of organic polymers was also secured. From 1975 to 1977, I was awarded more grants from SRC that enabled several visitors to be invited to Cardiff for collaborative work.

Chapter 10

The Search for Cosmic Life

Two years after I began my tenure at University College Cardiff, I succeeded in getting Fred appointed as an Honorary Professor. This meant that he would be a nominal member of my Department and be entitled to use the university as endorsement for his publications. In turn, Cardiff gained much kudos from this link with Fred Hoyle. During the period of 1975–77, Fred spent large chunks of his time in the United States, so that his appearances in Cardiff were rare. Even so, I usually managed to track him down and keep him abreast of our research developments on organic polymers.

Most of our efforts in Cardiff were now directed to securing optimal fits to the 8–13 μm emission feature in one particular astronomical source — the Trapezium region of the Orion Nebula. (See Figs. 2 and 3.) We began to regard this source as the touchstone of acceptable grain models. (Modelling of this source was particularly simple because it was optically thin in the infrared, with grains all emitting at a temperature within the range 150 to 300 K.) A most glaring inconsistency with silicate spectra showed up as a deficit of silicate emissivity over the wavelength interval from 8 to 9 μm. The difficulty appeared to be endemic to all types of silicates, amorphous and hydrated. Moon rock was also considered for comparison, as were silicates that had been irradiated in the laboratory with a view to simulating space conditions. But nothing seemed to work!

A trick was invented by astronomers who thought they had to stick with a silicate hypothesis. The normal logic of identification through spectroscopy was inverted. The astronomical spectrum of the Trapezium Nebula was used to infer the opacity properties of the emitting material. A hypothetical material with such opacity properties but which has no counterpart in the real world was named an "astronomical silicate". We considered this practice unsatisfactory in the extreme and openly deplored it as being a "cheat". The fact remains that no known silicate can account for the Trapezium data as well as the organic polymeric models we discussed in the last chapter. There was what seemed to us to be unequivocal evidence of organic polymers existing on a vast scale throughout the galaxy. Models involving co-polymers–mixed chains of formaldehyde and other molecules as well as polymer mixtures resembling tars — were leading inexorably in one direction — life. What if the interstellar grains that I had begun investigating in 1960 were indeed connected with biology, with life itself? This question, with all its profound implications, represented such an assault on conventional thinking that I felt a compulsion to involve Fred at this stage.

I began a correspondence with Fred early in 1976 by first suggesting that the polymeric grains in molecular clouds could represent the beginnings of a process that may lead to life, thus permitting life to originate and evolve on a much bigger scale than had hitherto been contemplated. Fred's first reactions were far more cautious than I had anticipated. This point is worth stressing because there is a general perception that he embarked on our joint projects rashly and uncritically. Nothing can be further from the truth. He was exceedingly critical of every radical proposition that was put to him at each stage in our collaboration. He played the role of devil's advocate until he was convinced that there were overwhelming arguments to support the radical proposition. And this is exactly how a true scientist should proceed.

After many exchanges of letters and papers, Fred and I agreed in August 1976 that organic polymers in grains could undergo a Darwinian-style prebiological or prebiotic evolution in grain-on-grain collisions during the collapse of a molecular cloud. Organic tarry grains tend to be sticky and grain clumps would form by particles colliding and sticking together. Such grain clumps would also trap other organic molecules from the gas, and chemical transformations, sometimes assisted by ultraviolet light, would take place in the condensed state. In our first paper referring obliquely to biology entitled "Primitive grain clumps and organic compounds in carbonaceous chondrites" (*Nature* **264**, 45–46, 1976) we wrote:

> "*The formation of simple amino acids (e.g. glycine) is expected to take place in dense molecular clouds which may well be the cradle of life.*"

Even such a tentative proposition was regarded as outrageous heresy in 1976, although now in 2004 it is regarded as obvious.

Towards the latter part of 1976 there were other events that were to have a bearing on our story at a much later stage in its development. Even at the risk diverting from the main thread of my argument I shall report one such event here if only to keep my record in an approximate chronological order.

Space exploration was gathering momentum, as well as the search for life in the solar system. There were two unmanned missions to Mars: Viking 1 arrived at the red planet on 20 July 1976 and Viking 2 on 3 September 1976. Each mission involved an "orbiter" that was set in motion around the planet and a "lander" that actually set down at a chosen spot. The two Viking landers arrived safely at their chosen destinations, equipped with apparatus to make *in situ* tests for the presence of microbial life. These experiments were of vital importance and in many ways much more explicit in detection of life than any subsequently contemplated.

In one experiment (designated LR) a nutrient broth (with an isotope label on the carbon) of the sort that is normally used to culture a wide range of terrestrial microorganisms was contained in a sterilised flask, and the Martian soil was robotically added to it. It was found that the nutrient was taken up by the soil and gases (CO_2) were expelled from the flask as would be expected if bacteria were present. In another experiment the soil sample was heated to 75°C for 3 hours before it was added to the nutrient. This led to a diminution of gas release by 90%, but significantly the reaction was not completely stopped. Since some bacterial and fungal spores could survive temperatures of 75°C, the result of the second experiment was also consistent with a biological explanation, especially as the activity recovered gradually to its former higher value as time went on. The bacterial explanation gained further support from a third result, obtained by heating the soil sufficiently to kill microbial life entirely, when all activity was found to stop. However, yet another experiment in the Viking package proved initially more difficult to reconcile with biology. This experiment designated GCMS, sought to analyse the organic content of Martian soil using a mass spectrometer. Here the results were disappointingly negative for organic matter, indicating that if such matter existed it was present only in very small quantities.

The fact that the LR experiment was decisively positive and the GCMS experiment was negative posed a difficulty for NASA. The outcome was indefinite, and this is the way it should have been presented to the public. Yet NASA elected in 1976/77 to announce that the Viking experiments did not support the presence of life, and their statement that Mars was an intractably lifeless planet was given a great deal of publicity in 1976. It was their view that some other non-biological explanation had to be sought and would eventually be found. This has not happened to date and it has to be conceded that the balance of evidence is still positive for Viking landers to have detected life. But even retrospectively, there still remains an enormous resistance to admit this, and modern pronouncements by NASA relate mostly to the presence of past life in an epoch when rivers flowed over the surface of the planet. The facts have been clouded over, perhaps in an attempt to hide the fact that they plumped for an incorrect theory. And when big government science and the media make errors together, neither is anxious to be seen correcting itself, a sufficient reason why nothing much that is good can come of public funded science done in the glare of publicity. Science is a quiet, reflective activity, which cannot flourish in modern egalitarian or totalitarian societies.

The principal investigator of the 1976 biology experiment on Viking, Gill Levin, is known to me, and has revealed many things that are not generally known to the public. For instance his studies of a sequence of pictures taken by the Viking cameras over the duration of a Martian year, showed subtle shades of green appearing on the tops of rocks in the spring. These receded in the winter, suggesting the growth of lichen-type microbial life. Levin's views about these findings, however, did not endear him to the administration of NASA, and he parted company from them shortly afterwards to pursue his investigations independently. And after several decades of experimentation it has turned out that no non-biological model is feasible for explaining the positive results of the LR experiment, and moreover the lack of free organic matter in quantity as revealed by the GCMS experiment can easily be explained on the basis of a slow turn-over rate of microorganisms to be expected under the relatively inhospitable conditions that prevail on Mars.

The Mars probe Odyssey was launched in April 2001 to orbit the red planet and map its surface for hydrogen, water and minerals. Named *Odyssey* after Arthur C. Clarke's blockbuster novel, the probe obtained pictures that showed clear evidence of heavy frost or snow in many locations including the Viking landing sites. Snow or frost deposits were found to be seasonal, pointing to some kind of water cycle. But still NASA was claiming that contemporary life was highly improbable, despite the fact that on many sites on Earth where life has been discovered — in antarctic ice and at depths of 8 km below the Earth's surface — there are unquestionable parallels with Mars. Even as recently as 2004, the spacecraft "Mars Express" has obtained traces of methane and oxygen in the atmosphere that together would normally be interpreted as indicating biological activity. NASA seems loath to publicise such findings, perhaps in the hope that if they present the case very slowly, they would then be able to claim exclusive priority for the discovery of extraterrestrial life. Arthur C. Clarke summarised the four stages of the way new ideas are accepted into mainstream, institutional science:

- these ideas are crazy, do not waste my time with them;
- these ideas are possible, but are of no importance;
- these ideas — we said they were true all along;
- these ideas — we thought of them first.

I shall now return to the story of the composition of the interstellar dust. I had been plagued for a while by the thought that graphite grains, for which idea I had by now acquired a degree of fame, could not offer a rational explanation of the 2175 Å interstellar absorption band. The requirement of spherical graphite grains, all of one radius, $0.02\,\mu$m, was difficult if not impossible to defend. The central wavelength of the absorption due to graphite shifted away from the astronomically observed wavelength for particles in the shape of flake or whiskers, or if the radii of spheres departed significantly from $0.02\,\mu$m. Whilst continuing to fine-tune agreements with the 8–30 μm astronomical spectra in the infrared waveband with various types of organic polymers, it occurred to me to look critically at their ultraviolet spectra as well. In the summer of 1976 I discovered that a significant class of organic materials that possess C=C double bonds in their structures have ultraviolet spectra that peak near the required wavelength. Moreover from the available laboratory measurements in the ultraviolet we could calculate that only 10% of the available carbon in space was needed in the form of this material to give a 2175 Å band of the observed strength. This was my next subject of correspondence with Fred. Within a short while he agreed to our publication in *Astrophysics and Space Science* (**47**, L9–L13, 1977), the authors being myself, Fred Hoyle and Kashi Nandy of the Royal Observatory Edinburgh who was involved in observation of the interstellar extinction curve. This publication carried the first exposition of the idea that the ultraviolet extinction band was due to complex organic molecules.

The year 1977 was a vintage year for our collaboration, with no less than 6 papers being published in *Nature*. We were moving steadily in the direction of astrobiology, possibly 20 years ahead of our competitors. It should also be noted that we were remarkably successful in airing most of our work in the so-called "high impact" journals. The campaign of outright censorship had not yet begun.

In the world at large outside science there were momentous events afoot: in Britain the Queen celebrated her Golden Jubilee, Space sensationalism glutted Hollywood with movies like "Star Wars" and "Close Encounters of the Third Kind", almost as though the populace at large was getting ready for the arrival of ET in some form! In my native country, Sri Lanka, Junius Jayawardene becomes Prime Minister.

In January 1977 Fred made his first long visit to Cardiff in his capacity as Honorary Professor. During this week-long stay, he resided with us in Lisvane, as he would do on numerous subsequent occasions over the next decade and a half. It was always a pleasure to entertain Fred, and as the years went by, we became increasingly relaxed in his company. Priya is a charming and peerless hostess, and her efforts, particularly her cooking, were greatly appreciated by Fred. In fact she later came to joke with him about her contributions to man and science. Our friends, associates, and many others, were inevitably keen to meet Fred when he was in Cardiff, so places at our table were in high demand! We would invite small groups of guests to dinner, and Fred was charming company on such occasions, and his anecdotes were varied and surprising. I remember that once he nonchalantly related how, when he had finished the script for *A for Andromeda*, he toured the reparatory theatre in search of a "boyish young girl" to play the lead role. He was arrested by a performance of an unknown actress, Julie Christie, who agreed to play the part. By the second series, however, she had acquired such a degree of fame, that she was too expensive to re-hire.

As far as our collaboration was concerned Fred's visit in January 1977 was mostly taken up with the task of pouring over vast tomes of atlases of laboratory infrared spectra of organic polymers. Our comparison to be made was with the spectrum of the so-called BN object in the Orion Nebula. (See Fig. 4.) One of the defining features of organic polymeric dust would be a $3.4\,\mu$m feature of C–H bonds. In astronomical sources studied so far this feature shows up only as very weak shoulder on the wings of a much more prominent $3\,\mu$m absorption band. The $3\,\mu$m band was interpreted at the time as being due to H_2O ice condensed on grains within the dense molecular cloud around the BN object. Looking at the best available astronomical spectra Fred and I were the first to spot the CH absorption effect quite clearly, albeit as a shoulder on the longwave wing of an "ice" band. The task we set ourselves now was to find an organic polymer with absorptions at 2.9–3.1, 3.4 and 8–13 μm that was able to produce a BN-like spectrum.

This was not easy for the reason that for most of the organics we looked at, the CH stretching band at $3.4\,\mu$m tended to be

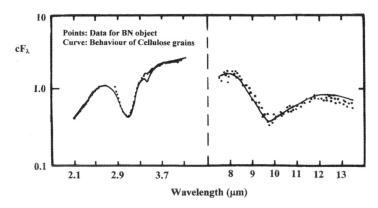

Fig. 4   The first argument for an interstellar biopolymer made in 1977. The infrared emission from the BN object in the Orion Nebula (points) matched to a model involving cellulose (curve).

much stronger than the OH stretching band at about $3\,\mu$m. We were, however, able to argue that there were both water-ice and organic polymers associated with the BN source. Because the mass absorption coefficient of water ice at $3.1\,\mu$m is a thousand times that of the $3.4\,\mu$m band for most organic polymers, a mass ratio of ice to organics of 1:1000 would produce the hint of a longwave shoulder in the $3.1\,\mu$m absorption band, exactly as it is observed in astronomy. In other words only a trace of water needs to be present in the dusty material around the BN object that had to be overwhelmingly organic. All these matters were sorted out during Fred's visit to Cardiff in January 1977.

We also began to look now at new ultraviolet spectra obtained for an extract of organic molecules from the Murchison meteorite. The data was supplied to us by A. Sakata in Tokyo, and we could immediately see here that the spectrum possessed a mid ultraviolet absorption feature near 2175 Å, very similar to the observed interstellar extinction feature to which we have referred earlier. We submitted a short letter to *Nature* making this point which confirmed our earlier contention of an organic carrier for the 2175 Å interstellar band. (A. Sakata *et al.*, *Nature* **266**, 241, 1977).

Our collaboration was rapidly gathering momentum. We entered a phase involving a brisk exchange of telephone calls, letters and graphs between Cockley Moor and Lisvane. We had eventually stumbled upon an organic material that captured our interest. It was an infrared spectrum of cellulose. The laboratory spectrum of cotton cellulose over the 2.5–30 μm waveband had shown even at a cursory glance most of the features required in order to explain astronomical spectra such as BN (see Fig. 4) and also the Trapezium nebula. The new agreements we obtained were somewhat better than those obtained earlier for polyformaldehyde as seen for instance in Fig. 3. Cellulose is of course the main component of the cell walls of plants and is by far the most abundant terrestrially occurring biopolymer. It is however logically at least the simplest biopolymer with an empirical formula $(H_2CO)_n$ just the same as for polyformaldehyde. It is a member of the most stable of a set of polymers known as the polysaccharides, which involves chains of various types of sugars, with a pyrolysis (heat destruction) temperature of 800 K. The advantage of cellulose is that it could exist in regions of relatively high temperature such as HII regions (ionised hydrogen), of which the Trapezium nebula in Orion is an example.

My own instinct at this stage was to consider interstellar polysaccharides as being derived from life and being indicative of fully-fledged microbial life in the Galaxy. Fred, however, was still inclined to tread more carefully. He conceded the existence of polysaccharides and similar molecules in interstellar space, but still sought non-biological or abiotic processes for their formation.

Both Fred and I spent a lot of time modelling a wide variety of astronomical observations over the 2–4 μm and 8–14 μm wavebands using the opacity measurements for cellulose. These efforts led to the publication of a series of papers (e.g. *Nature* **268**, 610–612, 1977; *Mon. not. R. Astron. Soc.* **181**, 51–55P, 1977). A particularly impressive fit with cellulose was calculated for the source OH26.5 + 0.6 where data was available over the very long wavelength interval from 2 to 40 μm (Fig. 5). Fred regarded this fit as the most decisive evidence for the validity of our point of view as it was in 1977.

The modelling of all these sources required a straightforward procedure known as a radiation transfer calculation which was a trivial job on the Cardiff University computer. But Fred insisted on checking

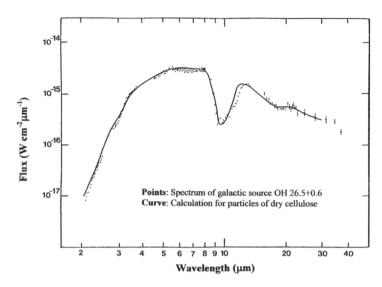

Fig. 5   The spectrum of the galactic infrared source OH26.5 + 0.6 over the very long waveband (2.5–40 μm) compared with a cellulose model (curve).

everything himself and he performed all his calculations on a simple hand-held programmable HP calculator that he carried everywhere. He maintained that being close to the logic of a calculation gave him a better insight into what was going on when it came to assessing the significance of the solutions that were obtained. Nowadays, school children use calculators and computers even for the simplest arithmetical operations, and this practice has undoubtedly led to a decline in the standards of numeracy, with multiplication tables being a thing of the past.

In parallel with our work on cellulose to model infrared spectra we also took another crack at the problem of understanding the interstellar ultraviolet absorption. This was an extension of the work on the Murchison meteorite extract that we referred to earlier. We argued that a class of double-ringed (bicyclic) aromatic compounds with the empirical formula $C_8H_6N_2$ (Quinozoline, Quinoxaline) provide an alternative explanation of the interstellar 2175 Å absorption band (*Nature* **270**, 323–324, 1977). A comparison of the astronomical data and the average absorption properties of these compounds is shown

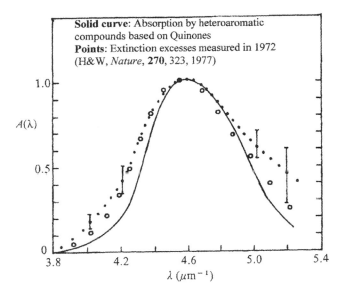

Fig. 6   The first argument for polyaromatic hydrocarbons in space in 1977. The average spectrum of compounds based on quinones (curve) compared with the 2200 Å interstellar absorption feature as it was known in the 1970's (points).

in Fig. 6. Our paper contains the first suggestion in the literature of interstellar aromatic compounds and accords unequivocal priority to Fred and myself for the idea of the existence of interstellar poly-aromatic hydrocarbons — a presence that is now taken for granted. This work was published in an accessible and credible journal, and so we found a lack of referencing to it inexcusable by any standards of moral propriety. We shall refer to similar matters in a later chapter.

To my mind the identification of polyaromatic hydrocarbons and cellulose-like polymers in interstellar space were tantamount to biology. Still preferring abiotic explanations Fred discussed a model involving mass flows from high mass O type stars, with mass loss rates amounting to about a hundredth of a solar mass per year. The outflowing matter then absorbs and re-emits the radiation from the star until an effective photosphere with a temperature (much cooler than the star) in the range 5000–10,000 K, forms at a certain distance. Beyond this point as the gas cools, one could argue that molecules form, including ring molecules and polysaccharide chains. This model was ingeniously conceived by Fred and worked out in great detail in an attempt still to justify an inorganic origin of biopolymers. Our joint paper on this subject was eventually published in *Nature* after a few unseemly squabbles with referees. This, incidentally was the first time we began to experience resistance from the conservative and hostile astronomical establishment. Seeing the direction in which we were moving, referee's reports of our papers were becoming increasingly more disdainful. There was a point when Fred's patience was tried by remarks that implied that the existence of organic polymers in space was impossible. They would all be destroyed by ultraviolet radiation, we were told, and indeed such comments were published by a number of distinguished astronomers.

At this point it should be put on record that two far-sighted individuals came to our rescue to ensure the rapid dissemination of our work. The first was Dr. C.W.L. Bevan, then Principal of University College, Cardiff, who chose to support our endeavours wholeheartedly in the belief that we were on the right track. The other was Zdenek Kopal who was the founding editor and Editor-in-Chief of the journal *Astrophysics and Space Science*, and at the time Professor of Astronomy at Manchester. Kopal offered us the opportunity of publishing our ideas in his journal, whilst Bevan agreed to the funding by University College Cardiff of a "Blue Preprint Series". The now extinct University College Cardiff Press was essentially placed at our disposal for publishing preprints as well as monographs on subjects of our choice. In this way our channels of communication with the scientific community were not interrupted as the result of the unexpected turn that our researches were to eventually take.

It was at about this time that we began to feel the need for a dedicated laboratory facility to obtain spectroscopic data as and when we required them. Whilst still exploring the role of interstellar polysaccharides we realised somewhat belatedly that the Biochemistry Department of University College Cardiff headed by Ken Dodgson had on its lecturing staff a leading expert on polysaccharides, Tony Olavesen. We wasted no time to seek his help to measure for us the spectra of a large number of different polysaccharides under conditions we considered were appropriate for making comparisons with astronomy. This work soon gave us confirmation of our earlier conclusions that were based only on the published spectra of cotton cellulose, and led to another paper that was published by *Nature* (F. Hoyle, A.H. Olavesen and N.C. Wickramasinghe, *Nature* **271**, 229–231, 1978).

But the end of our period of accommodation or toleration by the establishment was nigh. I felt we were walking into a kind of scientific exile. Even though we were still sticking firmly to prebiotic molecules in space a surge of resentment was beginning to surface. Fred's small hand-held programmable HP was being stretched beyond its limit for the ambitious calculations he was intending to do. Fred and I therefore decided to apply to the Science Research Council (SRC) for the purchase of a small computing facility to be housed at Cockley Moor for this purpose. The following reply came from S.T.G James of the SRC dated 4 November 1977:

> "*Dear Professor Hoyle,*
>
> *Your joint application with Professor Wickramasinghe for a grant a £10,540 in support of computational work on the identification of polysaccharides and related organic polymers in galactic infrared sources, was considered by Astronomy II Committee on 25 October.*
>
> *I regret to have to inform you that the Committee were not prepared to support the proposal outlined in your application. In arriving at this decision the Committee indicated that it was not at all clear that the problems mentioned in the application were ready for attack by computational methods...*".

This response was an indication that our work on interstellar prebiotic polymers was being firmly rejected by the astronomical establishment in 1977. In retrospect the irony here is that these ideas are now firmly in the mainstream, with little credit being recorded in our favour. The gross inequity of the rejection of our grant application made us take an unusual step. We wrote to the Prime Minister

James Callaghan, who was also an MP for Cardiff. A reply dated 7 February eventually came from the Secretary of the Department of Education and Science with the following pronouncement:

> *"It appears that this application has been properly assessed under the regular procedures of the SRC and that you have been given reasons for rejection. . . ".*

All this did not deter us from proceeding along the course upon which we had embarked. At this time Fred was firmly wedded to the idea of stellar mass flows and interstellar chemistry producing organic molecules. These were the complex structural building blocks of life that were supposed to become incorporated into the comets of the early solar system, within which an origin of life in the fashion of the Haldane–Oparin model took place. I was, however, personally inclined to flirt even more confidently with the radical idea of interstellar microbial life. Interstellar polymers being regarded as having a biological provenance.

Fred was in the US again when the next crucial step of our journey was taken. I received a letter from Dr. J. Brooks of the School of Chemistry at Bradford University including a spectrum of a complex organic biopolymer known as sporopollenin, which forms a major component of pollen and many spore walls. The spectrum had many of the features that were required to explain galactic infrared sources except for one fact. A 3.4 micron feature due to CH stretching was too prominent compared to the galactic sources we had seen. If, however, a thin layer of ice was condensed on particles comprised of sporopollenin, this difficulty would be rectified, thus making ice-coated bacterial spores a tenable model for interstellar dust. I promptly wrote to Fred in the United States soliciting his views and hopefully his support on this proposition. I soon followed this up with a draft of a joint paper for *Nature*. In a letter dated May 5 1977 written from Cornell University he wrote thus:

> *"Is the association of sporopollenin specific enough to support the final paragraph (speculating on the possibility of spores)? One might grant an interesting relation of the IR absorption*

obtained for the galactic centre or for galactic sources with the IR curve for sporopollenin, but the association may not require anything as complex biologically as sporopollenin itself...".

A few days later he reiterates the same point. In a letter dated May 9 also from Cornell he writes:

"My feeling, however, is that we cannot invert the situation toward a conclusion favouring a particular complex biological structure like sporopollenin:

Polymers → IR absorption	*O.K.*
IR absorption → Interstellar Biology"	*Not O.K.*

A drastically toned down version of the first draft I sent to Fred (that included references to interstellar biology) eventually appeared in October in *Nature* (Wickramasinghe, Hoyle, J. Books and G. Shaw, *Nature* **269**, 674–676, 1977).

Later that summer Fred was back home in Cumbria. Still using his hand held HP he was calculating spectra of a large number of galactic infrared sources with opacity data for polysaccharides. The results, in terms of the closeness of fits, were most impressive. We first issued our calculations as a preprint in our "Cardiff Blue Preprint Series", and later in two papers — a short version in *Nature* (*Nature* **268**, 610–612, 1977) and a fuller version in *Astrophysics and Space Science* (*Astrophys. Space Sci.* **53**, 489–505, 1978).

In July of this year, Phil Solomon, one of the pioneers in the detection of interstellar molecules using millimetre waves, visited me in Cardiff. He met with Fred both in Cardiff, and also in Mid-Wales for a workshop on Giant Molecular Clouds in the Galaxy that I had organised. The venue for the symposium was Gregynog Hall, a Conference Centre belonging to the University of Wales. With links dating back to the 12th century, Gregynog Hall is an imposing 19th century manor house set amidst 750 acres of gardens, woodland and farmland. It was bequeathed to the University in the 1960's by the Davies sisters, who had been avid art collectors, as well as generous patrons of the arts. Fred and Barbara, along with their two grandchildren, rented one of the many cottages in the grounds. Priya and I, and our two young children, did the same. The kids were all of a similar age and enjoyed having free reign to tear around the vast grounds together. The conference itself brought together a small group of scientists, and was hailed as a success. Fred and I had ample time between sessions to discuss our strategy for further development our ideas. Our joint presentation "Evidence of Interstellar Biochemicals", was published in *Giant Molecular Clouds in the Galaxy* (eds. P.M. Solomon and M.G. Edmunds, Pergamon Press, 1980).

In September 1977, we were abroad again, visiting the Astronomy Department of the University of Western Ontario, Canada. Fred was on his way to Caltech. Perhaps in need of a respite from the intensity of the life in space argument, Fred and I started pouring over a volume entitled "Cretaceous-Tertiary Extinctions and Possible Terrestrial and Extraterrestrial Causes" published in the previous year by the Canadian Museum of Natural History in Ottawa. Some 65 million years ago the dinosaurs and indeed all animals with body weights above 25 kg suddenly became extinct. We argued that this could be due to the interaction of the Earth with a cloud of porous cometary dust derived from the extended coma of a comet. The Earth's stratosphere will be dusted over in a way that two thirds of the light and incident energy from the sun will be blocked for several years whilst still permitting infrared radiation to leak out. The result would be semi-darkness for a decade, leading to the withering of foliage in trees and causing a severe interruption of food chains. Herbivorous creatures including dinosaurs would soon become extinct, and so would carnivores that feed on the herbivores. With rivers still continuing to run and some lakes remaining unfrozen, fresh water organisms would survive — their food chains depending on decaying vegetable matter, would take longer to disrupt than marine organisms dependent on phytoplankton. The seeds and nuts of land plants would also survive and small animals, including small mammals, living on nuts and seeds would also survive the dark and desolate years. We humans owe our descent through this ecological crisis to the survival of these small mammals. All these ideas were published in the form of a paper in *Astrophysics and Space Science* (*Astrophys. Space Sci.* **53**, 523–526, 1978).

Our ideas on the Cretaceous-Tertiary extinctions were similar (though not identical) to those of Alvarez and Alvarez that were published approximately 2 years after ours, and which have come to be more or less generally accepted. A direct hit by a comet seems to have occurred 65 million years ago, and the resulting crater has been discovered in the seabed of the Yuccatan Peninsula.

In addition to causing extinctions of species we also argued in our paper that cometary dusting over a more protracted period could trigger the onset of ice ages. This process was explored by us in greater detail in the late 1990's.

Chapter 11

Life from Comets and Pathogens from Space

Serendip is an ancient name for the island of Sri Lanka and was in use from the 4th Century AD. In the fairy-tale of Horace Walpole (1717–1797), *Three Princes of Serendip*, the heroes keep making delightful discoveries of things that they were not in quest of. This, according to the Oxford English Dictionary is the origin of the English word serendipity.

It is perhaps not a surprise that serendipitous events played a part in our collaboration at crucial stages of its development. This was certainly so for events that were to lead to a particular diversion that was to engage our attention almost obsessively for a full three decades. It is a curious tale that started with sniffles in the late summer of 1977, just prior to my trip to Canada. I had succumbed to an unseasonal bout of flu-like illness and this happened at a time when Fred and I were in a phase of brisk telephone exchanges over matters relating to the origin of life in comets. I was suddenly reminded of my mother's admonition in my childhood: "Don't go out in the rain or in evening mists or you'll get ill!" A similar belief is, of course, widely prevalent in the West. And in many cultures throughout the world comets have also been thought of as harbingers of pestilence and death. Could time-hallowed beliefs possibly have their basis in hard fact?

I telephoned Fred in Cockley Moor, on a depressingly grey afternoon in the late summer, with perhaps the most preposterous proposition I had ever made. Could the old wive's tales of diseases being connected with rain have possibly been right? Could viruses be present in comets, and could cometary viruses entering the Earth cause disease? Fred was caught unawares and I was greeted by a long silence at the other end, before he finally said he would think about it. I was indeed extremely surprised when he phoned back within hours of my call agreeing that this might be so. Fred was reminded of conversations he had had some years earlier with the Australian physicist E.G. (Taffy) Bowen. Bowen had discovered an amazing connection between freezing nuclei in rain clouds and the incidence of extraterrestrial particles.

What Taffy Bowen had showed was that there was a link between the frequency of freezing nuclei in tropospheric clouds and the occurrence of meteor showers. Meteor showers occur at regular times in the year whenever the Earth in its orbit crosses the trails of debris evaporated from short-period comets. If bacteria and viruses come in with cometary meteor showers they could, if they survive entry, act as freezing nuclei for rain. Raindrops laden with bacteria and viruses then become a distinct possibility. By analysing all the available data Bowen had reported in a paper in *Nature* (**177**, 1121, 1956) that as dust from meteor streams falls into the troposphere heavy falls of rain could be expected. This is found to happen some 30 days after the meteor dust first entered the very high atmosphere. It was discovered only much later that bacterial and fungal spores could indeed lower the freezing temperature of water and act as condensation nuclei for rain.

Because Fred was still hesitant in 1977 to accept the possibility of bacterial grains in interstellar space, his instant conversion to the idea of disease-causing bacteria and viruses coming from comets somewhat surprised me. We did not waste much time in issuing a preprint expounding our position, as we conceived it at the time, both on the origin of life in comets and on its consequences in regard to disease.

We began our exposition with the well-accepted premise that comets originated within our solar system as a by-product of the formation of the outer planets Uranus and Neptune. As I have previously discussed, a disc of gaseous material expelled from the equatorial regions of a fast rotating superluminous sun is the setting for the formation of all the planetary bodies. The inner planets condense as the disc cools to 1400 K thus removing metals and silicates from the expanding planetary disc. At the distances of the outer planets icy particles form which then coalesce into hundreds of billions of cometary bodies. This disc of comets remains immersed in the dense molecular cloud from the solar system condensed for hundreds of millions of years. Since this entire system undergoes random motions through the cloud, it would accrete through contact sweeping, tens of Earth masses of interstellar prebiotic molecules.

The aggregation of the outer planets took some 300 million years, after the inner planets had condensed into solid bodies. During this period cometary bodies would have been randomly deflected into elongated orbits that crossed the orbits of the inner planets. Several direct hits of the Earth by comets would have taken place early in its life and this led to the formation of the Earth's primitive atmosphere and oceans.

During close perihelion passages (closest approach to the Sun) of comets there would have been a tendency for selectively boiling off volatile materials in the nucleus. The surface temperature of the comet would oscillate between 300 K at perihelion and 100 K at aphelion (furthest distance to Sun) giving rise to periodic oscillations of melting and refreezing near the surface. We argued that such a process could lead over many millions of years to the elaboration of interstellar prebiotic molecules into more complex structures and eventually to life. A soft landing of a comet on the Earth about 4 billion years ago then could have started life. All this was to be the subject of our first book *Lifecloud: The Origin of Life in the Universe*, which was published by J.M. Dent and Sons in 1978.

But other attempts at starting life anew from prebiotic molecules, we argued, must continue on the surfaces of comets, particularly on "new" comets which have long periods of revolution around the sun. At irregular intervals the Earth must pick up debris from such comets in the form of micrometeorite showers; and at more regular and frequent intervals it picks up material from shorter period comets. The question naturally arises whether encounters with such material, containing fresh attempts at starting life, could sometimes have a deleterious effect on indigenous terrestrial biology. In answer to this question we began to argue that extraterrestrial encounters of this type may indeed lead to the injection of disease-causing bacteria and viruses on to the Earth.

By observing the pattern of appearance of diseases in the past we soon discovered that a strong *prima facie* case existed in support of our contention. The rather sudden appearance in the literature of references to particular diseases is significant in that it probably points to times of specific "invasions". Thus the first clear description of a disease resembling influenza is early in the 17th century AD. The common cold has no mention until about the 15th century AD. Descriptions of small pox and measles do not appear in a clearly recognisable form until about the 9th century AD. Furthermore, certain early plagues such as the plague of Athens of 429 BC, which is vividly detailed by the Greek historian Thucydides, do not seem to have an easily recognisable modern counterpart.

We noticed that epidemics and pandemics of fresh diseases, both in historical times as well as more recently, have almost without exception appeared suddenly and spread with phenomenal swiftness. The influenza pandemics of 1889–1890 and 1918–1919 both swept over vast areas of the globe in a matter of weeks. Such swiftness of spread, particularly in days prior to air travel, is difficult to understand if infection can pass only from person to person. Rather it is strongly suggestive of an extraterrestrial invasion over a global scale. We argued now that it is the primary cometary dust infection that is the most lethal, and that secondary person-to-person transmissions have a progressively reduced virulence, so resulting in a diminishing incidence of disease over a limited timescale.

On this picture, the pattern of incidence and propagation of any particular invasion is a somewhat complex matter. It depends, amongst other factors, upon the sizes of incoming micrometeoroids, the local physical characteristics of the atmosphere and on the distribution of global air currents. We might expect certain latitude belts on the Earth to be comparatively disease free, whilst others might be more prone to receiving space-borne pathogens. Also, depending on sizes of particles amongst other factors, some epidemics may be geographically localised while others may be global. In all cases any new epidemic must occur suddenly — when the Earth crosses the trail of infected cometary particles.

Fred and I became fully convinced that these ideas had to be substantively right. We both made a great effort to learn as much as we could about virology and infectious diseases, from text books as well as by talking to our medical colleagues at the University Hospital, particularly John Watkins, Professor of Medical Microbiology and Robert Mahler, Professor of Medicine. Fred's visits to Cardiff were now taken up with locating the right virologists, bacteriologists and historians who could enlighten us with facts about diseases past and present. We also made several visits together to the Central Virus Reference Laboratory in Colindale, London, and one memorable visit to meet Sir Christopher Andrewes (a virologist who played a role in isolating cold-type viruses) at the Common Cold Research Centre in Salisbury plains. Here we discovered that all attempts to infect volunteers with a common cold virus under controlled, epidemic-like conditions had been, up to that time, a failure.

We soon began to take a particular interest in one disease — influenza — because we discovered there were many puzzling aspects connected with its epidemiology (epidemic behaviour). Here was a disease that appeared to indicate the incidence of a virus or at least a trigger from the atmosphere. A distinguished epidemiologist Charles Creighton maintained as late as the final decade of the 19th century that influenza is not a transmissible disease. In his book *History of Epidemics in Britain* (Cambridge University Press, 1891) he discusses the influenza epidemics of 1833, 1837 and 1847, in which medical opinion held that populations living over considerable areas are affected almost simultaneously. Such evidence suggested to Creighton a "miasma" descending over the land rather than a disease which must spread itself from person to person. If one substitutes for "miasma" the phrase "viral invasion from space" it is a similar position to that which we arrived at in 1977. Creighton's hegemony, however, was short lived, and by the end of the 19th century, the concept of infectious disease caused by Earth-bound microorganisms had firmly taken root.

Strictly, the microbiological concept requires only that victims of the disease should acquire the virus from outside of themselves, which of course they would do if the infection came from the atmosphere. But such an idea seemed so much less plausible to scientific opinion than the concept of person-to-person transmission — this was not even considered as a hypothesis to be tested. It became an axiom.

Fred Hoyle and I became convinced that the transmission hypothesis could be tested if a new pandemic strain arose. Particles of viral size added to the Earth are stopped in the atmosphere at a height of about 30 km. Vertical descent of particles through the stratosphere and into the troposphere occurs mainly as a result of winter downdrafts that occur 6-months apart in the two hemispheres. This phenomenon offers a ready explanation of the fact that influenza and other respiratory viral diseases are distinctly seasonal in character.

It is commonly found that lingering mists in the winter season usher in a wave of flu-like disease. Since, as I have already said, bacteria and viruses can act as condensation nuclei around which water droplets form, this coincidence is not entirely unexpected. In situations where rain falls as large drops there is not much chance of direct inhalation of nucleating viruses, whereas misty weather provides the incoming virus with the best opportunity to become dispersed in aerosol form in a way that can easily be inhaled near ground level.

Among earlier evidence that pointed in this direction, the observations of Professor Magrassi in 1948 are worthy of note. The worldwide influenza epidemic of 1948 apparently first appeared in Sardinia. Magrassi, commenting on this epidemic, wrote:

> "We were able to verify the appearance of influenza in shepherd who were living for a long time alone, in open country, far from any inhabited centre; this occurred almost contemporaneously with the appearance of influenza in the nearest inhabited centres...".

One of the most striking features in this whole story is that the technology of human travel has had no effect whatsoever on the way that influenza spreads. If influenza is indeed spread by contact between people, one would expect the advent of air travel to have heralded great changes in the way the disease spreads across the world. Yet the spread of influenza in 1918, before air travel, was no faster, and no different from its spread in more recent times.

Probably the most disastrous influenza pandemic in recent history occurred in 1918–1919 and caused about 30 million deaths. After studying all the available information about the spread of influenza during this epidemic Dr. Louis Weinstein wrote:

> *"Although person-to-person spread occurred in local areas, the disease appeared on the same day in widely separated parts of the world on the one hand, but on the other hand, took days to weeks to spread relatively short distances. It was detected in Boston and Bombay on the same day, but took three weeks before it reached New York City, despite the fact that there was considerable travel between the two cities. It was present for the first time at Joliet in the State of Illinois four weeks after it was first detected in Chicago, the distance between those areas being only 38 miles..."*

As we were pondering on such matters serendipity intervened once again with a remarkable circumstance. The new pandemic we had talked about had become a reality. In November 1977 an outbreak of a strain of flu that had not been present in the human population for 20 years, was reported in the Far East region of the old Soviet Union. By the end of December the first cases were reported in Britain and by January the disease was rampant, most noticeably it was raging through the schools of England and Wales.

Our ideas on the emergence of life on comets and of a possible connection with plagues and pestilences had now advanced to a stage when I felt that Fred should try them out on an academic audience. Fred obligingly agreed to deliver a lecture entitled "Diseases from Outer Space" at Cardiff on January 18, 1978. Needless to say his lecture was a sell-out. There was only standing room in the main auditorium of Cardiff's largest lecture theatre at the time, the Sherman Theatre. His lecture, chaired incidentally by Principal Bevan, was greatly appreciated even though it stimulated controversy and a degree of hostility as well. The so-called Red Flu was around us everywhere, as was evident even in the coughing that was heard through Fred's lecture. During the days he spent in Cardiff, Fred and I developed a strategy to investigate this epidemic.

We saw this as an ideal opportunity for testing person-to-person transmission. School children under the age of 20 had not been exposed to the new virus in their lifetime, and so were all equally susceptible. We had the idea of using the school population as detectors of the new virus, rather in the same way that physicists use amplifying detectors to observe small fluxes of incident cosmic rays. Our first "experiment" was confined to schools, including boarding schools in South Wales and the South West region of England. We began with analyses of school absenteeism. For this purpose Priya and I did a mammoth circuit of schools within 30 miles of our house, and examined their attendance records. Our objective was to determine, for each individual school, the time dependence of absenteeism due to influenza during the epidemic, and also the overall attack rates that could be inferred. What surprised us most was the huge range in the attack rates, essentially 0% to over 80%, and this was determined only by the location of the school, indicating that the incidence of the virus was patchy on distance scales of a few kilometres or less. Data that Priya and I had collected for boarders in Howells School Llandaff, Cardiff and Atlantic College at St Donats gave a taste of more to come. There was clearly an effect connected with the houses where the children slept, some had high attack rates, others very low. Already, person-to-person transmission was beginning to look unlikely.

Fred and I published our preliminary findings in *New Scientist* (September 28, 1978) and started a more ambitious exploration of the school data throughout England and Wales. We circulated a questionnaire to all privately supported secondary schools seeking the following information:

1. Over-all attack rates amongst boarders and day pupils (separate data) during the recent influenza epidemic;

2. Day-to-day attack patterns as they are reflected in classroom absence and/or admissions to school sick bays;
3. The date of the peak of epidemic experienced in each school.

The results of our bigger survey only confirmed and strengthened those we had from the local schools. It is commonly stated that school boarders, members of the armed services in stations, and other closed communities, are highly susceptible to epidemic outbreaks of influenza. Replies to our questionnaire showed clearly that as far as school boarders are considered this is a myth. Our sample involved a total of more than 20,000 pupils with a total number of victims of some 8800. The distribution of attack rates in the schools showed that only three schools out of more than a hundred had the very high attack rates that have been claimed to be the norm.

If the virus responsible for the 8800 cases were passed from pupil to pupil, much more uniformity of behaviour would have been expected. We found evidence for great diversity, with a hint that the attack rate experienced by a particular school (or house within a school) depended on where it was located in relation to a general infall pattern of the virus. The details of this infall pattern are determined by local meteorological factors. The infall clearly displayed patchiness over a scale of tens of kilometers, the typical separation between the schools.

One particular school in our survey, Eton College, merits special attention. There were 1248 pupils distributed in a number of boarding houses and the total number of cases across the whole school was 441. The actual distribution of cases by house showed enormous heterogeneity. College house with a total population of 70 had only one case, compared with the expected value of 25 on the assumption of random distribution, in a person-to-person infection model. Here again we saw heterogeneity, but now on the scale of hundreds of metres. This entire distribution would be expected once in 10^{16} trials on the basis of person-to-person transmission. Clearly, if one looks objectively at all the facts, flu cannot be "catching" from person to person as our present-day scientific culture would have us to believe.

Further evidence against the standard dogmas of influenza transmission came from a study of influenza in Japan. The Japanese data was supplied by my friend Shin Yabushita, a contemporary of mine at Cambridge. Japan is remarkable in that several large areas have population densities in excess of 2000 per square km, whereas others have well under 200 per square km. Standards of monitoring and reporting disease are also uniformly efficient throughout most of Japan. The data showed extreme variability of attack rates between adjacent prefectures. Here again the patchiness of viral infall is over a distance scale of tens of kilometres, exactly as in the case of the Welsh and English schools we already saw.

We have already noted that the descent of the virus from the stratosphere to ground level depends on global circulation patterns of the atmosphere. This fact accounts for the otherwise mysterious phenomenon that epidemics of flu occur with a distinct seasonality in widely separated parts of the world.

Scientists who feel uncomfortable with the logical inferences drawn from a theory such as this often seek eye-catching one-line disproofs. For the case of influenza from space the often-stated disproof is that: "Viruses are host-specific, and so must have evolved in close proximity to the terrestrial species that they attack". In other words the critic says: "How could the incoming virus know ahead of its arrival the nature of the highly-evolved and specialised hosts that it may encounter?" Our answer is simple: "The virus could not of course anticipate us, but we, the host species, could anticipate the virus since we must have had a long and continuing exposure to viruses of a similar kind". It is also well known that viruses can on occasion add onto our genes and so viral DNA sequences serve as an invaluable store of evolutionary potential. Our genomes would be "made up" of cometary viruses according to this point of view. If the influenza virus, or one that is similar to it, forms part of our genetic heritage, then the so-called host specificity, or the apparent human-virus connection is instantly and elegantly explained.

According to our point of view, reservoirs of the causative agent for influenza are periodically resupplied at the very top of the Earth's atmosphere. Small particles, the sizes of viruses or smaller, tend to remain suspended high up in this region for long periods unless they are pulled down into the lower atmosphere. In high latitude countries, such breakthrough processes, where the upper and lower air becomes mixed, are seasonal and occur during the winter months. Thus a typical influenza season in a European country would occur between December and March. Frontal conditions with high wind, snow and rain effectively pull down viral pathogens close to ground level. The complex turbulence patterns of the lower air ultimately control the details of the attack at ground level, and determine why people at one place and at one time succumb, and why those in other places and at other times do not.

We also discovered in the literature that ozone measurements can be used to trace the mass movements of air in the stratosphere. Such measurements show a winter downdraft that is strongest over the latitude range from 40° to 60°. Taking advantage of this annual downdraft, individual viral particles incident on the atmosphere from space would therefore reach ground-level generally in temperate latitudes. Such locations on the globe would naturally emerge as places where upper respiratory infections are likely to be most prevalent, on the supposition of course that the Earth is smooth. The exceptionally high mountains of the Himalayas, rearing up through most of the height range to the stratosphere, introduce a large perturbation on the smooth condition, which may be expected to affect adversely this particular region of the Earth, especially regions lying downwind of the Himalayas, particularly China and South East Asia. In effect, the Himalayas are so high that they could act as a drain plug for most of the viruses incident on the atmosphere at latitude ∼30°N, the large population of China being inundated by this drainage effect, making China the quickest and worst affected region of the Earth. This could explain why new respiratory viruses such as Severe Acute Respiratory Syndrome (SARS) and new influenza viruses often make their first appearance in China. Concomitantly, other parts of the Earth at ∼30°N should be largely free of viral particles, unless it happens that such particles are incident as components within larger particles which fall fast under gravity.

A direct demonstration that the general winter downdraft in the stratosphere occurs strongly over the latitude range 40° to 60° was

given by M.I. Kalkstein (*Science* **137**, 645, 1962) in the last of
the series of atmospheric nuclear tests carried out in the middle of the
20th century. A radioactive tracer, Rh-102, was introduced into the
atmosphere at a height above 100 km and the fall-out of the tracer
was then measured year by year through airplane and balloon flights
at altitudes ~20 km. The tracer was found to take about a decade to
clear itself through repeated winter downdrafts, and this happened
mostly over the latitude belt 40–60 degrees.

The observed fall out patterns of a radioactive tracer agreed
closely with the well-known winter season of the viruses responsible
for the majority of upper respiratory infections in temperate latitudes. The time of a decade or so that was taken to clear the tracer
from the stratosphere also coincides with the average time of preva-
lence of any new influenza subtypes after it is first introduced. How
could all these facts be explained by the conventional ideas about
influenza? Fred Hoyle answered this question in his own inimitable
style in one of the preprints we put out at the time:

> "*Unfortunately so little has been understood of the mode of
> attack of the so-called infectious diseases that almost any form
> of hypothesis has come to be accepted in the past as an answer
> to questions of this sort. The truth is that, although the world
> may be extremely complex it is nevertheless extremely precise,
> with explanations every bit as clear-cut as that of the quantum
> mechanical analysis of the energy levels of the hydrogen atom
> being ultimately available for every phenomenon we observe*".

If one looks at the disc of the sun through specialised darkened glass
dark spots are often to be seen. These are vortices of gas on the
surface that are associated with strong magnetic fields and their
numbers vary enormously through a solar cycle which lasts about
11 years. I have referred earlier to my first scientific paper with Fred
in 1962 proposing a mechanism for periodic reversals of the sun's
polar magnetic field. This reversal is connected with the cumulative
effects of such sunspot activity. Sunspot numbers give a measure
of high-energy activity at the sun's surface, the peak numbers corresponding with frequent solar flares and the emissions of charged

particles that reach the Earth. Such activity on the sun is known to result in geomagnetic storms, ionospheric disturbances that interfere with radio communications, and most spectacularly the production of bright auroral displays (Northern Lights), the latter being caused by the streaming of charged particles from the sun moving along magnetic field lines.

A possible connection between peaks of sunspot activity and the times of influenza epidemics was first suggested by Edgar Hope-Simpson on the basis of data over a limited time span. We extended these comparisons over a longer time interval and found that a correlation of this kind did indeed hold. Peaks of solar activity will undoubtedly assist in the descent of charged molecular aggregates (including viruses) from the stratosphere to ground level. Thus according to our point of view serious influenza epidemics would follow such peaks, provided the culprit molecular aggregates were recently dispersed in the stratosphere from cometary meteor streams. With a more or less regular occurrence of such meteor showers, the limiting factor may be the intensity of solar activity, leading to coincidences between the timings of pandemics or major epidemics and sunspot peaks.

With all these considerations well in place on influenza as well as other epidemic diseases we now began to write our second book *Diseases from Space*, published in 1979 by J.M. Dent.

1. My father, who inspired me to start my journey, outside the Senate House, Cambridge, 1932.

2. With my brothers *en route* to England on a visit, 1946. Dayal, who figures in the story, is on the left.

3. Sunset at a beach near my home in Colombo, 1960.

4. With a group of Commonwealth Scholars on arrival in the UK, outside Houses of Parliament, 1960. (I am on the left.)

5. With Jayant Narlikar at Granchester, Cambs, 1961.

6. My trip to the Lake District with Fred, 1961.

7. Fred looks up to muse on the rain clouds, 1961.

8. Fred in animated conversation at a Varenna Summer School in 1961.

9. Mayo Greenberg has a captive audience in Troy, NY, 1965.

10. Enter Priya, Sri Lanka, 1966.

11. Dining on board ss Oriana, 1966.

12. Barbara Hoyle and Priya stop to drink the sap of a coconut, during Barbara's visit to Sri Lanka, 1969.

13. Barbara Hoyle, Viv Howes and Priya at "Land's End" Sri Lanka, 1969.

14. A family group with the Hoyles at Gregynog, 1977. *Left to Right*: Anil (my son), Liz (Fred's daughter), Fred, Geoff (Fred's son), Samantha (Fred's granddaughter), Nina Solomon, Jacqueline (Fred's granddaughter) and Kamala (my elder daughter).

15. My children and Fred's grandchildren with Matsuda. (Courtesy Anna Jones).

18. Priya signing copies of *Spicy and Delicious*, 1979.

16. Discussion at the black board, 1977.

19. With Fred and Bill Bevan at the launching of University College Cardiff Press, 1980.

17. With Fred in my office, Cardiff, 1978.

20. Fred and Bevan: Questions after a Public Lecture, 1980.

21. At the Arkansas Trial, with Priya and infant Janaki, 1982.

22. Presenting a copy of *Lifecloud* to President J. R. Jayawardene, President of Sri Lanka, 1989.

23. The Dyffryn Gardens Meeting on Alternative Cosmologies in 1989. *Left to Right*: Jayant, Geoff Burbidge, Fred, Chip Arp, and myself.

24. Discussions with Arthur C. Clarke in Sri Lanka, c 1995.

25. Jayant, Fred and Fred in bronze cast by Shiela Solomon, c 1997.

26. Fred looking again at *Space Travellers*, April, 2001.

Chapter 12

First Signs of Life

I had wondered for a while what astronomical discovery would encourage Fred to take the step from life emerging on a comet out of interstellar prebiotics to fully-fledged microbial life distributed throughout interstellar space. He had now been expounding the former thesis for nearly two years with conviction and much eloquence, and had he stayed with that our fortunes may have turned out differently. Our position of 1977, was after all the standard point of view of the scientific community in 2004, and we may now have been more openly acknowledged as its pioneers.

As a brief respite from our work on the influenza pandemic I decided upon impulse to take a closer look at the visual extinction curve of starlight, the way that starlight is extinguished by dust at visual wavelengths. This, it would be recalled, is where my researches into this whole subject began in 1961. There were many unresolved problems that had to be solved. Over the wavelength range from 7000 to 3000 Å, the extinction (dimming ratio on a logarithmic scale) was approximately inversely proportional to the wavelength, and this was the case in whatever direction one looked. Such an invariance of behaviour was difficult to reconcile with the grain models we had discussed so far involving mixtures of silicate or organic grains along with graphite and iron. In all these non-biological grain models the visual part of the extinction curve had to come mostly from the dielectric (non-absorbing) component of the mixture, and to get the correct shape of the extinction curve, grain radii had to be fairly sharply fixed. This condition could not be relaxed as long as one stayed with grain materials such as ice, organics or silicates which have visual refractive index values $n = 1.3, 1.5, 1.6$, respectively. In other words, the solutions obtained so far in all these cases are highly parameter sensitive, and therefore not very satisfactory.

I discovered that one way to relax the size constraint is to reduce the value of the refractive index, n, below 1.3. A value closer to $n = 1.15$ would be nearly optimal from this point of view, for the reason that the extinction efficiency of spheres with this refractive index can remain closely proportional to inverse wavelength over a wide range of wavelengths. But what material could possess such a low value of n? When I searched the relevant handbooks of physical constants it soon became obvious that, with the exception of solid hydrogen, there was no other homogeneous solid material that had the desired property. Solid hydrogen was of course ruled out because we had found earlier that it could not survive under normal interstellar conditions. It was at this point I recalled a collaboration I had been engaged in with Craig F. Bohren (C.F. Bohren and N.C. Wickramasinghe, *Astrophys. Space Sci.* **50**, 461–472, 1977) on the optical properties of heterogeneous grains. That is to say grains comprised of a mixture of two different particulate types welded together. I immediately turned to this earlier work, noting that a particle made of organic material with bulk refractive index $n = 1.5$ could have an average refractive index of 1.2 or less if it contained minute vacuum cavities. For instance an aerogel would have such a refractive index.

When I pointed this out to Fred his interest in the possibility of bacterial grains in space was immediately aroused. We got to work on discussing ways in which bacteria would freeze-dry in space leading to the production of vacuum cavities within them. We started with a vegetative bacterial cell with the following typical constitution: Organic material 20%, bound water 20%, free water 60%. Freeze-drying in a vacuum, such as in outer space, would maintain the cell wall in tact and also retain the interior organic content and bound water, while free water will escape and lead to the production of vacuum cavities. The average refractive index (by a straightforward averaging process) could then be shown to be

$$n = 0.2 \times 1.5 + 0.2 \times 1.3 + 0.6 \times 1.0 = 1.16.$$

For freeze-dried interstellar bacteria we concluded that it would be reasonable to assume an average refractive index in the range 1.15–1.16.

The next thing we needed to know was the size distribution that would be representative for actual bacteria; and for this, *Bergey's Manual of Determinative Bacteriology* was consulted. The best data we could find related to spore-forming bacteria, which gave the count of species in various ranges of sizes in the form of a histogram. By taking freeze-dried interstellar bacteria to have this particular size distribution that was appropriate for terrestrial spore forming bacteria, we had a situation totally different from anything we had experienced previously. For now there were no parameters at all to be fitted. The extinction behaviour of the entire ensemble of grains became immediately amenable to calculation using the computer program I had developed and used over many years. The result was staggering: we had discovered a perfect fit to the average interstellar extinction over the visual waveband with just the one assumption *that interstellar grains were freeze-dried bacteria.* Figure 7 shows this fit

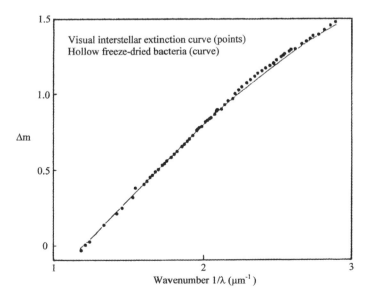

Fig. 7 Interstellar extinction observations over the visible waveband (points) compared with the theoretical extinction curve calculated for hollow freeze dried bacteria (curve).

and the solution, it should be stressed, is essentially parameter free. Referring to such matters in a public lecture Fred said:

> "*I have a particularly ferocious dislike of model calculations which achieve a tolerable correspondence with observation through assigning more-or-less arbitrary values to many parameters introduced by the investigator. Although in the past I strayed from the strict and narrow path of virtue by engaging in this self-deceiving practice, nowadays I refuse to spend five minutes on such parameter-fitting exercises. By a like token I am heavily impressed by agreements achieved by models without arbitrary parameters, and I am doubly impressed by agreements in which such calculations precede in time what the observations eventually turn out to be. In short, successful model calculations without free parameters that are predictive are for me doubly impressive.*"

Fred's adherence to this philosophy was evident throughout the four decades of my collaboration with him. I believe it was the extraordinary nature of this fit (Fig. 7) coming as it did after two decades of failure that won Fred over to the case for bacterial grains. And then there was no turning back. For the composition of interstellar dust we had progressed cautiously and in slow stages through a sequence of options: graphite, organic polymers, and then to complex biopolymers such as the polysaccharides. These organic polymeric particles that were in evidence everywhere in the galaxy had an average size comparable to a bacterium, and an average refractive index appropriate for a freeze-dried bacterium. Good fits to infrared, visual and infrared data were possible on the assumption of bacteria-like particles. Could all this be somehow explained without invoking biology? Of course this question has to be explored.

After weeks of fumbling through a sequence of ideas, all of which were proving to be woefully inadequate, we alighted on the most promising hypothesis. In the gigantic clouds of interstellar dust could we be witnessing no less than the dissemination of biology itself? Could the interstellar medium be choc-a-bloc, not simply with the building blocks of life arranged into prebiotic molecules or even pro-tocells, but with the end products of the living process itself? And this would then be required to happen on an unimaginably vast scale. At the end of a long run of frenzied telephone calls between Cardiff and Cockley Moor we decided that was it! Interstellar grains must surely be bacteria-albeit freeze-dried, perhaps mostly dead! At the very least this was a hypothesis that had to be explored.

Nobody takes lightly the prospect of walking into exile, albeit scientific exile, and moreover one that is self imposed. Yet by the Spring of 1977 it appeared to both of us that we had no option but to do so, carrying the heavy burden of not one but two scientific heresies. The heresy of diseases from space and now the heresy of microbial life in interstellar space. No sooner than we discovered the solution in Fig. 7 (which is further elaborated in Fig. 8) we wrote up our results, distributed them to colleagues in the form of a Blue Cardiff Preprint at the beginning of April 1979, and at the same time sent off a technical paper entitled "On the nature of interstellar grains" for publication in *Astrophysics and Space Science* (**66**, 77–90, 1979).

The immense power of bacterial replication is worth careful note at this point. Given appropriate conditions for replication, a typical doubling time for bacteria would be two to three hours. Continuing to supply nutrients, a single initial bacterium would generate some 2^{40} offspring in 4 days, yielding a culture with the size of a cube of sugar. Continuing for a further 4 days and the culture, now containing 2^{80} bacteria, would have the size of a village pond. Another 4 days and the resulting 2^{120} bacteria would have the scale of the Pacific Ocean. Yet another 4 days and the 2^{160} bacteria would be comparable in mass to a molecular cloud like the Orion Nebula. And 4 days more still for a total time since the beginning of 20 days, and the bacterial mass would be that of a million galaxies.[1] No abiotic process remotely matches this *potential* replication power of a biological template. Once the immense quantity of organic material in

---

[1] As the readers would know, this is not possible since a "continuous supply of nutrient" to such a massive colony is inconceivable!

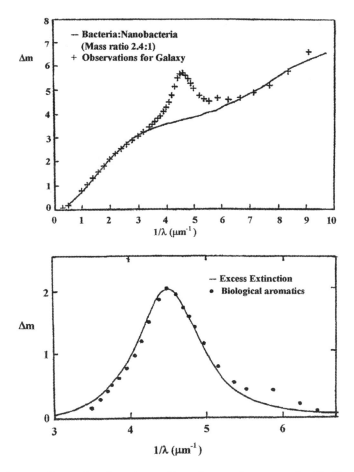

Fig. 8   Interstellar extinction observations (crosses) compared with the scattering behaviour of freeze-dried bacteria and nanobacteria (curve, upper panel). Lower panel shows how the excess extinction over the scattering background in the 3300 Å–1500 Å waveband is matched by the properties of biological aromatic molecules.

the interstellar material has been appreciated, a biological origin for it becomes a necessary conclusion.

But where are astronomical locations where conditions for replication of bacteria can be found? Certainly not in the cold depths of space, where microbes could merely remain in a freeze-dried dormant state. Planets like the Earth would provide too small a total mass of carbonaceous material to make any cosmic impact. It is therefore to comets we turned, arguing that comets are the sources of biological particles in interstellar clouds. An individual comet is a rather insubstantial object. But our solar system possesses so many of them, perhaps more than a hundred billion of them, that in total mass they equal the combined masses of the outer planets Uranus and Neptune, about 10^{29} grams. If all the dwarf stars in our galaxy are similarly endowed with comets, then the total mass of all the comets in our galaxy, with its 10^{11} dwarf stars, turns out to be some 10^{40} grams, which is just the amount of all the interstellar organic particles that are present in the dust clouds within the galaxy.

How would microorganisms be generated within comets, and then how could they get out of comets? We know as a matter of fact that comets do eject organic particles, typically at a rate of a million or more tons a day when they visit the inner regions of the solar system. We argued that comets when they are formed incorporate a interstellar bacterial particles, from which only a fraction 10^{-22} needs to retain viability for a regeneration process to operate. For at least a million years, at the time of their origin, comets have liquid cores due to radioactive heat sources such as ^{26}Al which are also incorporated within them. Within a very brief period as described above sequential doublings of viable microorganisms would lead to a sizable portion of the cometary core being converted into biomaterial. When the comets re-freeze this amplified microbial material is also frozen in, only to be released when they become periodically warmed up in the inner solar system. Some of this bacterial matter may reach the inner planets, which they can seed with life, some of it is expelled back into interstellar space.

Our point of view requires that bacteria must be space-hardy which recent research has shown is the case. On the whole microbiological research of the past 20 years has shown that bacteria and other microorganisms are indeed remarkably space-hardy (J. Postgate, *The Outer Reaches of Life*, Cambridge University Press, 1994). Microorganisms known as thermophiles and hyperthermophiles are present at temperatures above boiling point in oceanic thermal vents. Entire ecologies of microorganisms are present in the frozen wastes of the Antarctic ices. A formidable total mass of microbes exists in the depths of the Earth's crust, some 8 kilometres below the surface, greater than the biomass at the surface (T. Gold, *Proc. Natl. Acad. Sci.* **89**, 6045–6049, 1992). A species of phototropic sulfur bacterium has been recently recovered from the Black Sea that can perform photosynthesis at exceedingly low light levels, approaching near total darkness (J. Overmann, H. Cyoionka and N. Pfennig, *Limnol. Oceanogr.* **33**(1), 150–155, 1992) There are bacteria (e.g. *Deinococcus radiodurans*) that thrive the cores of nuclear reactors. Such bacteria perform the amazing feat of using an enzyme system to repair DNA damage, in cases where it is estimated that the DNA experienced as many as a million breaks in its helical structure.

There is scarcely any set of conditions prevailing on Earth, no matter how extreme, that is incapable of harbouring some type of microbial life. As for ultraviolet damage under space conditions, this is very easily shielded against. A carbonaceous coating of only a few microns thick provides essentially a total shielding against ultraviolet light, and there are several modern experiments that have demonstrated precisely that. Next, let us note that many types of microorganisms are not really killed by ultraviolet light, they are only deactivated. And this happens through a shifting of certain chemical bonds contained in the genetic structures of the organisms, without destroying the genetic arrangements themselves. And this permits the original properties to be recovered once the ultraviolet radiation has been shut off. Furthermore, we know that microorganisms that are normally sensitive to ultraviolet light can, through repeated exposures, be made just as insensitive as the more resistant kinds — yet another unearthly property.

All this was good news for our theory, but as with every good theory we had to be able to demonstrate its predictive capabilities. We had to make predictions that could be verified or falsified by future experiment or observation. This was yet a couple of years in the future. In the years 1979 and 1980 when these crucial steps were taken in our collaboration, the world outside was recording a few dramatic events. The Shah of Iran was driven into exile; there were referenda for home rule in Scotland and Wales, the Scots say yes, the Welsh say no; Margaret Thatcher becomes Britain's first woman Prime Minister. More relevant to our story the US spacecraft Voyager I returns the most dramatic pictures of Jupiter, Saturn and Uranus. Rings are discovered around both Jupiter and Uranus, leading us, Fred and myself, to speculate whether these contained bacterial particles.

On October 29th–31st 1980, a conference with the title "Comets and the Origin of Life" was organized by Cyril Ponnamperuma at the University of Maryland. Ponnamperuma, a Sri Lankan born chemist, always had his feet firmly planted in the opposite camp from us. He had made important laboratory studies on the abiotic synthesis of organic molecules, including sugars and nucleotide bases under conditions similar to those used in the famous Urey–Miller experiment of the 1950's. All such experiments are of course a very far cry from the generation of life, but they are often presented as being a significant step in that direction. Ponnamperuma, at the time was Director of the Laboratory of Chemical Evolution at the University of Maryland, and clearly was an opponent of the ideas being discussed by Fred and myself. I was, however, invited by him to present a joint paper with Fred entitled "Comets — a vehicle for panspermia," an invitation that I accepted with some trepidation.

Cyril had rounded up just about all of our potential adversaries, including J. Oro and of course J. Mayo Greenberg. Almost everyone at the meeting delivered a polemic denying any possibility of comets carrying life, or even life's building blocks in some instances. Greenberg made his first coherent claim to rival our 1974 ideas of polymeric grains with his own title "Chemical Evolution of Interstellar Dust — A source of prebiotic molecules". He was unfortunately a few years too late to steal any priority from Fred and myself. The conference volume, including our paper, was published by D. Reidel Co. in 1981 (*Comets and the Origin of Life* — ed. C. Ponnamperuma). Nothing that I heard at this meeting made me weaken my resolve to continue our search for an origin of life on a grand cosmic scale.

Chapter 13

Bacterial Dust Predictions Verified

The conference in Maryland gave me the first direct experience of the power of the opposition that was rallied against us. I formed the impression that nothing will be spared in an attempt to denigrate our work or to stifle its further progress. The hostility was getting steadily worse as the evidence grew in strength. Fred put this cogently in a piece he wrote for our joint book *From Grains to Bacteria* (University College Press, 1984):

> *"It is necessary to come now to a curious situation that we think will eventually be of interest to students of scientific methodology. The more precise the correspondences we calculated between our models and the observation, the greater was the measure of opposition we received from individuals, from journals and from funding agencies like SERC. The introduction of polysaccharides, because of their biological association apparently, became a signal for papers to be turned down by journals, and even for the most modest grant applications to be thrown back in our faces by SERC, an organisation which in a time span of no more than a decade and a half managed to go from a beginning of rich promise to one of the outstanding Gilbert-and-Sullivan operattas of the twentieth century ...".*

Such remarks coming from one who was not so long ago Chairman of a SERC committee, had a sense of cynicism as well as disillusionment. It was during the period 1979–1981 that we encountered the worst of the opposition that Fred describes. This entailed a succession of socially welcomed one line disproofs that came to be offered. For example it was often stated that organic matter could be expected to have many C–H linkages, with absorption due to stretching in the 3.3–3.5 μm wavelength range in the infrared region of the spectrum. Claims were published in refereed journals that this absorption had been looked for and had not been found. Therefore, it was argued, interstellar grains could not be of an organic nature as we had claimed. This was an argument which those with an anti-Copernican opposition to the thought that life might be cosmic may have found appealing, but unfortunately it turned out to be false.

The following quote from W.W. Duley and D.A. Williams published in *Nature* (**277**, 4 January, 1979) (one of many similar quotes) illustrates this state of mind:

> "*We conclude that no spectroscopic evidence exists to support the contention that much of interstellar dust consists of organic materials. While the presence of trace quantities of organic compounds on grains inside very dense clouds cannot be excluded by available data, the absence of any observation of a 3.3–3.4 μm absorption band even in objects with* $A_v = 50$ *mag strongly suggests that organic grains constitute at most a minor component of interstellar dust . . .*"

How wrong this has been because, in 2004, there is not the slightest dissent from the view that the bulk of interstellar dust is organic. The illusion for some in 1979 came from neglecting to take proper account of the band strength of the C–H linkages in actual biopolymers which we knew from the measurements of Tony Olavesen in Cardiff to be weak, with a mass absorption coefficient of not more than about $1000 \, \text{cm}^2 \, \text{g}^{-1}$. That is to say, the expected absorption in the 3.3–3.5 μm band could not exceed about 2% of the visual extinction. Such a small effect was not detectable even for the most highly reddened stars with the equipment and instruments available in 1979, so the claims made by David Williams and others were incorrect.

We had already shown that the C–H stretching organic band was present as a minor shoulder in the wings of 3 μm bands in galactic infrared sources, and the fits of polysaccharide models to such sources were exceedingly close. Also the extinction curve of starlight for a biological model matched the astronomical data with uncanny precision.

With the emergence of Cardiff as a new centre for astronomy, the Royal Astronomical Society (RAS) asked me to arrange their annual out of town meeting for 1980 in Cardiff. The then RAS President, Professor M.J. Seaton of University College London also requested Fred Hoyle to give an evening public lecture on this occasion. On April 15, 1980 Fred delivered his lecture with the title "The Relation of Biology to Astronomy" to a saturated hall at University College Cardiff, in which he presented an eloquent exposition of our position on the nature of interstellar grains. His frontal assault on conventional theories of biological evolution on the Earth did not win him much support, but the case against such theories was outspokenly presented. For instance:

> "*What may be the biggest biological myth of all holds that evolution by natural selection explains the origin of the phyla, classes and orders of plants and animals. There are certainly plenty of examples of minor evolutionary changes caused by natural selection, and on the evidence of these minor changes the major changes are assumed to be similarly caused. The assumption became dogma, and then in many people's eyes the dogma became fact...*"

This was a statement of intent that we were soon to take on the entire biological establishment in re-evaluating the evolution of life in a closed box setting, and opening up the process to the wider universe. This was to happen in the months that followed leading eventually to the publication of a book (*Evolution from Space*) and several short booklets.

I suspect that the majority of the audience in Cardiff, who were astronomers, did not take much interest in biology at this stage even though it was the dominion of astronomy that was being re-evaluated and enlarged. Thus he went on:

> "*Astronomers have become accustomed to thinking of the external Universe in the words of Macbeth, as being 'full of sound and fury, signifying nothing'. Can we seriously believe*

that anything as subtle as biology could have gained a toe-
hold in a world signifying nothing? I pondered this question
for a long time before arriving at a strange answer to it. If
the astronomer's world of fury is really in control, then the
prospects for biology would be poor. But what if it is really
biology which controls the astronomer's world?"

Unwittingly perhaps Fred laid the foundations for the modern discipline of astrobiology, a subject that is becoming increasingly popular
these days, with his concluding remarks on April 15, 1980:

"Microbiology may be said to have had its beginnings in the
nineteen-forties. A new world of the most astonishing com-
plexity began then to be revealed. In retrospect I find it remark-
able that microbiologists did not at once recognise that the
world into which they had penetrated had of necessity to be of
a cosmic order. I suspect that the cosmic quality of microbiol-
ogy will seem as obvious to future generations as the Sun being
the centre of our solar system seems obvious to the present
generation."

In the days immediately following Fred stayed with us as usual. We
took this opportunity to identify many of the loose ends of our theory
that had to be dealt with. The immediate question now was: what
further checks were there to be made for the thesis of interstellar
bacteria? Predictions of the behaviour of a bacterial model at infrared
wavelengths should be made, and these might then be looked for in
astronomy. But this required experimental work was preferable to be
done ahead of astronomical observations being made.

Here was our next instance of the intervention of serendipity. My
brother D.T. Wickramasinghe, (Dayal), Professor of Mathematics
at the Australian National University in Canberra, was also an
astronomer (trained at Fred Hoyle's IOTA!) and frequently used the
3.9 metre Anglo-Australian Telescope (AAT). The AAT, the completion of which was in large measure due to Fred Hoyle's untiring
efforts in the early 1970's, happened to be equipped with just the
right instruments to look for a signature of interstellar bacteria.

Shortly after the release of our April 1979 preprint on the scattering properties of bacteria, Dayal visited Cardiff to spend some time with our family. Dayal's visit happened to coincide with a time when Fred Hoyle was also in Cardiff. We naturally got talking about matters relating to interstellar bacteria. Dayal asked: "What do you think can be done at the telescope to prove or disprove your theory?" to which we promptly replied that he could use the infrared spectrometers on the AAT to look at infrared sources near the wavelength of 3.4 μm in greater detail than ever before. A very long path length through the galaxy was needed to have any hope of detecting such an effect unambiguously. The longest feasible path length through interstellar dust that existed within our own galaxy was defined by the distance from the Earth to the centre of the Galaxy. There were several sources of infrared radiation located near the galactic centre that could serve as search lights for interstellar bacteria. Dayal was doubtful that he would be allocated observing time if he applied for such time specifically to do this project. The general consensus then was that life in space could not be regarded as respectable science! Dayal overcame this difficulty, however. Although honesty is the best policy, it often pays handsomely to be economical with the truth in a world of dubious morality. The deceit involved applying for telescope time to do a quite different project, and then illicitly using part of the time to look for the signature of organic matter.

In February and April 1980, Dayal, collaborating with D.A. Allen, obtained the first spectra of a source known as GC-IRS7 which showed a broad absorption feature centred at about 3.4 μm (D.T. Wickramasinghe and D.A. Allen, *Nature* **287**, 518, 1980). When Dayal's published spectrum was examined we found that it agreed in a general way with the spectrum of a bacterium that we had found in the published literature. But at this time neither the wavelength definition of the astronomical spectrum nor the laboratory bacterial spectrum was good enough to make a strong case for interstellar bacteria. However, even from this early observation we were able to check that the overwhelming bulk of interstellar dust must have a complex organic composition, in flagrant contradiction with the statements of Duley and Williams. Hitherto the infrared data showing organic polymers in space had related only to localised dust clouds such as the Trapezium nebula. More general infrared absorption by bacteria-like grains was a possibility at that point, but now it was beyond any doubt.

It was precisely at this moment that Shirwan Al-Mufti, a practical man, the son of an Iraqi Army General, approached me to become a research student at Cardiff. Here was our chance to get the required laboratory work done. We approached Tony Olavesen at the Biochemistry Department at Cardiff and arranged for Al-Mufti to be given bench space and laboratory facilities in that Department to undertake spectroscopic studies of biological samples. The purchase of a modest amount of equipment that was needed was immediately authorised by Principal Bevan, and our experimental project got under way. Al-Mufti set about executing this task with military-style efficiency, and the experiments began to yield important results in the first few months of 1981.

Al-Mufti's experiments involved desiccating bacteria, such as the common organism *E. coli*, in an oven in the absence of air and measuring, as accurately as possible, the manner in which light at infrared wavelengths is absorbed. The normal technique for doing such a measurement involved embedding the bacteria in discs of compressed potassium bromide and shining a beam of infrared light through them. The standard techniques had to be adapted only slightly. There was the need to match the interstellar environment which involved desiccation, and the spectrometer that was used had to be calibrated with greater care than a chemist would normally exercise. When all this was done it turned out that a highly specific absorption pattern emerged over the 3.3–3.6 μm wavelength region, and this pattern was found to be independent of the type of microorganism that was looked at, (Fig. 9). Thus whether we looked at *E. coli* or dried yeast cells it did not matter. This invariance came as a great surprise. Our newly discovered invariant spectral signature was a property of the detailed way in which carbon and hydrogen linkages were distributed in biological systems.

The detailed structure of the pattern of absorption displayed in the upper panel of Fig. 10 is what was required to show up in

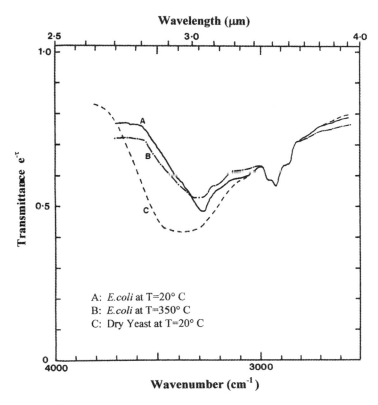

Fig. 9 Normalised transmittance properties of *E. coli* at 20 and 350 degrees Celsius and of dry yeast at 20 degrees Celsius showing an almost invariant absorption profile over the 3.3–3.7 μm waveband.

astronomy if our ideas were right. The original astronomical spectrum of GC-IRS7 of Dayal and Allen was not of high enough wavelength resolution to verify this prediction unequivocally. If astronomy turned up later with a totally different profile, then our model will have been falsified. Because we knew the exact amount of bacteria in the laboratory sample causing the absorption in Fig. 10 we could also determine how strong the absorption by bacterial dust should be at any particular wavelength. From our measurements it turned out that this absorption band was intrinsically weak. This means that a very long pathlength through interstellar dust was needed to get a strong positive signal confirming the presence of bacteria.

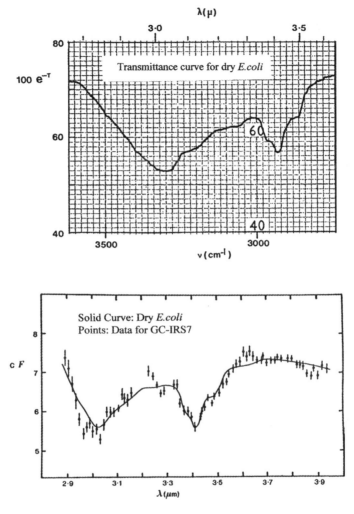

Fig. 10   Upper panel: Laboratory transmittance curve of dry *E. coli* measured by Shirwan Al-Mufti. Lower panel: Calculated behaviour of *E. coli* (curve) compared with the astronomical data for GG-IRS7 obtained by Dayal Wickramasinghe and David Allen (points with error bars) in 1981.

The observations that were to mark a crucial turning point in this entire story were carried out by Dayal and D.A. Allen at the AAT in May 1981. The new observations were of a far superior quality because a new generation of spectrometers were used. Dayal sent us his raw data by fax to compare with our new laboratory spectra which had been obtained just months earlier by Al-Mufti in March and April of the same year. After an hour or so of straightforward calculations we were able to overlay the astronomical spectrum over the detailed predictions of the bacterial model. This led to perhaps the most dramatic confirmation of the bacterial model of cosmic dust, as can be seen in the lower frame of Fig. 10. This for us was the best possible confirmation of our model, particularly because the experimental data in the comparison was obtained before the final astronomical observations became available. The agreement between a set of data points and a predicted curve as seen in Fig. 10 is normally regarded as a consistency check of the model on which the curve is based. Coming as it did after earlier fits of the same model to other sets of data, as we discussed earlier, the closeness of this particular fit would be hailed as a triumph of the model. But in our case, since the model of bacterial grains runs counter to a major paradigm in science, the situation was otherwise.

We were told by a number of chemists with experience of infrared spectroscopy that a curve like that of Fig. 10 could be obtained from non-biologically derived organic materials in many ways. Since by now we had examined without success literally hundreds of infrared spectra of organic compounds, we did not believe this claim. Consequently, we asked the chemists in question that an explicit example be produced. But it never was, with the exception perhaps that some expensive laboratory experiments, involving carefully controlled irradiation of inorganic mixtures, were claimed to yield undefined "organic residues" that may possess some of the desired properties.

We all to some degree tend to think when we run into an apparently absurd proposal, that any form of opposition to it will suffice. Because in the end what is absurd will be proved to be absurd, so that whatever we say in our opposition will eventually come out on top in the argument. Experience shows that when there are no good observations in favour of what seems absurd, this easily adopted policy is usually fairly safe. But in the face of good observations and in the face of many of them it is a highly questionable strategy.

So just how good is the agreement displayed in Fig. 10, particularly when it is taken with Fig. 7? By the early 1980's, when we attempted to answer this question, we had two decades of experience behind us in evaluating such correspondences. Expressed quite simply, we had never seen anything nearly so good. Yet even so was it all good enough to sustain a belief in such an apparently outlandish idea?

We recognised that to a person who had not followed the problem over the years the absorption characteristics of the interstellar grains in Fig. 10 might have seemed inadequate support for such a far-reaching hypothesis. And doubtlessly this was the way it appeared to many. But to us who had been involved over almost two decades it seemed otherwise, and we think it fair to add that time has supported our point of view here. Nobody among the critics of the 1980's has managed to find an alternative theory of the absorption characteristics of the grains to equal the success of the bacterial hypothesis.

And of the correspondences seen in Fig. 7 as well as Fig. 10 was not all that was there. Agreements with the data continued to

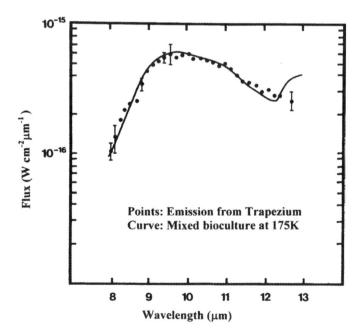

Fig. 11   Observed infrared flux from the Trapezium nebula (points) compared with the emission calculated for a mixed microbial culture including diatoms heated to 175 (curve).

emerge in whichever direction we cared to look. Our old friend the Trapezium spectrum, which was our starting point of organic dust models, now produced a perfect agreement with a biological model that combined purely carbonaceous microorganisms with a class of common algae called diatoms that included siliceous biopolymers. A microbial mix taken from a sample of water from the nearby River Taff in Cardiff resulted in the fit to the emission spectrum of the Trapezium spectrum over the entire 8–40 μm wavelength range. The correspondence is shown in Figs. 11 and 12. We referred in a previous chapter to the story surrounding the 2175 Å interstellar absorption feature and its incorrect assignment to graphite. No sooner than we had embarked on the organic polymer trail we were opting more decisively to attribute this absorption to the effect of aromatic carbon-ring structures.

From 1980 onwards infrared observations accumulated that also had a bearing on aromatic molecules (molecules involving hexagonal

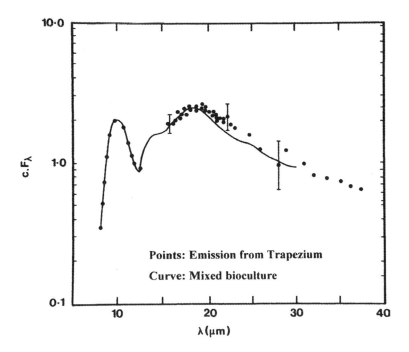

Fig. 12 Same as Fig. 10 but extended to longer infrared wavelengths.

carbon ring structures) in interstellar space. In the mid-1980's groups of astronomers both in the USA and France independently concluded that certain infrared emission bands occuring widely in the galaxy and in extragalactic sources are due to clusters of aromatic molecules. The molecules absorb ultraviolet starlight, get heated for very brief intervals of time, and re-emit radiation over certain infrared lines, including one at 3.28 μm. Needless to say, such molecules are part and parcel of biology, and their occurrence in interstellar space is readily understood as arising from the break-up of bacterial cells.

Fred and I showed much later that galactic infrared emissions at 3.28 μm and other infrared wavelengths, combined with extinction at 2175 Å can be explained on the basis of an ensemble of biologically generated organic molecules. (F. Hoyle and N.C. Wickramasinghe, *Astrophys. Space Sci.* **154**, 143–147, 1989; N.C. Wickramasinghe, F. Hoyle and T. Al-Jabory, *Astrophys. Space Sci.* **158**, 135–140, 1989.)

As was pointed out in an earlier chapter, even much earlier in 1962, the presence of aromatic molecules in space might have been inferred from the so-called diffuse interstellar absorption bands. It has been known for over half a century that some 20 or more diffuse absorption bands appear in the spectra of stars, the strongest being centred on the wavelength 4430 Å. Despite a sustained effort by scientists over many years no satisfactory inorganic explanation for these bands has emerged. I came across a possible solution at the conference in Troy New York to which I have already referred. The Chemist F.M. Johnson showed that a molecule related to chlorophyll — magnesium tetrabenzo porphyrin — has all the required spectral properties. Chlorophyll of course is an all important component of terrestrial biology — it is the green colouring substance of plants, the molecule responsible for photosynthesis, the process that lies at the very base of our entire ecosystem on the Earth.

Very recently we have unearthed yet another property of biological pigments such as chlorophylls, a property that clearly shows up in astronomy. Many biological pigments are known to fluoresce, in the fashion of pigments in glow worms. They absorb blue and ultraviolet radiation and fluoresce over a characteristic band in the red part of the spectrum. For some years astronomers have been detecting a broad emission feature of interstellar dust over the waveband 6000–7500 Å. Chloroplasts containing chlorophyll when they are cooled to temperatures appropriate to interstellar space fluoresce precisely over the same waveband. (See F. Hoyle and N.C. Wickramasinghe, *Astrophys. Space Sci.* **235**, 343–347, 1996.)

Chapter 14

Life on the Planets

Despite the mounting hostility towards us, we were able to carry on our work whilst maintaining a degree of cheer. This was due to several reasons. There was the fortunate circumstance that the research we were engaged in did not require much in the way of financial support — certainly not remotely like the funding our colleagues used to get from the public purse. Most importantly we had the support of Principal Bill Bevan who from the outset had an instinct that we were on the right track. We could count on him for all our modest financial needs. At a later stage, Bill Bevan introduced us to Gary Weston, Chairman of Associated Foods whose generous support for our work continued well into the 1990's. This latter support was useful, for instance, to provide Fred with his first fax machine to facilitate our communication, and for meeting the escalating costs of our telephone bills. Last but not least, we had the support of our wives. I know that Barbara encouraged Fred to fight in defence of his views as did Priya, who felt we should continue the struggle to win whatever it took!

We were now firmly committed to the view that an immensely powerful cosmic biology came to be overlaid on the Earth from the outside some 4 billion years ago, through the agency of comets. Other planetary bodies, within the solar system and elsewhere, must also be exposed to the same process. Wherever the broad range of the cosmic life system contains a form of life (genotype) that matches a local niche of a recipient planet, that form would succeed in establishing itself. In our view the entire spectrum of life, ranging from the humblest single-celled lifeforms to the higher animals must be introduced on a planet from the external cosmos. With this in mind we began to examine new data on the planets of our solar system obtained from Pioneer and Voyager spacecraft for tell tale signs of microbial life. We took as an indispensable condition for bacterial life the need to have access to liquid water.

Subject to this constraint we discovered what we thought were tentative signatures of bacterial life in the planets Venus, Jupiter and Saturn. Since Venus is exceedingly hot at ground level (about 450°C) it would be impossible for life to exist at the surface. Venus, however, has an extensive cloud cover and it is within these clouds that life may have taken root. Water is present in small quantities and in the higher atmosphere the temperature is low enough for water droplets to form. Moreover the clouds of Venus are in convective motion in the upper atmosphere, which ranges in height between 70 km to 45 km, with a corresponding temperature range of 75°C at the top to −25°C at the bottom. We argued that the survival of bacteria over the range of conditions in the upper atmosphere was possible, and that repeated variations of temperature in a circulating cloud system would tend to favour bacteria capable of forming sturdy spores. We argued for an atmospheric circulation of bacteria on Venus between the dry lower clouds and the wetter upper clouds where replication might take place. We discovered that Pioneer spacecraft data, including the presence of a rainbow in the upper clouds, could be interpreted as implying the presence of scattering particles that had properties appropriate to bacteria and bacterial spores.

These ideas have recently come into vogue. In 2002 Dirk Schulze-Maluch and Louis Irwin looked at data on Venus from the Russian Venera space missions and the US Pioneer Venus and Magellan probes. They discovered trademark signs of microbial life from studies of the chemical composition of Venus's atmosphere 30 miles above the surface. They expected to find high levels of carbon monoxide produced by sunlight but instead found hydrogen sulphide and sulphur dioxide, and carbonyl sulphide, a combination of gases normally not found together unless living organisms produce them. They conclude that microbes could be living in clouds 30 miles up in the

Venusian atmosphere, exactly in the manner we discussed 25 years earlier (*New Scientist*, 26 September 2002).

We had also argued for localised bacterial populations in Jupiter's atmosphere that might even have a controlling effect on its meteorology, including the persistence of the Great Red Spot. A kilometre-sized cometary object hitting Jupiter at high speed will be disintegrated into hot gas that would form a diffuse patch similar to the Great Red Spot. Such a region of the atmosphere would be rich in the inorganic nutrients needed for the replication of microorganisms. A large bacterial population could then be built up in this area and the possibility arises for a feedback interaction to be set up between the properties of the local bacterial population and the global meteorology of Jupiter as a whole. Additionally we presented a case for bacterial grains trapped in the rings of Jupiter, Saturn and Uranus, rings that were discovered in 1979 by the Voyager missions.

We had also come to regard the presence of methane and other organic compounds in any quantity on solar system bodies as being an indication of life. Essentially all the organics on the Earth today are either directly or indirectly due to biology. So it is likely to be for organic matter that is found in substantial quantity elsewhere in the Universe. Astronomers have been used to thinking of the presence of methane in the atmospheres of the four large outer planets as being the result of a thermodynamic trend of carbon compounds to change to methane at low temperatures in the presence of a great hydrogen excess. Yet in our model of the formation of the solar system (discussed earlier) there was no great hydrogen excess in the solar nebula: the bulk of the hydrogen and helium in the disc escaped at the periphery carrying away the excess angular momentum that led to the ejection of the disc in the first place.

If one considers a mixture of free hydrogen and carbon dioxide placed in a flask the separate gases will persist unchanged for an eternity. The thermodynamic trend to methane is essentially unobservable. It is in such situations that catalysts come into their own. Of all the catalysts in the Universe, bacteria are by far the most efficient. Methane producing bacteria (methanogens) exist precisely for speeding up the conversion of carbon dioxide and hydrogen to methane and water. If the transformation occurred inorganically there would be no niche for an entire kingdom of bacteria — the methanogens. So the outer planets, which are known to contain methane, in their atmospheres, must be teeming with methanogens according to our point of view. All these speculations were discussed at length in a Cardiff Blue Preprint entitled "On the Ubiquity of Bacteria" and were later published in abridged form in *Space Travellers: The Bringers of Life* (University College, Cardiff Press, 1981).

The next fellow traveller to accompany us in our journey was Hans Dieter Pflug from the Geological Institute of Justus Liebig Universtiy in Giessen, Germany. In 1979 Pflug had presented evidence for microbial fossils in the sedimentary rocks of South-West Greenland (the Isua Series). These rocks being dated at 3800 million years put the first appearance of life back by some 500 million years from previous estimates, thereby reducing the time available for the development of any primordial soup. In fact it turns out that before 3800 million years the Earth was subject to severe cometary bombardment, so that Pflug's microfossils could represent the first respite from these impacts and the first opportunity for life on Earth to survive.

Pflug discovered structures in the shape of fossilised cells occurring as colonies and individually in different stages of budding. His technique using thin sections of the rock seemed to be beyond reproach and free of contamination, but in view of the nature of his findings it was inevitable that controversy would ensue. Objectively, however, his case was water-tight. Spectroscopic studies showed organic molecules confirming the conclusion that these structures did indeed show evidence of early life. In one particular instance there was a cell that appeared to possess a nucleus — a eukaryotic cell resembling a yeast cell. This was of course contrary to the prevailing paradigm of biology which states that cells with nuclei came much later in the process of biological evolution on the Earth. Pflug's paper appeared in *Nature* (H.D. Pflug and H. Jaeschke-Boyer, *Nature* **280**, 483–486, 1975). Predictably several rebuttals were to follow claiming that Pfug's microfossils were not relics of biology but crystallographic artefacts. Similar arguments have continued to the present day.

Pflug contacted us in 1980 offering us information that was even more interesting than terrestrial microfossils. He claimed to find compelling new evidence for bacterial microfossils in carbonaceous meteorites. The historical background to this work is worth recalling before describing Pflug's new finds.

As the name implies, the carbonaceous meteorites, contain carbon in concentrations upwards of 2 percent by mass. In a fraction of such meteorites the carbon is known to be present as high molecular weight organic compounds. Although there is still some debate on the matter, it is generally held that at least one class of carbonaceous meteorite is of cometary origin. If one thinks of a comet containing an abundance of frozen microorganisms, repeated perihelion passages close to the sun could lead to the selective boiling off of volatiles, admitting the possibility of sedimentary accumulations of bacteria within a fast shrinking cometary body. We can thus regard carbonaceous chondrites (a type of meteorite) as being relic comets after their volatiles have been stripped.

Microfossils of bacteria in meteorites have been claimed as early as the 1930's, but the very earliest claims were quickly dismissed as being contaminants. The story did not end there, however, and the whole argument was revived in the early 1960's. The actors in the new drama included Harold Urey, who was personally known to Fred, and was one of the greatest geologists of the century. H. Urey together with G. Claus, B. Nagy and D.L. Europe examined the Orguel carbonaceous meteorite, which fell in France in 1864, microscopically as well as spectroscopically. They claimed to find evidence of organic structures that were similar to fossilised microorganisms, algae in particular. The evidence included electron micrograph pictures, which even showed substructure within these so-called "cells". Some of the structures resembled cell walls, cell nuclei, flagella-like structures, as well as constrictions in some elongated objects that suggested a process of cell division. These investigators, like their colleagues before them, became immediately vulnerable to attack by orthodox scientists.

With a powerful attack being launched by the most influential meteoritists of the day, the meteorite fossil claims of the 1960's became quickly silenced. One of the more serious criticisms that were made against these claims was that the meteorite structures included some clearly recognisable terrestrial contaminants such as rag-weed pollen. But the vast majority of structures ("organized elements") that were catalogued and described were not contaminants. Intimidated by the ferocious attack that was launched against them, Claus reneged under pressure, and Nagy retreated while continuing to hint in his writings that it *might be so*, rather in the style of Galileo's whispered "*E pur si muove*".

An alternative explanation was that these fossil-like structures were mineral grains which have acquired coatings of organic molecules by some non-biological process. The difficulty with this theory, however, is that the highly organized cell-like appearance of these structures would still remain a mystery. Terrestrial contaminations were a possibility, but this was also unlikely to be the correct explanation for most of the structures, because they have no modern terrestrial counterpart.

In 1980 Pflug reopened the whole issue of microbial fossils in carbonaceous meteorites. Pflug used techniques that were distinctly superior to those of Claus and his colleagues and found a profusion of cell-like structures comprised of organic matter in thin sections prepared from a sample of the Murchison meteorite which fell in Australia, about a hundred miles north of Melbourne on 28 September 1969. He showed these images to us and both Fred and I were convinced of their biological provenance. Pflug himself was a little nervous to publish the results, fearing for his career and anticipating the kind of reaction that was seen in the 1960's. We convinced him to present his work at the out-of-town meeting of the Royal Astronomical Society, held in 1980 in Cardiff, to which I have already referred.

The method adopted by Pflug was to dissolve-out the bulk of the minerals present in a thin section of the meteorite using hydrofluoric acid, doing so in a way that permits the insoluble carbonaceous residue to settle with its original structures in tact. It was then possible to examine the residue in an electron microscope without disturbing the system from outside. The patterns that emerged

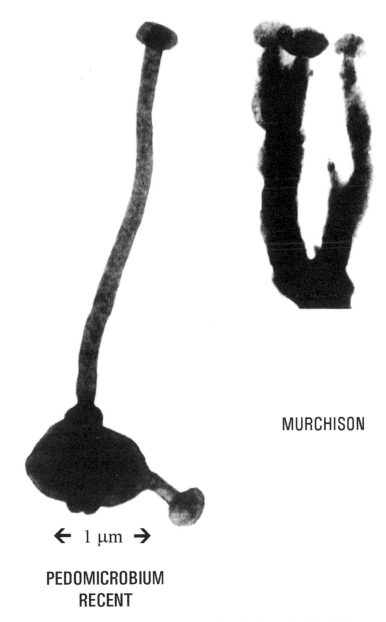

MURCHISON

← 1 μm →

PEDOMICROBIUM
RECENT

Fig. 13   One of Hans Pflug's many microfossils from the Murchison meteorite compared with recent microorganism Pedomicrobium.

were stunningly similar to certain types of terrestrial microorganisms. Scores of different morphologies turned up within the residues, many resembling known microbial species. An example is shown in Fig. 13. It would seem that contamination was excluded by virtue of the techniques used, so the sceptic has to turn to other explanations as disproof. No convincing non-biological alternative to explain all the features was readily to be found.

The so-called stereoisomerism of amino acids (the Left-handed and Right-handed forms) in meteorites is a tangled story. Biological proteins are made almost exclusively of the levorotary (L) form, whereas amino acids made synthetically as in a Urey–Miller type experiment are made of equal numbers of L and R forms for each amino acid. Many investigators claimed that meteoritic amino acids have equal L and R numbers and so could not have a biological connotation. However, during fossilization of organisms on Earth it is known that some switches from L to R do in fact take place, although equal L and R numbers may be hard to explain in this way. Pflug brought to our attention unpublished work of M.E. Engel and B. Nagy that indicated a slight preponderance of the L form in the biologically relevant amino acids of the Murchison meteorite. (This was published subsequently as M.E. Engel and B. Nagy, *Nature* **296**, 837, 1982.) Thus, as of 1981, the question of the biological origin of the meteoritic amino acids remained unresolved.

We kept in close touch with Pflug throughout the period 1980–1983, and on November 26th, 1981 invited him to deliver a public lecture in Cardiff with the title "Extraterrestrial life: New evidence of microfossils in the Murchison meteorite". The talk was introduced by Fred and the meeting chaired by Principal Bill Bevan. As always with any lecture event involving Fred we had a packed hall. The audience was stimulated as well as entertained, and the Earth Scientists left in a state of open-mouthed bewilderment.

Chapter 15

Evolution from Space

With the amount of evidence we now had for an organic composition of interstellar dust we found it exceedingly puzzling to understand the reluctance to accept even this relatively simple fact. Perhaps there was a perception that very much bigger issues were at stake. If the whole of Darwinian evolution was to come under scrutiny there would be a motive to turn away from even the simplest facts that pointed in such a direction. After all the victory of Darwinism over the Judeo Christian view of creation as exemplified in the Huxley–Wilberforce debate was a hard-won affair and the memory of the blood-letting must still linger in our collective consciousness. It is a victory to be cherished at all cost, and smaller truths may need to be sacrificed in the interests of larger perceived goals.

On Fred's visits to us we would often touch upon such matters. At times we were philosophical, at other times irate in dealing with the attitudes of our scientific colleagues. One afternoon, when we had tired of figuring out some astronomical problem, we decided to take a stroll to Caerphilly mountain, a few miles from where I lived. We had reached the top of the mountain when a dog darted across the footpath and snapped at Fred, sunk a tooth into his ankle and tore the bottom of a trouser leg before an embarrassed owner came to the rescue. Later that evening the owner came to our house to further apologise and offer compensation and remarked that this was the very first time his dog had behaved in this aggressive way. Fred surmised that his anger at the intransigence of our critics may be the cause, the dog sensing the adrenaline that he produced!

The next major project we undertook during the period 1980–1981 was an attempt to connect cosmic life, viruses and bacteria causing disease, with the evolution of life on the Earth. If life started on the Earth some 4 billion years ago with a comet bringing the first cosmic microorganisms, how did it evolve and diversify to produce the magnificent range of life forms we see today?

It is believed by neo-Darwinists that the full spectrum of life is the result of a primitive living system being sequentially copied billions upon billions of times. According to their theory, the accumulation of copying errors, sorted out by the processes of natural selection, the survival of the fittest, could account for both the rich variety of life and the steady upward progression of complexity and sophistication from a bacterium to man. This is perhaps a simple representation of Neo Darwinism, but it encapsulates its essential features. Is this enough to explain the available facts of biology?

When we began to examine the matter our answer turned out to be an emphatic no. Our mathematical objections to the theory were published in a booklet entitled *Why Neo-Darwinism Does Not Work* (University College Cardiff Press, 1982), and a more general critique of Neo-Darwinism in our book *Evolution from Space* (Dent, 1981).

In essence our basic argument was simple. Major evolutionary developments in biology require the generation of new high-grade information, and such information cannot arise from the closed box evolutionary arguments that are currently in vogue. The same difficulty that exists for the origin of life from its organic building blocks applies for every set of new genes needed for further evolutionary developments. One of the earliest arguments in this context concerned the origin of the set of enzymes needed for a primitive bacterium.

A typical enzyme is a chain with about 300 links, each link being an amino acid of which there are 20 different types used in biology. Detailed work on a number of particular enzymes has shown that about a third of the links must have an explicit amino acid from the 20 possibilities, while the remaining 200 links can have any amino acid taken from a subset of about four possibilities from the bag of 20.

This means that with a supply of all the amino acids supposedly given, the probability of a random linking of 300 of them yielding a particular enzyme can be calculated to be as little as 10^{-250}. The bacteria present on the Earth in its early days required about 2000 such enzymes, and the chance that a random shuffling of already-available amino acids happens to combine so as to yield all the required 2000 enzymes is about one in $10^{500,000}$.

Everybody must surely agree that a probability as small as this cannot be contemplated. So to a believer in the paradigm of the origin of a bacterium or indeed a set of genes in the warm little pond, there has to be a mistake in this argument. Although it is known that the bacteria present on the Earth, almost from the beginning, were ordinary bacteria, modern bacteria as one might say, it has been argued by some that the first organisms managed to be viable with considerably fewer than 2000 enzymes. A number of about 256 has been quoted and in this case the probability of origin of this severely sawn-down enzyme set is one in 10^{6900}, still not a bet one would advise a friend to take. For comparison, there are about 10^{79} atoms in the whole visible universe, in all the galaxies visible in the largest telescopes.

Such statistics convinced us beyond any shadow of doubt that life must be a truly cosmic phenomenon. The first origin of the magnificent edifice we recognise as life could not have begun in a warm little pond here on Earth, nor indeed in any single diminutive location in the cosmos. It must have required all the resources of the stars in a large part of the universe to originate, and thence its spread is easily achieved. Many attempts to convey this were made by Fred and myself in our writings and lectures by means of comparison with everyday situations. One such comparison was that if a population of several thousand people were to throw a pair of unbiased dice, the probability of everybody throwing two sixes is of the same order of difficulty as the origin of the set of enzymes needed for a primitive bacterium. Another comparison is that the origin of life from organic molecules on Earth is of a smaller order of improbability than a tornado blowing through a junk yard assembling a fully working Boeing 707!

A Journey with Fred Hoyle

The many attempts that have been made thus far to overcome this hurdle were in our view wholly unsuccessful. In a lecture delivered by Fred at the Royal Institution on 12 January 1982 he said thus:

> *"Some people are putting out statements that by appealing to a mysterious process called non-equilibrium thermodynamics the problem of finding the required explicit orderings of amino acids can somehow be solved. This is like saying to a person trying to throw a sequence of 5 million sixes that they would do better if there were to roll the dice a bit faster, whereas of course it would scarcely help at all even if the dice were thrown at the speed of light."*

To overcome these seemingly insurmountable obstacles for the origin of life there seemed to us to be two logical options:

Option 1: The alternative to the assembly of life by random, mindless processes in a finite universe is assembly through the intervention of some form of cosmic intelligence. Such a concept would be rejected outright by many scientists, although there is no purely logical reason for such a rejection. With our present technical knowledge human biochemists and geneticists could now perform what even 10 years ago would have been considered impossible feats of genetic manipulation. We could for instance splice bits of genes from one species into another, and even work out the possible outcomes of such splicings. It would not be too great a measure of extrapolation, or too great a license of imagination, to say that a cosmic intelligence that emerged naturally in the Universe may have designed and worked out all the logical consequences of our living system. If the Universe was of the standard Big Bang type with an age of no more than 16 billion years one might well be stuck with such an explanation, unless life itself can somehow beat the improbability factors we have just discussed. For instance, one might assert that the information content of life is deeply buried in the structure of matter at an yet undiscovered subatomic level.

Option 2: The first premise for the origin of life and all its genetic facets is the existence of a spatially infinite universe, a universe that ranges far beyond the largest telescopes. Then the very small chance of obtaining a replicative primitive cell will bear fruit somewhere and, when it does, exponential replication will cause an enormous number of the first cells to be produced. It is here that the immense replicative power of biology shows to great advantage. It generates enough copies of itself for a second highly improbable evolutionary event to occur in one of the profusion of offspring from the first cell. And so by an extension of the argument to the third improbable event. Indeed to a whole chain of improbable occurrences, which result at last in the enormous diversity of cells we have today, the cells that were already present at the formation of the Earth. This is option that Fred himself considered to be the most reasonable, and one that connected with his cosmological preferences.

On this view of the origin of life there would be little variation in the forms to which the process gives rise, at least so far as basic genes are concerned, over the whole of our galaxy. Or indeed over all nearby galaxies. The rest of the story concerns the many ways in which the same basic genes can combine to produce rich varieties of living forms from one environment to another, always remembering that because of the large numbers involved — large numbers of stars, large numbers of planets and large numbers of galaxies, the system can afford many failures. For instance the Earth, in its four and a half billion year history, would not have produced anything very noteworthy but for the chance events of the last half-billion years.

Our views on cosmic evolution connect also with the idea of disease-causing viruses coming from space. The question was asked by our critics: "How could a virus or bacterium coming from space relate to evolved life forms on the Earth?" The answer must be that higher organisms evolved in response to exposure to space borne viruses. Apart from causing diseases, viruses could on occasion add on to our genes, and so provide a store of future evolutionary potential. This would be the raison d'être for the persistence of viral diseases in evolved life forms such as ourselves. One might legitimately ask: if virus infections are bad for us why did the evolution of higher life not develop a strategy for excluding their ingress into our cells. Logically it seems easy enough for the greater information content of our cells to devise a way of blocking the effects of the much smaller information carried by a virus, and yet this has not happened in the long course of evolution. Could it be, we wondered, whether this "invitation" to viruses was retained for the explicit purpose of future evolution?

Chapter 16

Theories of Trial

Fred was far from being a religious man, not in the conventional sense at least. His position, as far as I could assess, was that *if* there were a cosmic creator it would be scarcely conceivable that any of the world's religions would have fully grasped either *His* intent or *His* plan. A degree of incompleteness in comprehending such matters must necessarily remain. I believe he kept an open mind, as I did, and regarded "creation", in some form, as being a valid intellectual position to hold in relation to the origin of life. He was also cynical of the ambivalent scientific attitude that prevailed in relation to this whole question: whilst it was considered untenable for "creation" to be used in connection with life, it was perfectly acceptable to contemplate that an entire Universe, with all its inherent laws, suddenly came into existence some 16 billion years ago, created to all intents and purposes. If Fred and I ever discussed "creation" or a "creator" we did so only as an abstract concept free of any specific Judeo–Christian implications.

In 1981 we had already published our book *Evolution from Space* which was receiving a great deal of media attention, particularly a chapter with the enigmatic heading "Convergence to God?" It was not surprising, therefore, that Fred was approached by the State Attorney of the State of Arkansas asking if he would give expert evidence for the State at a forthcoming Creation trial. On March 19, 1981, the Governor of Arkansas had signed into law an Act which stated: "Public schools within this State shall give balanced treatment to creation-science and to evolution-science." The US Federal

Government had challenged the constitutional validity of the Act, and the case being heard was the State of Arkansas versus the Federal Government. Fred was not able to oblige due to other commitments, and directed the request to me for consideration. I spoke at length to the Attorney who convinced me that all I was required to do was to defend the ideas we published in *Evolution from Space*. To be their expert witness I had to rebut the claim of the Federal Government that neo-Darwinian evolution was a proven fact. Although I was a little apprehensive of what I might be letting myself in for, I did not see an immediate reason for declining their invitation. After several long telephone conversations with Fred it was agreed that I should go to Arkansas and present a testimony that we would agree upon beforehand. I had religious friends and I respected peoples' freedom to hold such beliefs. I did not feel that their legitimate aspirations should be thwarted on scientific grounds that seemed to me to be insecure.

My wife Priya, my youngest daughter and I set out to Arkansas on a cold December day in 1981 through a snowbound airport at Heathrow. There were long delays due to snow, and I remember thinking many times that this was an ill omen and that we should turn back and return home. But we made the trip and eventually reached Arkansas just in time for the deposition and the trial. The case I presented essentially summarised my scientific beliefs. The following quotations are an extract of my testimony.

> "*The facts as we have them show clearly that life on Earth is derived from what appears to be an all pervasive galaxy-wide living system ... Life was derived from and continues to be driven by sources outside the Earth, in direct contradiction to the Darwinian theory that everybody is supposed to believe ...*
>
> *It is stated according to the theory that the accumulation of copying errors, sorted out by the process of natural selection, the survival of the fittest, could account both for the rich diversity of life and for the steady upward progression from bacterium to Man ... We agree that successive copying would*

*accumulate errors, but such errors on the average would lead
to a steady degradation of information... This conventional
wisdom, as it is called, is similar to the proposition that
the first page of Genesis copied billion upon billions of time
would eventually accumulate enough copying errors and hence
enough variety to produce not merely the entire Bible but all
the holdings of all the major libraries of the world... The
processes of mutation and natural selection can only produce
very minor effects in life as a kind of fine tuning of the whole
evolutionary process...*

*In our view every crucial new inheritable property that
appears in the course of the evolution of species must have
an external cosmic origin... We cannot accept that the genes
for producing great works of art or literature or music, or
developing skills in higher mathematics emerged from chance
mutations... If the Earth were sealed off from all sources of
external genes: bugs could replicate till doomsday, but they
would still only be bugs: and monkey colonies would also
reproduce but only to produce more monkeys. The Earth
would be a dull place indeed...*

*The notion of a creator placed outside the Universe poses
logical difficulties, and is not one to which I can easily sub-
scribe. My own philosophical preference is for an essentially
eternal, boundless Universe, wherein a creator of life may
somehow emerge in a natural way. My colleague, Sir Fred
Hoyle, has also expressed a similar preference. In the present
state of our knowledge about life and about the Universe, an
emphatic denial of some form of creation as an explanation
for the origin of life implies a blindness to fact and an arrogance that cannot be condoned."*

My own testimony which was consistent with my beliefs, and which
Fred wholeheartedly endorsed, is not a source of regret in itself.
The case was won against the State of Arkansas Education Board
whom I was supposed to be representing. In his summing up of the
judgement on 5 January 1982, Judge William R. Overton made the

following statement:

> "*In efforts to establish 'evidence' in support of creation sci-
> ence, the defendants (The State of Arkansas) relied upon
> the same false premise..., i.e., all evidence which criti-
> cized evolutionary theory was proof in support of creation
> science... While the statistical figures may be impressive evi-
> dence against the theory of chance chemical combinations as
> an explanation of origins, it requires a leap of faith to inter-
> pret those figures so as to support a complex doctrine which
> includes a sudden creation from nothing, a worldwide flood,
> separate ancestry of man and apes, and a young earth...*
>
> "*The defendants' argument would be more persuasive if,
> in fact, there were only two theories or ideas about the ori-
> gins of life and the world... Dr. Wickramasinghe testified at
> length in support of a theory that life on earth was 'seeded' by
> comets which delivered genetic material and perhaps organ-
> isms to the earth's surface from interstellar dust far outside
> the solar system... While Wickramasinghe's theory about the
> origins of life on earth has not received general acceptance
> within the scientific community, he has, at least, used scien-
> tific methodology to produce a theory of origins which meets
> the essential characteristics of science.*
>
> *The Court is at a loss to understand why Dr.
> Wickramasinghe was called in behalf of the defendants. Per-
> haps it was because he was generally critical of the theory
> of evolution and the scientific community, a tactic consis-
> tent with the strategy of the defense. Unfortunately for the
> defense, Dr. Wickramasinghe demonstrated that the simplis-
> tic approach of the two model analysis of the origins of life is
> false. Furthermore, he corroborated the plaintiffs' witnesses
> by concluding that 'no rational scientist' would believe the
> earth's geology could be explained by reference to a worldwide
> flood or that the earth was less than one million years old.*"

The repercussions of my court appearance unfortunately lasted for
several years. Although I had not compromised my beliefs (when

cross-examined, I often had to agree with the plaintiffs' claims), many scientists were angry at what they wrongly perceived as our attempt to give credibility to "creation science" which had come to be regarded as the antithesis to science. It was only after meeting "creation scientists" in Arkansas who believed in the literal truth of the Bible, including a belief in an Earth no older than 6000 years, that I began to doubt the wisdom of our decision. For many years following the trial, my family and I were plagued by death threats from several unknown extremist groups, and together with Fred I bore a heavy burden of ostracism for expressing our views in the Arkansas trial.

It was a great relief at this time to be able to escape from such troubles. From the beginning of 1980's I became involved in academic and scientific affairs in Sri Lanka in my capacity as advisor to President J.R. Jayawardene. In this connection I had to make frequent brief visits to the island, which were quite welcome. President J.R. Jayawardene was my father's contemporary at school and had sought me out through the extensive publicity I was receiving at the time for my collaboration with Fred. He had visited India and had been much impressed by the standards of scientific research that prevailed there. As newly appointed President he set himself the task of revitalising Sri Lankan science. He invited me in the summer of 1980 to work on a blueprint for an Institute of Fundamental Studies, based roughly on the model of the Tata Institute in India, with the difference that the President himself was to be the Chairman of its Board of Management. By January 1983 the Institute was set up and I was invited by "JR" (as he was called) to be its founding Director, whilst still retaining my position in Cardiff.

I went to Sri Lanka in the summer of that year with Priya and our three children and prepared for our longest stay in the country since our marriage in 1966. We rented an apartment in Colombo and made every effort to reintegrate with the social and academic scene. I had secured a substantial grant from the United Nations Development Programme (UNDP) to put the fledgling Institute on the world scientific map by organising an interdisciplinary international conference under its aegis. The aim of the conference was to introduce prominent scientists from around the world to the Sri Lankan scene in the hope that they would be able to forge links with researchers working in the island.

Once I had successfully got together the basic infrastructure of the Institute — rented a building with offices located just outside the heart of the city, hired a secretary, an accountant and other office staffs — I started work on planning the conference. I proceeded to do this by involving local academics in Sri Lanka whilst also involving Fred closely on decisions as to who we should invite. How much more interdisciplinary could one get than the areas of research we ourselves were currently engaged in? We thus found a rationale for including an extended session relating to our immediate research interests. Our list of invitees included Zdnek Kopal, Gustav Arrhenius (grandson of Svante Arrhenius), Hans Pflug, Bart Nagy (who was involved in the controversy over organised elements in meteorites), Keith Bigg (an atmospheric physicist who had found bacteria-like structures in the stratosphere), Sri Lankan expatriate scientists Cyril Ponnamperuma and Asoka Mendis, Phil Solomon, Arthur C. Clarke, Tom Gehrels, Jayant Narlikar and Arnold Wolfendale (then President of the Royal Astronomical Society).

The participants began to arrive a few days before the conference in December 1982 and were all accommodated at the Lanka Oberoi, a five star hotel in Colombo. The meeting itself was to take place at the Bandaranaike Memorial International Conference Hall (BMICH) a well-equipped auditorium and conference facility that had been gifted to Sri Lanka by the People's Republic of China. Fred arrived in Colombo from Sydney after an extended visit to Australia. The previous summer we had prepared a preprint entitled "Proofs that Life is Cosmic" which contained what we considered to be the arguments from many different disciplines, all pointing to the cosmic origins of life. We had decided to publish this document as a *Memoir of the Institute of Fundamental Studies, Sri Lanka, No.1*. Our manuscript was delivered to the Government Printer of Sri Lanka to be printed as a government publication on the instructions of the President himself. Within days, Fred and I found ourselves holed up together in the Printer's office — a dingy, Dickensian space — to correct the proofs. We were pleasantly surprised to find very few typos and the book soon saw the light of day.

The conference itself was a high profile national event in Sri Lanka with President Jayawardene delivering an opening address, followed by a keynote lecture given by Fred Hoyle. Fred gave as usual a brilliant exposition of our theories in the unusual circumstance in which a Head of State was in the audience.

One might have thought that the conjunction of talks by Pflug on the Murchison microfossils, Bigg on microbes in the upper atmosphere, Nagy on D/L ratios of amino acids in meteorites in the presence of Gustav Arrhenius (grandson of Arrhenius), may have struck a chord of consonance. But this was not to be. Hans Pflug cautiously presented his slides as he had done on several occasions in the past, and stated the barest of facts without making any inferences. Likewise Bigg presented his intriguing pictures of particles resembling bacteria in the atmosphere, shyly and with minimal commentary. No sooner than these presentations were completed, Gustav Arrhenius then set upon both them with vengeance, claiming that all such finds had to be interpreted as non-biological artefacts. It seemed that he was determined to turn his back on grandfather Svante's old ideas of panspermia that he considered to be improper. There was clearly no way of winning him over, no matter how strong the arguments in our favour might turn out to be. People do indeed see what they want to see and don't see what they don't.

Bartholomew Nagy's case was different. He knew that he had discovered something profoundly important, first with the microfossils and later with the D/L ratios of amino acids in meteorites. In his formal presentation in Colombo his delivery was strident and straightforward, but when Fred and I met him in the hotel bar and engaged him in conversation it was evident that he was an exceedingly frightened man. As Fred put it *"like a rabbit who was being hunted"*.

The sessions that dealt with the question of cosmic origins of life were concluded without any resolutions of the main issues that were raised. Arnold Wolfendale (then President of the Royal Astronomical Society) who was a silent observer of these sessions assured us that he would arrange a discussion meeting at the RAS to continue the debate in London.

Apart from the arguments and conflicts over the Life issue, the conference was enjoyed by everyone. The traditionally lavish Sri Lankan hospitality seemed to surpass itself and there was plenty of time for socialising and relaxing. On the final weekend, a fleet of black Mercedes cars provided by the Foreign Ministry, arrived at the hotel to take participants on a 3-day cultural tour of the island.

After the formal part of the conference was over, President Jayawardene asked me if I would bring Fred over to President's House, first for lunch and later for a discussion to his office. The question raised at our meeting was what should be done with the new Institute after the inaugural conference was over. It had become clear at this stage that I would not be able to make a long-term commitment as Director of the IFS in view of the research work I continued to be engaged in. Fred suggested the name of C.J. Eliezer, his old Cambridge friend and student of Dirac, who was at the time a Professor in an Australian University. Jayawardene was visibly disturbed by this suggestion but Fred could not understand why. I had later to explain to him that Eliezer was a Tamil, and Jayawardene was fearful of any links that he might have with the so-called Tamil Tiger movement (LTTE) that was threatening to destroy the integrity of the island. All Tamil intellectuals working abroad were regrettably on a suspect list as far the Government was concerned.

One of the highlights of Fred's brief stay in Sri Lanka was a visit to meet the legendary science fiction writer Arthur C. Clarke. Arthur Clarke had adopted Sri Lanka as his home since 1956 to pursue a passion for deep sea diving and also to write his long series of books including *2001: A Space Odyssey* in delightful surroundings and reflective solitude. Besides his prolific output in science fiction Clarke is also known as the inventor of the idea of the telecommunication satellite. Way back in October 1945 he wrote an article in *Wireless World* that was to change the world forever. He pointed out that a network of satellites in geostationary orbit (Clarke orbit) at a distance of 35786 km above the Earth's surface could serve to give worldwide telecommunication coverage, overcoming the problem of the curvature of the Earth faced by the old telecommunication masts. His ingenious idea is now widely exploited throughout the telecommunication industry, and thanks to Arthur we can make instant contact in sound and pictures across the world.

I first met Arthur in 1962 in an airplane on a journey from London to Colombo and have kept in touch with him ever since. Whenever I go to Sri Lanka I always visit him to talk about astronomy and space and to exchange news. He is a witty and scintillating conversationalist who is never shy of voicing an iconoclastic point of view, and I am always entertained in his company. Arthur was always strongly supportive of my work with Fred, and as our ideas evolved towards cosmic life his support grew stronger. He always had an instinct that these ideas *must* be right, and this was a great source of encouragement to me. Fred and Arthur also shared a publisher and had many common interests, both in science and science fiction. So it seemed appropriate that they should meet.

Whilst we were chatting in Arthur's air-conditioned study in Barnes Place, Fred's eyes alighted on a copy of *Diseases from Space* that was on a shelf in the library. Arthur then made an extremely beguiling comment. He said that he had recently been visited by someone high up in the CIA who had remarked that "they" had evidence to support our view that bacteria come from space. Since then we discovered that in the 1960's, NASA had supported a series of balloon flights into the stratosphere to heights above 40 km and had recovered viable microorganisms that could be cultured by relatively simple means. Their results showed that there were 0.01–0.1 viable bacteria per cubic metre of air, and that the density appeared to increase with height. Because it is exceedingly unlikely that bacteria could be lofted to such heights and in such large numbers, the conclusion that they came from space must surely have drifted even momentarily into the heads of the experimenters — an idea totally alien to the belief system that prevailed. Of course in the 1960's the conditions under which such experiments were conducted could have left room for the objection of possible contamination. But the findings may have worried NASA nevertheless and they dealt with the situation expeditiously by withdrawing support for further flights.

We shall return to similar experiments carried out by ISRO in the year 2001 in the last chapter of this book. But in the mean time Arthur Clarke's personal stand in the "Life from Space" debate was made quite clear to us in Colombo in December 1983. He was, of course, firmly on our side.

When all our visitors had left I spent a few more months in Sri Lanka trying to find ways of making the Institute of Fundamental Studies a long-term success. With local jealousies, amongst other factors, this goal was turning out to be more difficult than I had expected. Our time in Sri Lanka effectively ended in July 1983 when the most savage communal riots in the country's history unexpectedly broke out. I was in audience with President Jayawardene at the time when news broke of looting and arson in the South of Colombo, and our meeting was abruptly terminated. The riots evidently began as retaliation for a Tamil Tiger ambush of an army patrol in the north of the country leaving 13 soldiers dead. The Sinhalese retaliated violently and riots continued throughout the island for several weeks.

Ever since this time the Tamil Tigers, who are seeking to establish a separate Tamil state in the north, have carried out sporadic acts of violence and suicide bombings mainly directed upon Government targets. They were responsible for the murders of several prominent politicians, including Rajiv Gandhi, the Indian Premier, President Premadasa (who succeeded J.R. Jayawardene as President) and several ministers of state in the Government of Sri Lanka. This was not a country which one could have felt comfortable to work in.

By 1983 Cardiff Astronomy was advancing in diverse ways. There was a noteworthy development under Bernard Schutz on Relativistic Astrophysics that led eventually to a major group devoted to the search for gravitational waves — a prediction of Einstein's theory of relativity. In our own particular areas of research, which were by now getting increasingly distant from the rest, we had several students, mostly from Iraq, working alongside Shirwan Al-Mufti on various aspects of the biological grain thesis. Niama Jabir was doing more detailed work on theoretical modelling of the interstellar extinction curve (N.L. Jabir, F. Hoyle and N.C. Wickramasinghe, *Astrophys. Space Sci.* **91**, 327–344, 1983) and Laith Karim was studying spectra recorded by the International Ultraviolet Explorer Space-craft (1981–1982). Karim was accessing this data from the Ruther-ford Appleton Laboratory to search for details in the extinction curve around 2800 Å for stars which were known to have relatively weak 2175 Å absorption features. This was of interest because DNA or RNA, if it existed in any quantity in free form or as viruses or viroids, would exhibit an absorption band centred at 2600–2800 Å. We did not have great expectations of finding such a band because it is a relatively very weak band and would tend to be submerged in the wings of the 2175 Å feature. Moreover, the effect had not shown up in bacteria studied by Al-Mufti in the laboratory, spec-tra that were found to have an absorption that peaked at about 2200 Å (F. Hoyle, N.C. Wickramasinghe and S. Al-Mufti, *Astrophys. Space Sci.* **111**, 65–78, 1985). Karim, however, was able to find IUE spectra where the 2200 Å was weak, in which with an appropri-ate comparison star, a hint of a 2800 Å absorption feature was seen. Looking over his work at the time we did not see any obvious prob-lems with the analysis, so both Fred and I permitted the appearance under joint authorship of a paper reporting this provisional discovery (L.M. Karim, F. Hoyle and N.C. Wickramasinghe, *Astrophys. Space Sci.* **94**, 223–229, 1983).

At the present stage of our progress towards panspermia any tech-nical error we might commit could all to easily become a hostage to fortune. Unsubstantiated opinions, on the other hand, were easier to cope with. It was only after Karim's work was published that we realised that there might be a problem with the data that was being used. An examination of the IUE documentation belatedly revealed that the spectra over the wavelength range of interest suffered from a problem known as "saturation". Under such circumstances no con-clusion about the spectra was possible. This unfortunate glitch meant that we had to disregard any effect such as a 2800 Å absorption band in our stars, although this did not pose any threat to our overall model of the grains. But it was an error that our adversaries could seize upon.

After the heat of Colombo conference, the next testing ground for our ideas was a discussion meeting of the Royal Astronomical

Society which took place on 11 November 1983. As promised to us in
Colombo, the meeting was initiated by Arnold Wolfendale. The discussion had the title: "Are interstellar grains bacteria?" Apart from
myself and Fred there was Hans Pflug, Max Wallis and Phil Solomon
on one side of the debate, and Mayo Greenberg, Harry Kroto, and
Doug Whittet on the other. After Fred and I had presented our evidence in support of an affirmative answer to the question at issue,
the others set out to argue the case against. In our view, the case
against and the rebuttals were largely polemical.

Quite predictably Mayo Greenberg pounced on the work of Karim
that was mentioned earlier, implying that if we got one thing wrong
everything that we said had to be dismissed. Doug Whittet raised
several objections to the bacterial grain model based on availability of
constituent atoms in interstellar space. The various atomic species —
carbon, oxygen, nitrogen etc. — present in the interstellar gas and
in the dust must together add up to what we know to be overall cosmic abundances. Those elements that go to make up the dust must
therefore be depleted from the gas phase. Whittet's first argument
was that recent estimates of carbon and oxygen depletions from the
interstellar gas were inadequate to allow for the carbon and oxygen
that were needed to be tied up as bacterial grains. This argument
was shown by Phil Solomon to be insecure and probably wrong. More
recent studies have shown that Phil was correct and the measured
carbon and oxygen depletions are indeed consistent with the bacterial
model of dust. Whittet's next point was that the interstellar abundance of phosphorus (present in DNA) was inadequate to support
the bacterial model — in other words there was not enough phosphorus around. This was also shown by us to be wrong. If one takes
solar abundances as being strictly correct for interstellar gas, and if
about 2/3 of a percent of the dry weight of bacteria is taken to be
DNA we could face a phosphorus deficit by a factor of between 5 and
10. But not all the interstellar bacteria can be assumed to be viable,
and nutrient-starved bacteria are known to have phosphorus deficits.

David Williams provided new data on hydrogenated films of
amorphous carbon and argued that this material too could be used
to explain the 2.9–3.5 μm spectrum of the galactic centre source

GC-IRS7. On closer inspection we did not think the fits were good enough to compete with the bacterial model. Finally, Harry Kroto (now Professor Sir Harry Kroto, Nobel Laureate for discoveries connected with C_{60}) presented the standard chemist's point of view. Nothing that is definite about precise chemical compositions can be inferred from IR spectra, it was argued. Although our fit of a bacterial model to the spectrum of interstellar dust cannot be denied, a chemist's contention is that a combination of organic absorbing groups could have arisen inorganically in just the right proportions to mimic a bacterial spectrum. That was not an argument we heard for the first time, and our answer has been already documented in this book.

Chapter 17

A Fossil Controversy

It would be difficult to continue my story without reference to a strange development in 1983 when Fred was cheated of a Nobel Prize. There is little doubt, even in the minds of his arch enemies, that his work on the origin of the elements in the 1940's and 1950's constitutes a monumental contribution to science. I have mentioned earlier that theory of nucleogenesis (the synthesis of elements in the hot interiors of stars) was an outstanding scientific landmark of the 1950's, and Fred's role as leader and pioneer of this entire venture is beyond question. Fred's early calculations had shown that in order for carbon to be produced in adequate quantities in stars the nucleus of the carbon atom had to possess an excited state, and precisely this level was later discovered in the laboratory by Willy Fowler and a Bob Whaling. In the further development of the theory he collaborated with Willy Fowler and with Geoffrey and Margaret Burbidge in the mid-1950's. In their classic paper B^{2}FH the four authors published a comprehensive account of stellar nucleosynthesis that remained a cardinal influence in astronomy over many decades.

The 1983 Nobel Prize for Physics was awarded to William A. Fowler and Subramanyam Chandrasekhar for contributions to nucleogenesis. Why Fred Hoyle was excluded in this award remains an inexplicable mystery. It is ironical that some months before the award was announced one of Fred's grand daughters had asked Willy Fowler (who was a family friend of the Hoyles) to write an article for a school magazine. In it Fowler stated that Hoyle was the pioneer of all this work and indeed the main driving force.

There have been, of course, other instances of breaches of justice in regard to acknowledgement of pioneering work on a lesser and a greater scale. The inherent weaknesses of human nature, fraught with jealousies and prejudices, often interferes with objective judgement, and Fred would have been the first to recognise this fact. Fred was also well aware that his own place in the history of science was secure, so that the opinions of politicised academies of science were largely irrelevant in a longer-term perspective of things.

I believe, however, the Nobel incident affected Fred. And although he rarely talked about it, he never talked again to his old friend Willy either. One perceivable effect of the episode was that he was even more vituperative in his attacks on the scientific establishment. There have been many speculations as to the reasons for Fred's exclusion for a prize for his own work. One curious speculation came from John Maddox, Editor of Nature, who surmised that the reason for the exclusion was Fred's involvement in panspermia, a theory that the Swedish Academy did not wish to endorse.

Despite such setbacks, our efforts in the next few years were focussed towards fine-tuning our theory of panspermia as well as dealing with criticisms that were levelled against it. There was a growing consensus that interstellar grains were of a complex organic nature, and this situation we regarded to be a victory. But the identification of interstellar grains with bacteria was still vigorously challenged in many quarters. As described in the last chapter it is our claimed uniqueness of spectral fit to the galactic infrared source GC-IRS7 that came under the closest scrutiny. Our line of defence has always been that the fit proposed was not just for positions of a few absorption peaks over the relevant infrared waveband, but for the entire opacity function $\tau(\lambda)$ over essentially an infinity of wavelengths. The latter requirement was clearly far more stringent, and it was precisely for this reason that properly calibrated laboratory experiments were specially designed. Whether there were other combinations of functional chemical groups in an abiotic system that did the job equally well was always open to question. If an appropriate mixture of such functional groups can be obtained non-biologically, one is faced with the dilemma of explaining how it can occur with the same relative proportions with infallible accuracy throughout the galaxy.

The challenge of finding such a mixture to explain the spectrum of GC-IRS7 was considered so important that several laboratories in the United States and elsewhere began to devote their energies to this task. Such attempts were at best only partially successful.

Fred took pride in his Yorkshire roots, and as a rather forthright Yorkshireman he tended to be impatient when dealing with his critics, particularly if he felt sure that he was right and they were wrong. His responses to false criticisms were often far more aggressive than they might have been, and this situation may have worsened after the conduct of the Swedish Academy. In one instance he commented on a paper by M.H. Moore and B. Donn (*Astrophys. J.* **257**, L47, 1982) which had claimed that an organic residue extracted after irradiating a mixture of inorganic ice possessed properties which could rival our bacterial model in the 3–4 μm infrared waveband. Their laboratory spectroscopic data were displayed in their paper in a way that made it difficult to verify their claim. In a Cardiff preprint under our joint authorship entitled "From NASA with love" he wrote thus:

> "*When any organisation is first puffed-up with a gross over-supply of funds that are subsequently reduced to a moderate over-supply, the organisation always finds itself consuming its entire resources in overheads. This is true even when the supply remains at hundreds of millions, or billions, of dollars a year. NASA today is reduced to a situation in which its overheads are so demanding that it can no longer afford a straightforward sheet of paper... in which the researchers were obliged to publish their spectra so shrunken in scale that you would put any one of them on a decent-sized postage stamp... At this stage we could take note of the statement of Moore and Donn: 'In the 3.4 μm region, the spectrum of our laboratory-synthesized residue matches closely that of E. coli'... A sad situation indeed, for here we have the once-opulent NASA so reduced in circumstances that it can no*

*longer manage to send its researchers on a desperately-needed
visit to the nearest oculist...*"

In other publications Mayo Greenberg and his colleagues had claimed
that a variant of the processes used by Bert Donn and Moore could
lead to trace quantities of a complex polymer, that he called "yellow
stuff" for a better description. The spectrum of this material was also
claimed as another rival to our bacterial spectrum. The repeatability
of all such experiments to produce exactly the same composition of
end product always worried us, and this we thought to be a short-
coming of the abiotic solutions that were being offered. We were keen
to test it out but this was not possible. Our spectrum of a desiccated
bacterial cell, however, was everywhere reproducible by the simple
process of biological replication. In a comment addressed to Mayo
Greenberg, Fred wrote:

*"If Professor Greenberg would be good enough to provide us
with a milligramme of his yellow stuff, we shall in return offer
him a bucketful of horsedung!"*

The horsedung being replete with *E. coli* of course.

In my journey with Fred we sometimes took diversions that led us
through treacherous paths. With hindsight some of these excursions
may better have been avoided. One such diversion, which extended
over 3 years, was concerned with the famous fossil of *Archaeopteryx*.
When we were thinking about evolution, *Archaeopteryx* appeared a
somewhat discrepant oddity in the fossil record. It has been considered to be a link between reptiles and birds and its relevance to the
Neo-Darwinian theory of evolution is obvious. When it was discovered it was hailed as one of the long sought after missing links in the
fossil record. The fossil, ostensibly of a small reptile with exquisitely
well preserved feather impressions, was discovered in a 160 million
year old limestone deposit from the Jurassic period. The discovery
was made in 1877 by Ernst Habelein, a German doctor, in a quarry
at Solnhofen, some 40 miles south Nurenberg. A few years earlier the
same doctor had provided the palaeontological community with a single feather impression on limestone also of the same age, and from the

same quarry. The limestone slab containing the full *Archaeopteryx* fossil, as well as its counter slab (containing the mirror impression of the fossil) were sold to the British Museum, who have prized it as one of their most precious possessions. Little wonder then that even the slightest attempt to challenge its authenticity would provoke a furore.

In September 1984 Fred receives a letter, sent to him c/o the Royal Society, by Leo M Spetner in Rehovot, Israel. Spetner introduced himself as a friend of the distinguished Israeli physicist Cyril Domb with whom Fred had worked closely during the Second World War (1941–1945) on the development of Naval radar. I believe it was Fred's respect for Cyril Domb that made him take Spetner's letter seriously in the first instance. Spetner wrote:

> "*For several years I have had a strong suspicion that the Archaeopteryx fossil is not genuine... I suspect that the fossils were fabricated by starting with a genuine fossil of a flying reptile and altering it to make it appear as if it originally had feathers...*"

Fred passed this letter on to me, and we were both naturally intrigued by the suggestion. We promptly headed to our nearest libraries (the internet was not available yet) to learn all we could about the history of this fossil. It appeared that fossil forgery was quite commonplace in those days, so for museums to be sold forgeries was by no means an absolute impossibility. There were at least three *Archaeopteryx* related fossils on record from the same source: the single feather (just referred to), the specimen at the British Museum, and another similar specimen in Germany. If the one fossil turned out to be a forgery, the likelihood was that they all were.

A few weeks later Spetner sent us a detailed manuscript in which he summarised his concerns about the authenticity of *Archaeopteryx*, and we felt at the time that he had a *prima facie* case for his thesis. We met Spetner shortly afterwards when he visited Cardiff with his wife, and both Fred and I formed the opinion that he was an honest man — an orthodox Jew who lived by the Book. In fact he turned out to be so orthodox that when my wife Priya entertained him to dinner she discovered that he did not eat any food that was cooked in a non-Jewish home. We had to provide him with raw carrots and apples, which was evidently all he was permitted to eat.

Without indicating what our motives were, we now approached Dr. A.J. Charig at the British Museum for permission to photograph the fossil. Permission was duly granted. In the afternoon of 18 December 1984 we went over to London with the Physics Department's photographer R.S. Watkins and took hundred of pictures of both the slab and the counterslab under various lighting and exposure conditions. When we studied the pictures we saw many features that convinced us that Spetner's case was one that had to be taken seriously. The perfect feather impressions looked as though they were impressed on a thinly applied overlayer (limestone cement) that looked markedly different in texture from the rest of the fossil. We also found that the slab and the counterslab did not match in certain crucial areas. Playing the role of amateur detectives we poured over our pictures for hours on end, and after several months we were sufficiently convinced to go into print. We published our pictures with accompanying comments raising some doubts about the authenticity of the fossil in several articles written for the *British Journal of Photography*. (R.S. Watkins *et al*, *BJP*, Volume 132, Issues of March 8, March 29 and April 26, pages, 264, 358, 468, 1985). The Editor of the British Journal of Photography, Mr. Crawley, issued press releases highlighting the articles, and as a consequence the whole affair received far more media publicity than we might have desired.

Fred's interest on the many puzzles that seemed to be associated with our December 1984 photographs prompted many telephone calls and a couple of visits to Cardiff. Our studies of the pictures as well as our delving into the history of the fossil occupied several months and led to the publication of perhaps our most controversial book, *Archaeopteryx: The primordial bird — A case of fossil forgery* (Christopher Davies Publishers, Swansea, 1986). Just when I thought the fuss was all over, in the late Spring of 1985 I receive a transatlantic phone call from Fred who was visiting the Museum of Natural History in Washington DC. He had discovered that this Museum had a cast of the main slab of the London Archaeopteryx fossil, but not of the counterslab. This may have been quite innocent, but Fred was inclined to think it was not. He asked me to arrange another photographic session to resolve some issues connected with what he saw. I arranged this for 2.30 pm, 23 May 1985, but when Fred and I arrived at the Museum we found a hostile reception awaiting us. The message was simple: they had had enough and would not permit any further access. In fact the Museum went to the trouble of mounting a public exhibition to set out their case against the forgery claim. Finally they thought a line could be drawn under an ugly interlude in their history when they published a rebuttal in a high-impact scientific journal (A.J.F. Charig *et al*, *Science* **232**, 622, 1986). But all this was not enough to convince either Fred or Spetner. Eventually Spetner managed to secure a few minute samples of the fossil, one from the winged area, another from outside. He carried out scanning electron microscope studies and chemical analyses in Israel and arrived at the conclusion that there were differences between the "suspect" areas and the rest of the fossil. However, the smallness of the samples that were examined still left room for considerable doubt.

The upshot of all this was that nothing was decisively resolved. We did not convince our opponents as we had set out to do, and we also lost many friends! The prudence of taking on so powerful an institution like the British Museum must on retrospect be called to question. Particularly since the outcome of our intended inquiry, whichever way it went, would have had no bearing whatsoever on the bigger issues at hand. Notwithstanding this setback, our progress continued unerringly in the direction of cosmic life. Don Brownlee had begun his programme of collecting cometary dust particles using high-flying U2 aircraft that swept through the lower stratosphere at a height of 15 km (J.P. Bradley, D.E. Brownlee and P. Fraundorf, *Science* **223**, 56, 1984). The method employed was a "flypaper technique" where a sticky plate swept through large volumes of air and captured aerosols as they struck the surface at high speed. Fragile structures like clumps of bacteria or volatile dust would have unfortunately been destroyed by this procedure. What was recovered were mostly porous siliceous clumps with some embedded organics, but occasionally entire organic structures were found buried within them. Terrestrial particles were easily separated from particles of cometary origin by studying isotope ratios. (Several key isotope ratios such as $^{12}C/^{13}C$ are different in comets and terrestrial material). When we looked at the published pictures we found at least one clear case of an embedded bacterium-like organic structure with minute embedded magnetite substructures. This structure also turned out to be uncannily similar to a well-recognised fossilised iron-oxidising bacterium in the Earth's sediments dated at 2000 million years. Since the latter was found by Hans Pflug, we got together with him and published this comparison in a paper entitled: "An object within a particle of extraterrestrial origin compared with an object of presumed terrestrial origin" (F. Hoyle, N.C. Wickramasinghe and H.D. Pflug, *Astrophys. Space Sci.* **113**, 209–210, 1985) (Fig. 14). This, in our opinion, was the first strong indication that cometary particles with a biological provenance are still entering the Earth's atmosphere.

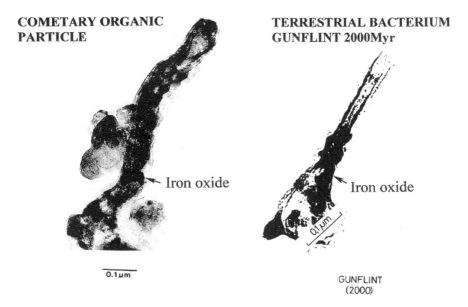

Fig. 14   Comparison between an organic particle collected in the lower stratosphere and an iron oxidizing bacterium in 2000 million year old terrestrial sediments. (F. Hoyle, N.C. Wickramasinghe and H.D. Pflug, *Astrophys. Space Science* **113**, 209–210, 1985.)

Very recent studies of similar stratospheric particles by C. Floss and his colleagues (*Science* **303**, 1355–1358, 2004) have shown the presence of heteroaromatic organic compounds with anomalous carbon and nitrogen isotope compositions attributed to a cometary origin. It is ironic that the authors claim that this constitutes evidence for comets seeding the Earth with the complex building blocks of life, exactly the situation we had maintained in 1977.

During 1984 and 1985, both Fred and I, on separate occasions, visited the NASA Marshall Space Flight Centre at the invitation of Richard B. Hoover who is now (2004) the Head of Astrobiology at this Centre. In 1985, Hoover's astrobiological interests were embryonic, and I suspect it was his interaction with us that made him move further towards the position he now holds. Richard and his wife Miriam had studied diatoms for many years, and were beginning to become intrigued by some of their bizarre properties. Diatoms are a group of golden-brown algae that have intricately woven silica shells. They are the most dominant microbial life form in terrestrial ice ecosystems such as the Antarctic, and all together they constitute a major component of all marine phytoplankton. Their unearthly properties that the Hoovers pointed out to us include an ability to survive very long periods of desiccation, an ability in some cases to live in total darkness, as well as to endure ionising radiation. In the latter context we learnt that many diatom species are capable of living in environments that contain extremely high concentrations of normally lethal radioisotopes such as americium and strontium. Diatoms thrive in highly radioactive waste ponds including the infamous U-pond. Moreover, they do not merely live here, they actually concentrate radioactive isotopes from an environment! The final surprise was that diatoms appear abruptly in the fossil record 112 million years ago during the late Cretaceous period. This to us was a clear indication that they came from space at this time.

In a joint paper with the Hoovers, Fred and I argued for diatom habitats in ice-water interfaces in comets as well as in the multi-cracked surface domains of the Jovian satellite Europa (Richard B. Hoover *et al*, *Earth, Moon and Planets* **35**, 19–45, 1986). This paper we believe contains the first attempt to identify the characteristic orange coloration of the cracks of Europa as arising from biological pigments. Very recently these ideas have been taken up by Brad Dalton of NASA's Ames Research Centre. He has argued that visible and infrared spectra of some pigmented extremophilic bacteria can explain the observations of Europa (see *New Scientist*, 11 December 2001). Again we might have been nearly two decades ahead of our time.

Chapter 18

Comet Halley and its Legacy

Fred's visits to Cardiff were always major family events. During his stay with us he would find time to discuss matters that were far removed from science and as my children grew older they too came to appreciate his rich and diverse company. Fred had an unerring interest in classical music and on some mornings (he was an early riser), I would find him in the living room listening intently to Beethoven's Fifth Symphony or Mozart's Death Requiem – music that I recall hearing on so many occasions blaring out from a gramophone at 1 Clarkson Close. He had grown up in the midst of music, his mother being a gifted piano teacher who had studied at the Royal College of Music, and he himself a paid chorister at a local church. Our home in Cardiff too tended to be filled with music as Priya and our three children share a passion for music. Whenever my elder daughter played the piano, Fred would stop whatever he was doing and praise her talents. When at a later date a publisher suggested that she edits our next joint book, Fred readily agreed saying that he would happily entrust such a job to anyone who played the piano so sensitively and so well! The book, however, never materialised.

An amusing incident took place when his birthday (June 24) happened to fall during one of his visits. Priya and I had organised a dinner party to celebrate the event at the "The Walnut Tree", near Abergavenny. Situated at the foot of the Skirrid Mountain and frequented by Elizabeth David, it was arguably the best restaurant in Wales. We had invited a few other astronomers and friends to join us and the occasion turned out to be most memorable. At the end of the evening the waiter brought me the bill to pay. I produced my visa card as I always do on such occasions, and to my horror was informed that they accept only cash or cheques. This moment of embarrassment was aggravated by a request to write down my name and address to be shown to the proprietor for appropriate action. Within minutes an ecstatic Franco Taruschino rushes to our table and hugs Priya! Evidently he knew Priya from a recent cookery book she had published. Fred Hoyle and all the other astronomers at our table were unknown quantities as far as he was concerned. His excitement at seeing Priya was so great that he went back in and woke up his young daughter to introduce her. All's well that ends well — after a round of complimentary liquors, we were sent home as friends of VIP Priya, and asked to post a cheque whenever we had the time!

We were now occupied with work for two books, *Living Comets* (F. Hoyle and N.C. Wickramasinghe, University College, Cardiff Press, 1985) and *Viruses from Space* (F. Hoyle, C. Wickramasinghe and J. Watkins, University College, Cardiff Press, 1986). For *Living Comets* we considered all aspects of comets that seemed to have a bearing on life, in particular the question of radioactive heating of comet cores that would permit bacterial replication to occur in the early history of the solar system.

In *Viruses from Space*, we proceeded to update our earlier arguments and also collaborated with a General Practitioner Dr. John Watkins who kindly provided us with data from his family practice in Newport. Looking through his case notes dating from 1970, Watkins identified 16 pairs of twins with ages ranging from six months to 14 years to determine how they succumbed to influenza during epidemics. Of the 118 instances when one twin was diagnosed with acute upper respiratory tract infection (presumed influenza during epidemics) the other twin was found to succumb only in 28 instances. The implied cross-infection rate was only 24%, very close to the attack rate that prevailed in the populace at large. A transmission probability as low as 0.24 would, at any rate, be quite insufficient to explain the facts relating to most influenza epidemics. In another project Watkins confirmed an earlier result that had been obtained by a Cirencester GP, Dr. Edgar Hope-Simpson, from his family practice data. During 1968/69 and 1969/70 Hope-Simpson considered a set of families where at least one member reported with influenza-type illness. In this group he looked at the incidence of subsequent cases on days 1, 2, 3, 4 etc. after the index case. From this data he found that the probability of a second member succumbing to influenza was no more than the attack rate in the community at large. Thus being a member of an "infected" household did not appear to increase the risk significantly. All this confirmed to us that in order to explain the pattern of influenza epidemics some trigger — perhaps a biochemical trigger — if not the entire virus must fall through a turbulent atmosphere and reach the ground in an exceedingly patchy distribution. Whether one succumbed or not during an epidemic depended mainly upon one's location in relation to a general infall pattern.

The next high point in our journey was connected with the return to perihelion of Halley's comet in 1986. This was the first time that a comet was being studied by scientists since the beginning of the space age. From as early as 1982 a programme of international cooperation to investigate this comet came into full swing, the objective being to coordinate ground-based observations, satellite-based studies, and space probe analysis on a worldwide basis. No less than 5 spacecrafts dedicated to the study of Comet Halley were launched during 1985, the rendezvous dates being all clustered around early March 1986, about one month after the comet's closest approach to the sun.

In the immediate run-up to these events Fred and I met to discuss what observations might be likely according to our present point of view. What predictions might we possibly make? Our deliberations led us to conclude that organic/biologic comets of the kind we envisage would have exceedingly black surfaces. This is due to the development of a highly porous crust of polymerised organic particles that can permit vigorous outgassing only when the crust comes to ruptured. We put all our arguments in the form of a preprint entitled "Some Predictions on the Nature of Comet Halley" dated 1 March 1986 (Cardiff Series, No 121) which came to be published much later in *Earth, Moon and Planets* (**36**, 289–293, 1986). This was only twelve days before the encounter, and our priority would have gone unrecorded had it not been for the fortunate circumstance that the *London Times* picked up on it and reported its contents (*The Times*, March 12, 1986).

On the night of March 13, 1986 we watched our television screens with nervous anticipation as Giotto's cameras began to approach within 500 km of the comet's nucleus. The fears that the spacecraft might be badly damaged and even destroyed by impacts with cometary dust were proved to be wrong, and the equipment functioned well throughout the encounter. The cameras were expecting to photograph a bright snowfield scene on the nucleus consistent with the then fashionable Whipple dirty snowball model of comets. In the event the television pictures transmitted world-wide on 13 March proved to be a disappointment. The cameras had their apertures shut down to a minimum and trained to find the brightest spot in the field. As a consequence, very little of any interest was immediately captured on camera. The much publicised Giotto images of the nucleus of Comet Halley were obtained only after a great deal of image processing. The stark conclusion to be drawn from the Giotto imaging was the revelation of a cometary nucleus that was amazingly black. It was described at the time as being "blacker than the blackest coal . . . the lowest albedo of any surface in the solar system . . . ." Naturally we jumped for joy! As far as we were aware at the time we were the only scientists who made a prediction of this kind, a prediction that was a natural consequence of our organic/biologic model of comets. Fred and I regarded this development as yet another decisive triumph of our point of view. More triumphs were soon to follow.

A few days after the Giotto rendezvous, infrared observations of the comet were made by Dayal Wickramasinghe and David Allen using the 154 inch Anglo-Australian Telescope (*IUA Circular* No. 4205, 1986). On March 31, 1986 they discovered a strong emission from heated organic dust over the 2 to 4 μm waveband. As noted earlier basic structures of organic molecules involving CH linkages absorb and emit radiation over the 3.3–3.5 μm infrared waveband, and for any assembly of complex organic molecules such as in a bacterium, this absorption is broad and takes on a highly distinctive

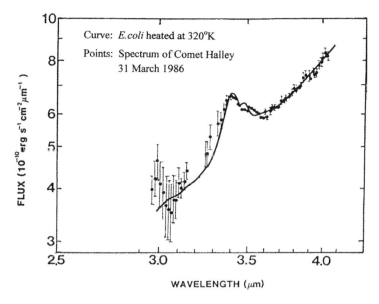

Fig. 15 Comparison of the spectrum of Comet Halley (points) obtained by Dayal Wickramasinghe and David Allen with a microbial model (curve).

profile. The Comet Halley observations by Dayal and David Allen were found to be identical to the expected behaviour of desiccated bacteria heated to 320 K (Fig. 15). Another victory for our model! Later analysis of data obtained from mass spectrometers aboard Giotto also showed a composition of the break-up fragments of dust as they struck the detector to be similar to bacterial degradation products.

The Halley observations, in our view, clearly disproved the fashionable Whipple's "dirty snowball" theory of comets. The theory dies hard, however, with variants of it still in vogue with the claim that Whipple was still mostly right, except that there was more dirt (organic dirt) than snow! It could not be denied that water existed in comets in the form of ice, but great quantities of organic particles indistinguishable from bacteria are embedded within the ice. This conclusion was unavoidable unless one chose to ignore the new facts. (D.T. Wickramasinghe, F. Hoyle, N.C. Wickramasinghe and S. Al-Mufti, *Astrophys. Space Sci.* **36**, 295–299, 1986). With such clear-cut documentation of our priorities for organic comet models we were obviously annoyed at the total lack of any reference to our work when the new data came to be discussed. We took particular exception to several papers that appeared in a special supplement of *Nature* on 15 May 1986 devoted entirely to the Halley results. These papers essentially congratulated various investigators for predicting a dark organic comet. It seemed to us particularly invidious that our old adversary Mayo Greenberg was presented as the hero of the day, attributing the credit solely to him for dark organic comets. The citation of an article by Greenberg dating to 1979 did not supersede our priority for the earlier work on this subject.

We were sufficiently incensed now to issue a preprint with the title "On Deliberate Misreferencing as a Tool of Science Policy" (Cardiff Series, 125, 1 June, 1986) setting the record straight in relation to our clearly established priorities, and with sarcastic comments directed at the journal *Nature*. The document was printed on one sheet resembling a *Nature* reprint in the journal's own house style. The following quotes would suffice to convey the gist of our contention.

> *"Despite our admiration of Professor Greenberg's sibylline achievements, we think that most people require predictions to be announced publicly in advance of events. Otherwise there would be no point at all in making predictions. No ordinary bookmaker, for example, pays out money to would-be clients who seek to back a winner two months after a race has been run. In science, however, bookmakers gladly pay out money even years after a race has been run, especially it seems if the client happens to be Professor Greenberg.*
>
> *In the special supplement to the issue of Nature for 15 May there are two examples, both concerning the likely organic nature of the bulk of cometary dust. The communication "Composition of Comet Halley dust Particles from Giotto observations" (J. Kissel et al, p. 336) attributes the suggestion that cometary dust might be organic to an article by Greenberg dated 1979, while the communication "Composition of Comet Halley from dust particles from Vega observations" also references Greenberg, through an article dated 1982. Our own first paper on organic material appeared in*

1975.... Our views by 1979 had indeed become so widely known as to justify the word "notorious"...

Misrepresentation would not be in question if our views from 1975 onwards have been wholly wrong. It is just because we have sometimes been right that misreferencing has arisen....

Our opposition to the Darwinian theory lies, we believe, at the root of the matter. At a microbial level, where only single base-pair changes of DNA are involved, the Darwinian theory is correct. But at the macrolevel involving multiple base-pair changes, the theory is wrong... Evolution that is Darwinian proceeds mostly at the lowest level of all, the level of varieties, just as it was conceived to do already in the 1850's. Beyond what was then apparent, the theory does not do much. But even if all this was not clearly demonstrable it would be usual in science, in physical science at any rate, for us to be granted the right to our own opinion, instead of being exposed to an orchestrated campaign of calumny and misreferencing...."

There is no law that misreferencing shall not be used as an instrument of science policy. Nor is there a law outside the courts which even requires even an approximation to the truth to be spoken or written... Yet it should be remembered that cultures of high quality are more fragile than they appear at first sight. While they were still operative there seemed no limit to the productions of the Elizabethan dramatists, the Florentine painters and the Viennese musicians. But all are gone now and no amount of effort, desire or money will bring them back. There is no more likely cause of a similar decline and virtual disappearance of productive science it seems to us than a calculated disrespect for the truth...."

Such provocative words were not ignored by *Nature*. A response by John Maddox appeared in the issue of *Nature* dated 19 June 1986 in the form of an article entitled "When reference means deference". Referring to the direction that our work on grains had taken, he

Chapter 19

Alternative Cosmologies

I should mention that the academic year 1987/1988 turned out to be traumatic for reasons unconnected with our work. University College Cardiff was deemed to be in financial difficulty by the Government and Principal Bill Bevan was forced to resign. Furthermore, University College Cardiff and the neighbouring University of Wales Institute of Science and Technology was forced to form a merged University institution that came into legal existence in 1988 under the name University of Wales College of Cardiff, later to be called Cardiff University. In the process of merger individual departments in the two constituent colleges had also to merge. There were 4 mathematics departments at University College Cardiff and one at the University of Wales Institute of Science and Technology that were combined into a single School of Mathematics. The large contingent of astronomers in my old Department of Applied Mathematics and Astronomy were now "displaced persons" and were forced to join a Department of Physics, that later became the Department of Physics and Astronomy. I had a choice: I could have gone over to physics or stayed with mathematics. Fred advised me to stay in the School of Mathematics, which I did. His argument was that there would always be a call for mathematics, whereas if proposed new courses in astronomy somehow failed to catch on, my position in the new outfit may not be so secure.

So after 1988 my "group" within the School of Mathematics had dwindled to four: myself, Fred Hoyle, Max Wallis, Shriwan Al-Mufti and 2 research students. I also now ceased to be Head of Department, through a manoeuvre that was scarcely legal, with the result that I was given an unprecedented workload of teaching. Fred of course still continued to come to Cardiff from time to time. Needless to say I received much encouragement from his visits, particularly to feel that we were confronting the really important problems of the Universe, against the backdrop of which my own squabbles with the University could be viewed in proper perspective. Fred's visits to Cardiff in the late 1980's and early 1990's tended to be more relaxed than before. We felt now that most of the hard work had been done. A conceptual framework for a grand theory of cosmic life was fully in place, and its predictions were being borne out in observations from several disciplines. Interstellar dust and cometary dust were found to possess exactly the properties we had predicted, and the oldest life on the Earth was pushed back to a time when intense cometary bombardment was taking place. We had the strongest indication that comets seeded the planet with life some 4 billion years ago. The discoveries of microbial life enduring the most extreme conditions were suggesting an alien context for such properties, and opening possibilities of microbial habitats in a wide range of bodies in the solar system. A wrong theory does not come up repeatedly with such an indefatigable series of successes. Sooner or later a contradiction turns up and the theory has to be abandoned. This has not happened in our case. Why then is there such deep-rooted hostility to our ideas? Perhaps such ideas went against the grain of an essentially geocentric scientific culture?

We talked at length about the possible causes of the opposition we faced, often despairing for future generations to whom the search for objective truth would become an ever more distant goal. Whilst the real source of the antagonism remained a mystery, the lack of support from our home institution Cardiff was undoubtedly a contributory factor. Following the departure of Bevan and the dissolution of my department, it had become fashionable to dismiss our ideas as insane, and the message gradually filtered out to the outside world.

A major shift in attitude was apparent in other University institutions as well. A golden age when philosophers and academics held sway had given way to an age of hard headed accountants. Money was all that mattered. The search for truth was subjugated by an over-powering greed for accumulating research funds and political power. Nor were these unsavoury developments confined to the cloisters of academia. Indeed it appeared that Universities were merely responding to major changes that were taking place in the world at large. Margaret Thatcher, whose government introduced the idea of a market economy for universities, was now in her third term of office as Prime Minister. People were reckoned to be less important than dubiously conceived objectives within institutions, including Universities. Intolerance of all sorts was on the rise. In 1989 Salman Rushdie publishes his novel *The Satanic Verses*, and the Iranian leader Ayatollah Khomeini orders muslims to execute him. In Sri Lanka conflicts between Tamil separatists and the Government flared sporadically. In China racial attacks on black students early in 1989 were followed by the historic protest and massacre of dissidents in Tiananmen Square. Even in the UK the incidence of racial attacks were noticeably on the increase, with the Police and authorities often turning a blind eye. My wife and I continued to have our fair share of "go home Paki" comments in Cardiff, and similar written threats were delivered to me at the University by persons who could not be identified. I felt that I had suddenly become a victim of racial abuse and discrimination within my own University, and this was sometimes heartbreaking.

Despite all these problems Fred and I continued to pursue our own researches in whichever direction that new data directed. And new data did indeed come our way at a brisk pace. A discovery of a $3.28\,\mu$m emission feature in the diffuse radiation emitted by the Galaxy confirmed that aromatic molecules of some kind were exceedingly common on a galactic scale. We argued that the infrared emissions not just at $3.28\,\mu$m but over discrete set of wavelengths — 3.28, 6.2, 7.7, 8.6, 11.3 μm — must arise from the absorption of ultraviolet starlight by the same molecular system that degrades this energy into the infrared. We had shown much earlier that the 2175 Å extinction of starlight may be due to biological aromatic molecules, and it seemed natural then to connect the two phenomena. Thus we developed a unified theory of infrared emission and ultraviolet extinction by the same ensemble of aromatic molecules. A currently fashionable non-biological aromatic molecule was coronene ($C_{24}H_{12}$), and it was easy to demonstrate that this type of molecule was nowhere near as good as biological aromatics. (F. Hoyle and N.C. Wickramasinghe, *Astrophys. Space Sci.* **154**, 143–147, 1989; N.C. Wickramasinghe, F. Hoyle and T. Al-Jubory, *Astrophys. Space Sci.* **158**, 135–140, 1989.)

Returning to our earlier interest in exploring possibilities for the cosmic microwave background, we turned our attention next to iron whiskers. It is known that the element iron is produced in supernovae, and we could easily show that in an expanding supernova envelope containing newly synthesised nuclides ^{56}Ni and ^{56}Fe, iron particles would eventually condense from the gas when the temperature fell to $1000\,K$. Since laboratory studies showed that iron vapour has a tendency to condense into slender long whiskers, we argued now that a similar process would occur in supernovae. Iron whiskers of diameter $0.02\,\mu m$ and lengths of the order of $1\,mm$ formed in this way are expelled into space. We computed the absorption properties of these whiskers using standard formulae and showed that they could have higher opacities at mm wavelengths than in the visual waveband. This is of course an excellent condition for modelling the cosmic microwave background — the light from distant galaxies will not be blocked, while microwave thermalisation can occur. We discovered also that iron whiskers will experience very high radiation pressure forces and could be expelled from galaxies and clusters of galaxies at high speed. (F. Hoyle and N.C. Wickramasinghe, *Astrophys. Space Sci.* **147**, 245–256, 1988.) Extragalactic iron whiskers are just what we needed to rescue the beleaguered Steady State Cosmology.

Fred and I both attended the 22nd ESLAB Symposium that was held in the delightful Spanish town of Salamanca between 7 and 9 December 1988. We presented three joint papers here, two on our theory of interstellar grains and one on a model of the cosmic microwave background based on long iron whiskers. Apart from the presence of the predictably hostile Greenberg contingent, we felt there was a mellowing of attitude towards us, compared to our experiences both in Colombo and at the RAS a few years earlier. There was still a long way yet to conceding a biological grain model, but people were at least willing to listen to the arguments and to extract what they thought was its acceptable essence. Organic dust everywhere in the cosmos, including in the comets, had become the order of the day.

Our new microwave background theory might have provoked hostility, but it did not. Again peopled listened with tacit interest. In fact many of the participants used dictaphones to record our lectures. The paper describing the new model was published in the conference proceedings (F. Hoyle and N.C. Wickramasinghe, *ESA-SP*-290 489–495, 1989). It was fortunate that the new supernova SN187A came to be discussed at this meeting, and the first signs of dust condensation were becoming evident (W.P.S. Meikle, *ESA-SP*-290 329–337, 1989). Firm evidence in favour of iron whiskers around SN1987A was, however, to become available only much later (N.C. Wickramasinghe and A.N. Wickramasinghe, *Astrophys. Space Sci.* **200**, 145–150, 1993).

In the world outside, momentous changes were under way. Two such changes are worthy of note. The Berlin Wall that was erected in August 1961, when East Germany sealed off the border between East and West Berlin, was triumphantly and dramatically brought down in November 1989. The ugliest symbol of a divided Europe disappeared overnight and with it the Cold War had come to an end. The map of Europe had to be redrawn.

And in South Africa, even more spectacular events were taking place. For over half a century the white South African government, under the control of the Afrikaner National Party, had pursued a policy of apartheid, which involved a total racial segregation — interracial marriages and mixed-race sporting events being outlawed. In the 1970's the government had established tribal "homelands" in the poorest parts of the country. Blacks needed passes to work outside these "homelands". In most instances, only single persons or married men received passes so that when workers left the homelands, they had to leave their families behind. Racially segregated worker "townships" became established outside all the major cities, and the blacks lived under conditions of unspeakable squalor, police brutality and discrimination. The situation was economically so favourable for the White settlers that the prospect of change seemed only a distant dream in the mid 1980's. In 1989, under the Presidency of F.W. de Klerk, changes came swiftly and unexpectedly. Although he knew he was going against the mainstream wishes of the South African whites, growing international pressure pushed de Klerk to work towards dismantling apartheid. And against all the odds he succeeded. In 1990, he released Nelson Mandela from prison, after over a quarter of century of incarceration, and started negotiating with him and the African National Congress (ANC) on the transfer of political power.

Quite a different kind of development that attracted our attention in the winter of 1989 was an outbreak of influenza throughout the UK. It was described as the worst influenza epidemic for 12 years, and hospitals were inundated with cases of complications from the flu. Fred and I decide to undertake our second Hoyle–Wickramasinghe Influenza Survey by sending out questionnaires to all independent schools in the UK, and by visiting particular schools in our immediate neighbourhood. Priya and I went in person to discover what happened in a place not far from Cardiff. We discovered that among the earliest to succumb in the new outbreak were the inhabitants of the sleepy little village of Gowerton, near Swansea. Attendance registers at the local school showed a sudden rise of absences on 27 November, the same day when a local publican, his wife and their child all came down with flu. Furthermore it was documented that the entire village became shrouded in a persistent low-lying mist starting about two days before the outbreak.

Although the influenza type known as H3N2 was the dominant subtype found to be involved in the epidemic, a whole host of other airborne viruses appears to have been in circulation at the same time. The incidence patterns of these viruses were consistent with an atmospheric fall out model, and inconsistent with direct person-to-person spread. All the data we collected appeared to corroborate our findings from the 1977/78 pandemic — even the epidemic patterns at Eton College. At this stage we were surprised to receive an invitation from the Royal Society of Medicine to write a review paper of our findings about influenza. Our article appeared as: "Influenza — evidence against contagion: discussion paper" (F. Hoyle and N.C. Wickramasinghe, *J. Roy. Soc. Med.* **83**, 258, 1990).

Our work in areas relating to cosmic life continued for the next half decade at least. Apart from presentations at international conferences and numerous public lectures that we gave, we had also begun writing a technical monograph, *The Theory of Cosmic Grains* (Kluwer Academic Publishers, 1991). We also started to think more explicitly about the relationship of life to cosmology. The Universe had to be of such a kind that the superastronomically improbable origin of life occurred. The use of the existence of life as an indicator of a hitherto unknown property of the Universe, is known nowadays as the Weak Anthropic Principle. This principle was pioneered by Fred in the 1950's by examining a condition that is needed to produce adequate amounts of the elements carbon and oxygen in stars. I have already mentioned that to achieve this Fred deduced that the nucleus of ^{12}C must possess an excited state close to 7.65 MeV above ground level. No such state was known to exist at the time when this deduction was made, but a state with exactly the predicted energy was discovered shortly afterwards.

In order to overcome the superastronomical information hurdle of life (to which I have referred earlier), Fred began to turn again to Steady-State Cosmology. In our preprint dated March 1991 "The Universe and Life: Deductions from the Weak Anthropic Principle" which was published later in *Astrophysics and Space Science* (F. Hoyle and N.C. Wickramasinghe, *Astrophys. Space Sci.* **265**, 89–102, 1999), we argued that access to superastronomical masses of carbonaceous matter is to be needed to get life started in the Universe. This leads to a sharp distinction between Big Bang and Steady-State cosmologies. In the standard Big Bang cosmology the total mass of carbonaceous matter available for life is a mere 10^{40} g, and the time available is limited to 15–18 billion years. In a steady-state or quasi-steady-state universe, the situation is dramatically different. In a quasi-steady-state model of the universe discussed by Fred in collaboration with Geoff Burbidge and Jayant Narlikar (F. Hoyle, G. Burbidge and J.V. Narlikar, *A different approach to cosmology*, Cambridge University Press, 2000), the universe expands exponentially on a timescale of 1000 billion years, while undergoing oscillations on a timescale of 50 billion years which continue for an eternity. New matter is created at the beginning of each oscillation when the density of the universe is highest. It is this matter that is processed in stars during each cycle and turned into carbonaceous material that is available for biological processing. If one starts with a biological message in some lifeless part of such a universe at a particular place at a particular time, and if the message can be copied and distributed, then in the Quasi-Steady-State Cosmology we estimated, that after a hundred billion Earth-ages, the message would have spread through $10^{90,000,000}$ grammes of carbonaceous material. This means that there is now a superastronomical mass of carbonaceous matter within which the superastronomically improbable event of the origin of life could occur. Thus life may arise in a non-living universe in about a hundred billion Earth-ages, provided the Universe is in a Steady-State or a Quasi-Steady-State. All this suggests that Fred's thoughts were never far divorced from cosmology, even when we were pondering the seemingly different problem of the origin of life.

In the summer of 1989 Jayant Narlikar contacted me to ask if I might be able to host a small cosmology workshop in which Fred could discuss the current state of cosmology with his closest collaborators. We first considered Gregynog Hall in mid-Wales as a possible venue for the meeting but decided against it because of the logistics of transporting participants to and from airports. Between 25 and 29 September 1989, Fred Hoyle, Geoffrey Burbidge, Harlton Arp, Jayant Narlikar and I met at a small residential conference centre — Dyffryn Gardens — just outside Cardiff city. We discussed the many lines of evidence that all appeared to go against the standard Big Bang model of the universe.

By now the evidence accumulated by Harlton Arp and Geoff Burbidge concerning QSO's with large redshifts being physically linked to galaxies of low redshift seemed to put paid to the assumption that high redshifts imply cosmological distances. The insecure nature of many of the other assumptions in the conventional Big Bang theory were also discussed and led this group to propose an alternative cosmology that was more consistent with the facts. My own contribution was mostly directed towards interpreting the cosmic microwave background. This has been the strongest piece of evidence that had been cited in support of a hot Big Bang universe for over two decades. At the time of the Cardiff meeting new measurements made by the COBE satellite were pointing to a microwave background that had a black body spectrum to a very high degree of approximation, and an isotropy on angular scales down to a few arc minutes. All this was being further adduced as support for the Big Bang cosmological model. Fred was quick to point out that the new COBE observations posed a problem for the Big Bang models. An early Black Body spectrum must surely come to be distorted by subsequent events — the condensation of galaxies and clusters of galaxies in the primordial universe must surely leave a mark on the isotropy of the background.

Our alternative point of view was that the cosmic microwave background was the final end product of the thermalisation of the energy produced by the conversion of hydrogen to helium in stars. The thermalisation is a multi-stage process in our view: starlight energy is first converted into the infrared by the normal dust in galaxies, then the infrared in turn is degraded into microwaves by absorption and re-emission by millimetre long iron whiskers.

Our deliberations at the meeting were written up in the form of an article and sent to John Maddox, Editor of *Nature*. It is still a little bewildering how Maddox was persuaded to publish such a devastating attack on conventional cosmology. The paper appeared with the title "The Extragalactic Universe: an alternative view" in the issue of *Nature* of 30 August 1990 (H.C. Arp., G. Burbidge, F. Hoyle and N.C. Wickramasinghe, *Nature* **346**, 807–812, 1990). Referring to the implied fallacy of the conventional explanation of the cosmic microwave background we wrote thus:

> "*The commonsense inference from the Planckian nature of the spectrum of the microwave background and from the smoothness of the background is that, so far as microwaves are concerned, we are living in a fog and that fog is relatively local. A man who falls asleep on the top of a mountain and who wakes in a fog does not think he is looking at the origin of the Universe. He thinks he is in a fog.*"

Our alternative view of the cosmic microwave background could not have been put more succinctly.

Chapter 20

The Last Decade

We did not stall in any of our various projects during the decade 1990–2000, nor did we feel we had come to the end of the road. Barbara's health was giving cause for concern throughout this period so Fred was finding it increasingly difficult to spend time away from home. Home for the Hoyles was now in Bournemouth, where they had moved from the challenging climes of the Lake District, partly owing to Barbara's health. This meant we saw less of him in Cardiff although our interaction and collaborations continued, by telephone and fax.

Fred's book, *Ice*, published in 1981 (F. Hoyle: *Ice*, Hutchinson & Co. Lond, 1981), had introduced the question of ice ages being mediated by cometary dusting. To examine this question more thoroughly we needed an accurate determination of the behaviour of ice particles in the stratosphere. The requirement now was to calculate the fraction of incident sunlight that was absorbed and scattered in directions that did not reach the Earth. This part of the incident sunlight is therefore lost to the process of heating the Earth. I performed the relevant mathematical analysis for the geometry of the Earth using standard theories of light scattering by spherical grains, and sent this to Fred in the autumn of 1990. Fred checked every step of my calculation and our joint paper "Back-scattering of sunlight by ice grains in the mesosphere" was published in *Earth, Moon and Planets* (F. Hoyle and N.C. Wickramasinghe, *Earth, Moon and Planets* **52**, 161–170, 1991), and paved the way to other related projects. Several shorter contributions from us followed, all of which stressed the sensitivity of the Earth's climate to stratospheric particle loading, particles that were derived from either terrestrial or extraterrestrial sources (e.g., Hoyle and Wickramasinghe, *Nature* **350**, 467, 1991).

Throughout much of the decade we continued to pursue various consequences of stratospheric dusting. In 1996, we collaborated with Bill Napier and Victor Clube to explore the effects of the break up of a giant comet, leading to fragments in Earth-crossing orbits, and to recurrent episodes of impacts and cometary dusting on the Earth. We argued for a connection between such cometary events, periodic glaciations as well as episodes of mass-extinctions in the geological record. We also suggested that the entire history of human civilization, over the past 10,000 years, after the end of the last ice age, bears witness to a record of repeated episodes of assaults from the skies (S.V.M. Clube, F. Hoyle, W.M. Napier and N.C. Wickramasinghe, *Astrophys. Space Sci.* **245**, 43–60, 1996). In a later paper, Fred and I also worked out a more specific analysis of the possible connection between cometary events and ice ages (F. Hoyle and N.C. Wickramasinghe, *Astrophys. Space Sci.* **275**, 367–376, 2001).

The modelling of ice ages took us in a somewhat esoteric direction to examine more carefully an idea that Fred had touched upon in a preprint some years earlier. The connection might sound bizarre, but our studies of ice ages led us to the origin of race prejudice among humans. This issue had come recently to the fore in view of a much publicised report of a British government inquiry into the gratuitous murder of Stephen Lawrence (a black youngster) in South East London, in April 1993. The London Metropolitan Police had refused to prosecute the white youths who murdered Lawrence, and the family launched a private prosecution that led to a Government inquiry under the chairmanship of Sir William Macpherson. The Macpherson report, published in March 1999, found the Metropolitan Police to be "institutionally racist", a phenomenon that was said to exist also in other institutions.

How, one might ask, could such a strong emotional response to skin colour prevail at a time in our civilization when we pride ourselves as being "enlightened"? When Fred and I first discussed this,

we soon agreed that there must be a powerful biological imperative for racism to persist, not just in Britain, but throughout much
of the modern world. We published our speculations on this subject in 1999 in the *Journal of Scientific Exploration* (**13**, 681–684,
1999). The theory is very simple, if somewhat Lamarckian in its
implications.

It is an undeniable fact that human evolution over the past
2 million years has led to the emergence of two broadly distinct
groups of people with regard to skin colour, one fair the other dark.
The lighter skinned group now occupies countries in northern latitudes that were on the borders of glaciers during ice ages, and
the darker skinned group mainly inhabit temperate and equatorial
regions. The difference in skin colour between these groups hinges
on a variable efficiency to produce the pigment melanin. Melanin is
produced in a special group of cells known as the melanocytes that
are located at the base of the skin, and although the density of such
cells remains more or less invariable, the efficiency of melanin expression is highly variable. Many genes appear to be involved in melanin
production, and the overall situation for melanin expression (that is
to say for being black) is strongly dominant.

The present-day situation for maintaining selective pressures for
melanin expression rests on a razor's edge between two competing
effects. On the one hand melanin as a pigment protects the base of
the skin from being damaged by ultraviolet radiation from the Sun
which in extreme instances leads to carcinomas of the skin. On the
other hand, an adequate penetration through the skin of ultraviolet radiation with wavelength shortward of 3130 Å is needed for the
production of Vitamin D. Whilst the latter requirement is not too
relevant in the present day with high levels of nutrition generally, it
would have been a strong selective factor for survival in harsher prehistoric times. With lower levels of dietary acquisition of Vitamin D,
the lack of an adequate absorption of sunlight would lead to the
crippling disease of rickets. In this disease severe bone deformities
result from the lack of Vitamin D, a substance that plays a crucial role in the absorption of calcium from food. The correct level
of pigment expression depends on the available ultraviolet light at

any given location at a given time. The balance is between rickets causing skeletal deformities with a consequent lower fecundity, and excessive sunburn radiation leading to skin cancers with attendant high levels of mortality.

The higher incidence of skin cancers in lighter skinned Caucasian migrants in the tropics is well attested. Likewise a high incidence of rickets has been recorded in black and Asian populations living in northern latitude countries before the large-scale introduction of vitamin supplements into staple foods. At the beginning of the 20th century, rickets was reported to affect 90% of black infants in New York. Even as recently as the 1970's high rates of incidence of rickets have been recorded in children of Asian immigrants living in Britain, as for instance in a Glasgow based study. The intensity of sunlight available to Asian immigrants in countries like the UK is obviously mismatched to their expressed pigment level, but routine food fortifications and vitamin pills now generally compensate for this deficiency. Needless to say such dietary supplements were unavailable in prehistoric times.

Throughout the Pleistocene epoch when human evolution occurred, the Earth was locked in an ice age that lasted for nearly 2.5 million years. There were warmer interglacial remissions interspersed throughout this time, each one lasting for about 10,000 years, and the total duration of all such warm periods making up just 10 per cent of the entire Pleistocene era. The Earth emerged from the last ice age approximately 11,000 years ago.

During ice ages the average temperature of the Earth's surface was about 10 degrees Celsius colder than today, and ice sheets were about three times as extensive as they are now. White skinned Nordic tribes living close to the edge of rugged windswept ice sheets under grey skies would have been eking out a precarious existence, grabbing whatever food could be gathered and utilising every photon of ultraviolet from the sun in order to stay alive and free of rickets. For them survival was crucially contingent upon having their genes for melanin suppressed. For people living in the tropics, however, the drier ice-age conditions with less cloud cover than at present would have made for a remorseless flood of ultraviolet radiation to fall on their skins. Survival for them was contingent on the fullest expression of their melanin genes, being as black they possibly could be.

In the course of random migrations from the south, the white populations living at the edge of ice sheets would have been at risk through matings with people possessing darker skins. Black–white matings would have tended to produce offspring with darker skins and thus more prone to rickets. Fewer of these malformed children would reach reproductive age, so black–white matings posed a real extinction threat to the white races. Under such circumstances the emergence of mating prohibitions and colour prejudice would be a natural outcome. The prejudice would become deeply ingrained in social traditions, language, mythology and religion. Thus the depiction of Satan as a black figure cannot be regarded as accidental, nor can the association of evil generally with blackness. The strong emotions manifest in modern racism could be understood, but not forgiven, in these terms.

Returning from this extended diversion back to our mainstream activities, we examined more carefully the theory behind our earlier computations of cross sections for iron whiskers. It would be recalled that these calculations were crucial for non-cosmological explanations of the cosmic microwave background. One deficiency of our earlier work that began to worry us was an assumption we made relating to the electrical conductivity of iron at low temperatures and low frequencies. We had taken a frequency independent value of conductivity of 10^{18} s^{-1} somewhat arbitrarily. Was there a better way to deal with this matter? In attempting to answer this question we came upon a way of using a well-attested theory of metals known as the Drude theory to work out the low temperature dielectric function of iron as a function of frequency. The only input that was now required was the DC conductivity of iron which of course was well known. Our calculations which broke new ground were published in 1994 (Wickramasinghe and Hoyle, *Astrophys. Space Sci.* **213**, 143–154, 1994) and the formulae set out here were used in all our subsequent calculations involving iron whiskers.

The dust in interstellar space continued to turn up many surprises even after more than three decades. Keeping abreast with the latest developments in this field, we noticed a remarkable observation that had gone almost unnoticed. Dust in a wide range of astronomical situations — reflection nebulae, planetary nebulae, HII region, high latitude galactic cirrus and in the extended halo of the external galaxy M82 all showed a broad emission feature over the waveband 6000–7000 Å. Astronomers were trying to identify this feature as being an effect of the so-called PAH molecules in interstellar space. But the fits for inorganic PAH's that had been considered in relation to this data remained poor. We discovered that a much closer fit could be obtained if one considered a fluorescence phenomenon that is well-known for many biological pigments. A galaxy like M82 appeared to be glowing in a pigment similar to that present in glow-worms! (F. Hoyle and N.C. Wickramasinghe, *Astrophys. Space Sci.* **235**, 343–347, 1996)

The giant Comet Hale-Bopp made its début in the Spring of 1997. It was certainly the brightest comet that had been seen for some time. The comet was estimated to have a nucleus of some 40 km in diameter and an orbital period, as it came in from the outskirts of the solar system, of about 4200 years. It brightened steeply as it came into perihelion on April 1 with a tail extending over 10–30 degrees, and remained a spectacular object for most of March and April 1997. Many types of molecules including organic molecules were discovered spectroscopically in the coma of Comet Hale-Bopp, and the infrared spectrum over the wavelength range 2.5–45 μm range was measured by the European Space Agency's Infrared Space Observatory (ISO) satellite when the comet was at a distance of 2.9 Astronomical Units from the sun. This spectrum is displayed by the jagged curve of Fig. 16. The dashed curve shows a spectrum we calculated for a model involving approximately 90% by mass of a bioculture including diatoms, and 10% in the form of pure olivine dust. Although an inorganic olivine component is needed to fit the positions of the peaks near 10 μm, this alone is hopeless for explaining the full range of the data. A dominant contribution from an organic (biologic) component is required, and any larger contribution from olivine than about 10 percent is inconsistent with the data. We published our results first

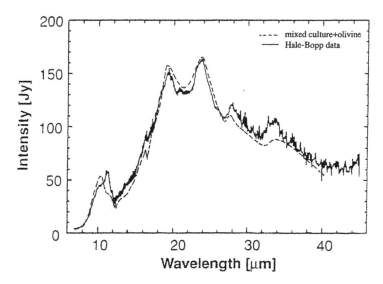

Fig. 16   Fit of the spectrum of Comet Hale-Bopp to a mixture of microorganisms and crystalline olivine.

in the new internet journal *Natural Science* of May 1997, and later in *Astrophysics and Space Science* (**268**, 379–383, 1999).

A further indication of cometary biology was the discovery that comets show considerable activity when they are at relatively great distances from the Sun, far beyond the orbit of Jupiter. This was observed for comet Halley after its 1986 perihelion when it had retreated to a distance of some 6–10 Astronomical Units from the Sun. Similar sporadic outbursts of activity had been known to be occurring in the case of comet Schwassmann–Wachman I, which has a period of about 15 years and an orbit that lies outside the boundaries of Jupiter and Saturn. Now in the case of the new comet Hale-Bopp it was reported that before it came to perihelion (in August, September and October of 1995), it was producing extensive dust and carbon monoxide halos. We analysed this data and concluded that a strong case for biological outgassing could be made. Inorganic comets would not be expected to explode in the manner that Comet Hale-Bopp had done in the cold depths of space. But biological activity simmering beneath a frozen crust and intermittently provoked by meteorite impacts could lead to the build-up pockets of high pressure gas that periodically explode, releasing gas and dust particles (N.C. Wickramasinghe, F. Hoyle and D. Lloyd, *Astrophys. Space Sci.* **240**, 161–165, 1996).

Panspermia theories came into sharp focus in August 1996 following the announcement of a possible detection of microbial fossils in a Martian meteorite. The meteorite ALH84001 — a piece of Martian rock ejected by a cometary impact — was studied by David S. McKay and a team of scientists and was found to contain complex organic molecules associated with μm-sized carbonate globules (D.S. McKay *et al*, *Science* **273**, 924, 1996). The team made the startling claim that the organics were likely to be generated biologically, and moreover that structures such as are shown in Fig. 17 are most likely to represent bacterial fossils. The headline news prompted by this work that "we are all descended from Martians" — provoked a storm of controversy that continues to the present day.

Although that claim itself has since been challenged, the impact of the initial announcement has not diminished in the intervening years. Astrobiology has suddenly emerged as a new scientific discipline and

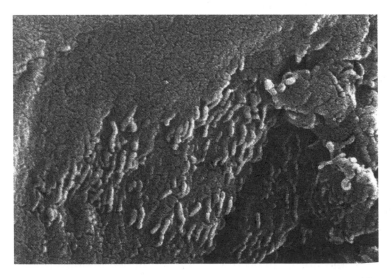

Fig. 17   Putative microbial fossils in the Martian meteorite ALH84001 (Courtesy NASA).

several international organisations including NASA have expressed their commitment to research in this general area. Connecting an impending paradigm shift in a roundabout way to Mars was a decision that was politically astute. For the concept of life on Mars has been filtering slowly into the public's consciousness since at least 1898 when H.G. Well's novel, *The War of the Worlds* was first published and introduced the frightening fiction that Martians were threatening to invade the Earth.

The Mars meteorite ALH84001 has shown beyond any doubt that complex organic structures, and by inference even microbial cells, could be transferred in a viable form from one planetary body to another. Planetary panspermia or transpermia as this concept has recently come to be known, is not by any means a new theory. It was discussed by Lord Kelvin well over a century ago. In his presidential address to the 1881 meeting of the British Association, Kelvin drew the following remarkable picture:

"When two great masses come into collision in space, it is certain that a large part of each is melted, but it seems also quite certain that in many cases a large quantity of debris must be shot forth in all directions, much of which may have experienced no greater violence than individual pieces of rock experience in a landslip or in blasting by gunpowder. Should the time when this earth comes into collision with another body, comparable in dimensions to itself, be when it is still clothed as at present with vegetation, many great and small fragments carrying seeds of living plants and animals would undoubtedly be scattered through space. Hence, and because we all confidently believe that there are at present, and have been from time immemorial, many worlds of life besides our own, we must regard it as probable in the highest degree that there are countless seed-bearing meteoric stones moving about through space. If at the present instant no life existed upon the earth, one such stone falling upon it might, by what we blindly call natural causes, lead to its becoming covered with vegetation."

Thus the ideas that have recently come to the forefront of scientific discussion were in circulation over 123 years ago. Such interplanetary transfers of life as described by Kelvin are possible of course, but in our view they represent a relatively unimportant route for exchange of life on a cosmic scale. Moreover they fail to address the all-important question of how life began in the solar system in the first place. According to the ideas we have discussed earlier, a far better option is to have Mars, Earth and every other habitable planetary body infected with the same cometary source of life — a source of life that is derived from an even bigger system. As we have already noted, comets impact all planetary bodies, and so cometary panspermia must surely remain the principal route for the transference of cosmic life.

It has been widely claimed that interstellar panspermia is unlikely because of the hazards of ultraviolet light and ionising radiation that has to be faced by iterant bacteria (C. Mileikowsky, *et al*, *Icarus* **145**, 391, 2000). A typical transit time between amplification sites in the galaxy could take a few million years, and the fraction of survivors needed for our theory to be valid is less than 10^{-22}. Firstly, we have argued that ultraviolet radiation is very easily protected against: only a thin layer of carbonaceous coating around a bacterium provides almost complete shielding. Ionising radiation could pose a bigger threat, but survivors are still inevitable, at least in the cases where entire comets could be transported from one planetary system harbouring life to another nascent system. Even individual bacteria or clumps of bacteria could withstand the doses of ionising radiation received during transit. Experiments suggesting otherwise are based on large fluxes of ionising radiation delivered in seconds or minutes, whereas in the interstellar medium a trickle of such radiation is incident over millions of years. The two situations could be dramatically different and not directly analogous as is normally assumed. It is well known that the oxidising effects of free radicals, particularly the hydroxyl radical, cause over 90% of DNA damage. So reducing the water content (from which hydroxyl is derived) can drastically diminish the lethal effects of ionising radiation. Moreover, irradiation in an inert atmosphere or vacuum such as exist in interstellar space, would also reduce potential damage. Low temperatures also go in the same direction by immobilising and preventing the diffusion of free radicals. For all these reasons it is fair to surmise that there is a considerable uncertainty as to the effects of cosmic radiation on interplanetary or interstellar bacteria. A low flux of ionising radiation delivered over astronomical timescales to dormant freeze-dried bacteria (in the absence of H_2O and air) would perhaps bear no comparison with equivalent doses on vegetative cultures in the laboratory (N.C. Wickramasinghe and J.T. Wickramasinghe, *Astrophys. Space Sci.* **286**, 453, 2003).

Direct proof of the survival of bacteria exposed to radiation environments in the near Earth environment has also been demonstrated using NASA's Long Exposure facility (G. Hornek, *et al, Adv. Space Res.* **14**, 41, 1994). Viable cultures of bacteria have been recovered from ice drills going back 500,000 years, from isolates in amber over 25–40 million years (R.J. Cano and M. Borucki, *Science* **268**, 1060, 1995) and from 120 million year old material (C.L. Greenblatt *et al, Microbial Ecology* **38**, 58, 1999). Similarly viable bacteria were recovered in salt crystals from a New Mexico salt mine dated at 250 million years (R.H. Vreeland, W.D. Rosenzweig and D. Powers, *Nature* **407**, 897–900, 2000). The present day dose rate of ionising radiation on the Earth arising from natural radioactivity is in the range 0.1–1 rad per year. Well-attested recoveries of dormant bacteria/spores after 100 million years imply tolerance to ionising radiation with total doses in the range ∼10–100 million rads. All the indications are that a large enough fraction survives to ensure the operation of panspermia, even for "naked" bacteria or bacterial clumps.

The theory that we developed throughout our journey requires life to have been introduced to Earth for the first time by comets some 4 billion years ago. But that process could not have stopped at a distant time in the past. Comets are still with us, and the Earth is entwined in the debris shed by comets. We know that at the present time some 100 tonnes of cometary material reaches our planet on a daily basis. One might then ask: What evidence is there of living particles, microbes, coming in with this influx of debris? Much of the infalling cometary debris would of course be in the form of millimetre-sized or larger particles that burn up as meteors on entry. But a significant fraction of infalling cometary material will be of sizes that will enable them to travel safely through the atmosphere, and this, according to our ideas, must include clumps of bacteria, including nanobacteria and viruses, freshly released from cometary surfaces.

I have previously discussed our theory of the 1980s that bombardments from space could lead to pathogenic interactions with higher life forms. And our interest in this process never ceased. In December 2000, during the epidemic of BSE that was raging through farms in the UK, we wrote the following letter to the *Independent*:

"THE CAUSE OF BSE

SIR — Diseases of plants and animals have a long history of mysterious appearances, like mysterious characters that appear inexplicably on stage in a play, without any satisfactory explanation being offered as to where they have come from. An example some years ago was the lethal respiratory disease that suddenly hit the grey seals in the remote Siberian Lake Baikal.

As the remarkable complexity of genetic systems comes increasingly to light it should be obvious that life on the Earth is far too intricate to have evolved here in isolation from the rest of the Universe. It is because life here is a part of a far vaster system that it is so complex.

The connection comes in our view from material of cometary origin being incident on the Earth in considerable amount. Recent studies have shown that much of the material of escaping from comets is in the form of organic particles that cannot be distinguished from biomaterial. The mass input to the Earth is estimated to be several tens of tonnes of cometary stuff per day, sufficient if it was all in the form of bacteria to give a daily incidence of several hundred thousand bacteria per square metre of area. For the most part the material proves to be harmless. It simply washes away. But in rare cases a connection may occur and if a connection escalates, mostly due to fortuitous circumstances, a new disease is born."

Small particles of bacterial and viral sizes descend through the Earth's stratosphere mostly during the winter months, and in our opinion it was the nearly unique English practise of out-wintering cattle that explains why BSE hit English farms more severely than elsewhere. English farmers move cattle frequently from field to field, maximising their chance of picking up any pathogen that may fall from the air onto the grass.

Once a causative agent (genetic fragment or piece of infective protein) got into a few cattle man took a hand, by grinding up infected animals and including them in feed for more cattle. In retrospect this may look a foolish thing to have done, but without knowing what was going on it is roughly comprehensible on economic grounds.

We live nowadays in a blame culture, egged on relentlessly by television. Somebody, we are constantly being told, has to be held responsible for BSE, when according to our point of view there was no culprit, not unless blame be equated with ignorance. Indeed the political authorities, by banning the inclusion of infected portions of cattle in cattle feed, may be said to have acted both quickly and responsibly.

Whether they should also have banned any use of cattle products in medical vaccines remains another question with disturbing possibilities.

Prof. Sir Fred Hoyle
Prof. Chandra Wickramasinghe"

In February 1999 the Stardust Mission to Comet Wild 2 was launched with the aim of conducting *in situ* experiments as well as collecting samples of cometary dust. The rendezvous and collection took place in January 2004. The collection was executed using an aerogel block to gently break the speed of falling cometary particles, but even so it is not expected that any microbes could survive the impact on the aerogel. When the material is finally returned to Earth in 2006, fragments of organic structures bearing tell-tale signs of life may be all we can expect to find.

The most promising method of detecting incoming cometary microorganisms is the use of sterile collection systems sent on balloons into the high stratosphere. Above the tropopause, which is 18 km in the tropics and 10 km in temperature latitudes, aerosols of 1–$10\,\mu$m in size, including bacterial clumps, could not stay for more than a very short timescale, weeks or less. They would quickly fall under gravity. If small amounts of bacteria from the Earth's surface get lofted on rare occasions to great heights, for example after a volcanic eruption, they would quickly fall. Above 40 km you would not expect to find any terrestrial bacteria at all in normal times, so if significant quantities of stratospheric bacteria are discovered, this would provide *prima facie* evidence of panspermia.

For many years Fred and I tried to convince organisations that had the capacity to carry out such an experiment that this was a project of potential value. The responses we received were consistently discouraging except in one instance. Fred had visited Jayant Narlikar at the Tata Institute on several occasions during the 1980's and had suggested that they attempted to collect cometary dust in the stratosphere using sterile equipment carried on balloons. The Tata Institute Balloon launching facility had a distinguished track record since the 1950's when they were engaged in pioneering work on the detection of cosmic rays. Evidently, Fred was given a polite hearing by the Indian scientists but the general impression at the time was that the experiment was not feasible because collection procedures that were sufficiently aseptic were not available. Methods of detecting small amounts of DNA using amplification techniques for example, were also not developed at the time. By the late 1990's the situation had significantly changed. Under the leadership of Jayant Narlikar a team of physicists and biologists proposed an experiment of the kind suggested by Fred Hoyle and myself to the Indian Space Research Organisation (ISRO). Funding was finally approved in the year 2000.

The object of the experiment was to collect stratospheric air aseptically, and to examine it in the laboratory for signs of life. The collection part of the project was as follows. A number of specially manufactured sterilized stainless steel cylinders were evacuated to almost zero pressures and fitted with valves that could be open and shut on ground telecommand. An assembly of such cylinders was suspended in a chamber of liquid neon to keep them at cryogenic temperatures, and the entire payload was launched from the TATA Institute Balloon launching facility in Hyderabad, India on 21 January, 2001. As the valves of the cylinders were opened at predetermined heights, ambient air rushed in to fill the vacuum, building up very high pressures within the cylinders. The valves were shut after a prescribed length of time, the cylinders hermetically sealed and parachuted down to the ground.

A set of cylinders was transported to Cardiff in February 2001, but due to the bureaucratic difficulties I had been experiencing with the authorities at Cardiff Univesity, the samples were not extracted from them until April 2001. Fred was kept fully appraised of developments of the balloon experiment and he gave us valuable advice at every stage. Priya and I saw Fred for the last time on 21 February 2001 in Bournemouth. The first question he asked me was "What have you found in the balloon samples?" My answer was that we were still waiting to appoint an assistant to do the work — a situation he found difficult to comprehend.

Eventually I managed to secure a minimal extent of support from Cardiff University and the short-lived services of a research assistant. The cylinders were opened and the collected stratospheric air made to flow through sterile membrane filters in a contaminant free environment. Any bacteria or clumps of bacteria present in the stratospheric air sample would then be collected on these filters. The analysis was first conducted for us by the microbiology department in Cardiff, and later investigations at Sheffield University were led by Milton Wainwright.

The first phase of this investigation was completed in July 2001 and we had unambiguous evidence for the presence of clumps of living cells in air samples from as high as 41 kilometres, well above the local tropopause (16 km), above which no micron sized aerosols from lower down would normally be expected to transported. The detection was made using electron microscope images (see Fig. 18), and by using a fluorescent dye known as Cyanine that is only taken

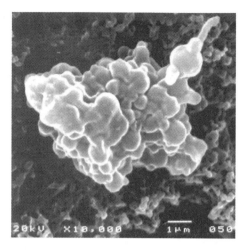

Fig. 18   A cluster of putative microorganisms collected from the stratosphere imaged using a scanning electron microscope.

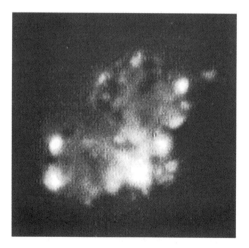

Fig. 19   A cluster of viable microorganisms from 41 kilometres, identified using a fluorescent dye taken up only by live cells.

up by the membranes of living cells. When the isolate treated with the dye was examined under a special kind of microscope the picture on the Fig. 19 was obtained. DNA was also detected in these clumps of cells using yet another fluorescence technique.

The variation with height of the density of such cells indicated strongly that the clumps of bacterial cells are falling from space. The input of such biological material was provisionally estimated to be between 1/3 tonne to 1 tonne per day over the entire planet. If this amount of organic material was in the form of bacteria, the annual transfer of bacteria is 10^{21}. These results were presented by me on behalf of our team at the Instruments, Methods, and Missions for Astrobiology IV session of the SPIE Meeting in San Diego at the end of July 2001. The paper was entitled: "The detection of living cells in stratospheric samples" by Melanie J. Harris, N.C. Wickramasinghe, David Lloyd, J.V. Narlikar, P. Rajaratnam, M.P. Turner, S. Al-Mufti, M.K. Wallis, S. Ramadurai and F. Hoyle, *Proc SPIE* **4495**, 192, 2002. This is sadly the last paper that Fred co-authored. The presentation made international headlines. Doubts about contamination were naturally raised by sceptics with a geocentric worldview, but our initial results have since received extensive confirmation in the work carried out by Milton Wainwright in Sheffield (M. Wainwright, N.C. Wickramasinghe, J.V. Narlikar and P. Rajaratnam, *FEMS Microbiology Letters*, **218**, 161, 2003; M. Wainwright, N.C. Wickramasinghe, J.V. Narlikar, P. Rajaratnam and J. Perkins, *Int. J. Astrobiol*, **3**, 13, 2004).

Science is certainly progressing towards vindicating the point of view that Fred Hoyle and I had developed over several decades. I have recently explored an alternative route to panspermia which compliments our previous work. (W.M. Napier, *Mon. not. R. Astron. Soc.* **348**, 46, 2004; M.K. Wallis and N.C. Wickramasinghe, *Mon. not. R. Astron. Soc.* **348**, 52, 2004.) Just as comets and asteroids impacting the Earth, after it was populated with life, could lead to extinctions of species, they could also splash material laden with life back into space. A fraction of this life-bearing material survives the ejection process and can actually escape from the solar system. The solar system (including the Earth) revolves around the centre of the Galaxy once in 240 million years, and the life-bearing material ejected from Earth would have periodic access at close quarters to hundreds of millions of nascent cometary and planetary systems. Such newly forming planetary systems would then be recipients of Earth life in the form of viable microorganisms. And since the Earth cannot be the sole centre of life, this same dissemination process would happen for every other life-bearing planetary system.

Space exploration of comets could be said to have barely got under way in the year 2004. From the time of the Giotto probe of Halley's comet in 1986, standard dogmas about comets continued to be revised. And the trend has been unerringly in the direction from inorganic comets to organic and life-bearing comets, just as we had first proposed in 1979. The most recent Rosetta mission to comet 67P/Churyumov-Gerasimenko (of which the author is a science team member) was launched in February 2004 with a landing on the comet's surface scheduled for 2014. The latest data, however, came from the Stardust Spacecraft's encounter with Comet Wild 2 which took place in January 2004 at a distance from the comet of just 236 kilometres (147 miles). The Principal Investigator of the Stardust Mission, Donald Brownlee is reported as saying:

> *"We thought Comet Wild 2 would be like a dirty, black, fluffy snowball. Instead it was mind-boggling to see the diverse landscape in the first pictures — including spires, pits and craters, which must be supported by a cohesive surface..."*

Stardust images of Comet Wild 2 showed pinnacles towering to heights of 100 metres and craters plunging to depths of more than 150 metres. The entire comet is only about five kilometres across, yet one of the largest craters is itself a kilometre across. Comet Wild 2 was imaged at close quarters (Fig. 20) and showed features that could not be further removed from the old "dirty snowball model", one that had acquired a hallowed status in cometary theories of the 20th century. Most puzzling of all are the dozens of extended jets — particularly jets that emanated from the unlit dark side of the comet that faced away from the sun. Comet Wild 2 has provided startling evidence of a seething hot cauldron of organic material bubbling beneath a frozen crust. Weak spots on the crust become ruptured from time to time, venting the high pressure fluid beneath thus giving rise to jets. The new data is strikingly consistent with our biological

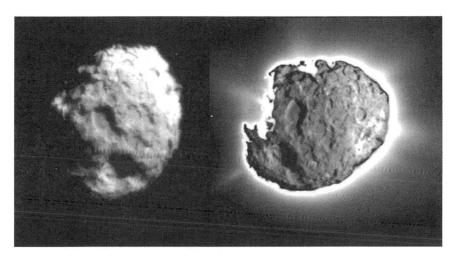

Fig. 20 Images of Comet Wild 2 taken in January 2004 with cameras aboard STARDUST (Courtesy, NASA, JPL).

model of a comet. Evidence supporting cosmic life seems now to be inescapable.

Over half a millennium ago, the Polish astronomer Nicolaus Copernicus (1473–1543) dethroned the Earth from its privileged position at the centre of the cosmos. The Earth, however, continued, well into the 20th century, to occupy an exalted status as the centre of life. Life was regarded as the result of terrestrial evolutionary processes occurring independently of the external Universe. Throughout our journey, Fred and I have sought to challenge this position using evidence derived from many different fields of science. Life on Earth cannot be regarded as being isolated from the rest of the galaxy. We are part of a truly enormous cosmic gene pool. Nothing of great innovative significance in biology ever happened on the Earth. The Earth was simply a receiving station, a building site for the incomparably magnificent edifice of cosmic life. When prejudice against the concept of cosmic life has run its course, our ideas will come to be regarded as self-evident truths. The cosmic quality of life will seem as obvious to future generations as the Sun being at the centre of our solar system is obvious to us today.

The sad news of Fred Hoyle's death at the age of 86 on 21st August 2001 marked the end of an era of 20th century science, a pursuit that was dictated solely by intellectual curiosity to discover what the world was really like. Sociological constraints were subjugated and the man-made boundaries between the various disciplines of science were broken. As Fred Hoyle would often say, the Universe itself has no respect whatsoever for such artificial boundaries. His own monumental work on nucleogenesis in the 1940's and 1950's, introducing nuclear physics to astronomy, epitomised this principle. So also did our own joint work that I have described in this book that lifted the barriers between biology and astronomy and gave birth to the burgeoning new science of astrobiology.

My journey with Fred Hoyle over nearly four decades was always filled with action — adventures into uncharted and sometimes dangerous terrain, and the excitement of new discoveries. If I were given another chance, I would gladly follow the same path again.

Where our journey ends
Another must begin
In some distant corner of the Universe
Following inexorably
The selfsame path.

Bibliography

Technical Papers

1. "A note on the origin of the Sun's polar field", F. Hoyle and N.C. Wickramasinghe, *Mon. not. R. Astron. Soc.*, **123**, 51, 1962.
2. "On graphite particles as interstellar grains", F. Hoyle and N.C. Wickramasinghe, *Mon. not. R. Astron. Soc.*, **124**, 417, 1962.
3. "On the deficiency in the ultraviolet fluxes from early type stars", F. Hoyle and N.C. Wickramasinghe, *Mon. not. R. Astron. Soc.*, **126**, 401, 1963.
4. "Impurities in interstellar grains", F. Hoyle and N.C. Wickramasinghe, *Nature*, **214**, 969, 1967.
5. "Condensation of the planets", F. Hoyle and N.C. Wickramasinghe, *Nature*, **217**, 415, 1968.
6. "Solid hydrogen and the microwave background", F. Hoyle, N.C. Wickramasinghe and V.C. Reddish, *Nature*, **218**, 1124, 1968.
7. "Condensation of dust in galactic explosions", F. Hoyle and N.C. Wickramasinghe, *Nature*, **218**, 1127, 1968.
8. "Interstellar grains", F. Hoyle and N.C. Wickramasinghe, *Nature*, **223**, 459, 1969.
9. "Dust in supernova explosions", F. Hoyle and N.C. Wickramasinghe, *Nature*, **226**, 62, 1970.
10. "Radio waves from grains in HII regions", F. Hoyle and N.C. Wickramasinghe, *Nature*, **227**, 473, 1970.
11. "Primitive grain clumps and organic compounds in carbonaceous chondrites", F. Hoyle and N.C. Wickramasinghe, *Nature*, **264**, 45, 1976.
12. "Organic molecules in interstellar dust: a possible spectral signature at 2200 Å?", N.C. Wickramasinghe, F. Hoyle and K. Nandy, *Astrophys. Space Sci.*, **47**, L1, 1977.

13. "Polysaccharides and the infrared spectrum of OH26.5 + 0.6", F. Hoyle and N.C. Wickramasinghe, *Mon. not. R. Astron. Soc.*, **181**, 51P, 1977.

14. "Spectroscopic evidence for interstellar grain clumps in meteoritic inclusions", A. Sakata, N. Nakagawa, T. Iguchi, S. Isobe, M. Morimoto, F. Hoyle and N.C. Wickramasinghe, *Nature*, **266**, 241, 1977.

15. "Polysaccharides and the infrared spectra of galactic sources", F. Hoyle and N.C. Wickramasinghe, *Nature*, **268**, 610, 1977.

16. "Prebiotic polymers and infrared spectra of galactic sources", N.C. Wickramasinghe, F. Hoyle, J. Brooks and G. Shaw, *Nature*, **269**, 674, 1977.

17. "Identification of the 2200 Å interstellar absorption feature", F. Hoyle and N.C. Wickramasinghe, *Nature*, **270**, 323, 1977.

18. "Origin and nature of carbonaceous material in the galaxy", F. Hoyle and N.C. Wickramasinghe, *Nature*, **270**, 701, 1977.

19. "Identification of interstellar polysaccharides and related hydrocarbons", F. Hoyle, N.C. Wickramasinghe and A.H. Olavesen, *Nature*, **271**, 229, 1978.

20. "Calculations of infrared fluxes from galactic sources for a polysaccharide grain model", F. Hoyle and N.C. Wickramasinghe, *Astrophys. Space Sci.*, **53**, 489, 1978.

21. "Comets, ice ages and ecological catastrophes", F. Hoyle and N.C. Wickramasinghe, *Astrophys. Space Sci.*, **53**, 523, 1978.

22. "Biochemical chromophores and the interstellar extinction at ultraviolet wavelengths", F. Hoyle and N.C. Wickramasinghe, *Astrophys. Space Sci.*, **65**, 241, 1979.

23. "On the nature of interstellar grains", F. Hoyle and N.C. Wickramasinghe, *Astrophys. Space Sci.*, **66**, 77, 1979.

24. "The identification of the 3 micron spectral feature in galactic infrared sources", F. Hoyle and N.C. Wickramasinghe, *Astrophys. Space Sci.*, **68,** 499, 1980.

25. "Organic grains in space", F. Hoyle and N.C. Wickramasinghe, *Astrophys. Space Sci.*, **69**, 511, 1980.

26. "Organic material and the 1.5–4 micron spectra of galactic sources", F. Hoyle and N.C. Wickramasinghe, *Astrophys. Space Sci.*, **72**, 183, 1980.

27. "Dry polysaccharides and the infrared spectrum of OH26.5 + 0.6", F. Hoyle and N.C. Wickramasinghe, *Astrophys. Space Sci.*, **72**, 247, 1980.

28. "Evidence for interstellar biochemicals", F. Hoyle and N.C. Wickramasinghe, in *Giant Molecular Clouds in the Galaxy*, (eds.) P.M. Solomon and M.G. Edmunds (Pergamon, 1980).

29. "Why Neo-Darwinism does not work", F. Hoyle and C. Wickramasinghe (University College, Cardiff Press, 1982).

30. "Comets — a vehicle for panspermia", F. Hoyle and N.C. Wickramasinghe (ed.) C. Ponnamperuma (D. Reidel Publishing Co., 1981).

31. "Infrared spectroscopy of micro-organisms near 3.4 microns in relation to geology and astronomy", F. Hoyle, N.C. Wickramasinghe, S. Al-Mufti and A.H. Olavesen, *Astrophys. Space Sci.*, **81**, 489, 1982.

32. "Infrared spectroscopy over the 2.9–3.9 micron waveband in biochemistry and astronomy", F. Hoyle, N.C. Wickramasinghe, S. Al-Mufti, A.H. Olavesen and D.T. Wickramasinghe, *Astrophys. Space Sci.*, **83**, 405–409, 1982.

33. "Interstellar absorptions at $\lambda = 3.3$ and 3.3 microns", S. Al-Mufti, A.H. Olavesen, F. Hoyle and N.C. Wickramasinghe, *Astrophys. Space Sci.*, **84**, 259, 1982.

34. "Organo-siliceous biomolecules and the infrared spectrum of the Trapezium nebula", F. Hoyle, N.C. Wickramasinghe and S. Al-Mufti, *Astrophys. Space Sci.*, **86**, 63, 1982.

35. "A model for interstellar extinction", F. Hoyle and N.C. Wickramasinghe, *Astrophys. Space Sci.*, **86**, 321, 1982.

36. "The infrared spectrum of interstellar dust", F. Hoyle, N.C. Wickramasinghe and S. Al-Mufti, *Astrophys. Space Sci.*, **86**, 341, 1982.

37. "On the optical properties of bacterial grains, I", N.L. Jabir, F. Hoyle and N.C. Wickramasinghe, *Astrophys. Space Sci.*, **91**, 327, 1983.

38. "Interstellar proteins and the discovery of a new absorption feature at $\lambda = 2800\,\text{Å}$", L.M. Karim, F. Hoyle and N.C. Wickramasinghe, *Astrophys. Space Sci.*, **94**, 223, 1983.

39. "The ultraviolet absorbance spectrum of coliform bacteria and its relationship to astronomy", F. Hoyle, N.C. Wickramasinghe, E.R. Jansz and P.M. Jayatissa, *Astrophys. Space Sci.*, **95**, 227, 1983.

40. "Organic grains in the Taurus interstellar clouds", F. Hoyle and N.C. Wickramasinghe, *Nature*, **305**, 161, 1983.

41. "Bacterial life in space", F. Hoyle and N.C. Wickramasinghe, *Nature*, **306**, 1983.

42. "The spectroscopic identification of interstellar grains", F. Hoyle, N.C. Wickramasinghe and S. Al-Mufti, *Astrophys. Space Sci.*, **98**, 343, 1984.

43. "Proofs that life is cosmic", F. Hoyle and N.C. Wickramasinghe, *Mem. Inst. Fund. Studies*, Sri Lanka, No. 1, 1983.

44. "2.8–3.6 micron spectra of micro-organisms with varying H_2O ice content", F. Hoyle, N.C. Wickramasinghe and N.L. Jabir, *Astrophys. Space Sci.*, **92**, 439, 1983.

45. "The extinction of starlight at wavelengths near 2200 Å", F. Hoyle, N.C. Wickramasinghe and N.L. Jabir, *Astrophys. Space Sci.*, **92**, 433, 1983.

46. "The radiation of microwaves and infrared by slender graphite needles", F. Hoyle, J.V. Narlikar and N.C. Wickramasinghe, *Astrophys. Space Sci.*, **103**, 371, 1984.

47. "The ultraviolet absorbance of presumably interstellar bacteria and related matters", F. Hoyle, N.C. Wickramasinghe and S. Al-Mufti, *Astrophys. Space Sci.*, **111**, 65, 1985.

48. "An object within a particle of extraterrestrial origin compared with an object of presumed terrestrial origin", F. Hoyle, N.C. Wickramasinghe and H.D. Pflug, *Astrophys. Space Sci.*, **113**, 209, 1985.

49. "On the nature of dust grains in the comae of Comets Cernis and Bowell", F. Hoyle, N.C. Wickramasinghe and M.K. Wallis, *Earth, Moon and Planets*, **33**, 179, 1985.

50. "Legionnaires' disease: Seeking a wider cause", F. Hoyle, N.C. Wickramasinghe and J. Watkins, *The Lancet*, 25 May 1985, p. 1216.

51. "Archaeopteryx — a photographic study", R.S. Watkins, F. Hoyle, N.C. Wickramasinghe, J. Watkins, R. Rabilizirov and L.M. Spetner, *British J. Photography* (8 March) **132**, 264, 1985.

52. "Archaeopteryx — a further comment", R.S. Watkins, F. Hoyle, N.C. Wickramasinghe, J. Watkins, R. Rabilizirov and L.M. Spetner, *British J. Photography* (March 29) **132**, 358, 1985.

53. "Archaeopteryx — further evidence", R.S. Watkins, F. Hoyle, N.C. Wickramasinghe, J. Watkins, R. Rabilizirov and L.M. Spetner, *British J. Photography* (April 26) **132**, 468, 1985.

54. "Archaeopteryx — problems arise, and a motive", F. Hoyle and N.C.Wickramasinghe, *British J. Photography* (June 21) **132**, 693, 1985.

55. "The availability of phosphorous in the bacterial model of the interstellar grains", F. Hoyle and N.C. Wickramasinghe, *Astrophys. Space Sci.*, **103**, 189, 1984.

56. "The properties of large particles in the zodiacal cloud and in the interstellar medium and their relation to recent IRAS observations", F. Hoyle and N.C. Wickramasinghe, *Astrophys. Space Sci.*, **107**, 223, 1984.

57. "From grains to bacteria", F. Hoyle and N.C. Wickramasinghe (University College, Cardiff Press, 1984).

58. "Living Comets", F. Hoyle and N.C. Wickramasinghe (University College, Cardiff Press, 1985).

59. "*Viruses from Space*", F. Hoyle and N.C. Wickramasinghe (University College, Cardiff Press, 1986).

60. "On the nature of the interstellar grains", *Q. Jl. R. A. S.*, **27**, 21, 1986.

61. "On the nature of the particles causing the 2200 Å peak in the extinction of starlight", F. Hoyle and N.C. Wickramasinghe, *Astrophys. Space Sci.*, **122**, 181, 1986.

62. "The measurement of the absorption properties of dry microorganisms and its relationship to astronomy", F. Hoyle, N.C. Wickramasinghe and S. Al-Mufti, *Astrophys. Space Sci.*, **113**, 413, 1985.

63. "The viability with respect to temperature of micro-organisms incident on the Earth's atmosphere", F. Hoyle, N.C. Wickramasinghe and S. Al-Mufti, *Earth, Moon and Planets*, **35**, 79, 1986.

64. "Diatoms on Earth, Comets, Europa and in interstellar space", R.B. Hoover, F. Hoyle, N.C. Wickramasinghe, M.J. Hoover and S. Al-Mufti, *Earth, Moon and Planets*, **35**, 19, 1986.

65. "The effects of irregularities of internal structure in determining the ultraviolet extinction properties of interstellar grains", F. Hoyle, N.C. Wickramasinghe, S. Al-Mufti and L.M. Karim *Astrophys. Space Sci.*, **114**, 303, 1985 .

66. "The case for interstellar micro-organisms", F. Hoyle, N.C. Wickramasinghe and S. Al-Mufti, *Astrophys. Space Sci.*, **110**, 401, 1985.

67. "Some evidence against the authenticity of Archaeopteryx Lithographica", F. Hoyle, N.C. Wickramasinghe, L.M. Spetner and M. Magaritz, *Bild der Wissenschaft* **5**, 51, 1988.

68. "Interstellar extinction by organic grain clumps", F. Hoyle and N.C. Wickramasinghe, *Astrophys. Space Sci.*, **140**, 191, 1988.

69. "Polymeric complexes in comets and in space", F. Hoyle and N.C. Wickramasinghe, *Astrophys. Space Sci.*, **141**, 177, 1988.

70. "Cosmic Life Force", F. Hoyle and N.C. Wickramasinghe (J.M. Dent, 1988).

71. "A diatom model of dust in the Trapezium nebula", Q. Majeed, N.C. Wickramasinghe, F. Hoyle and S. Al-Mufti, *Astrophys. Space Sci.*, **140**, 205, 1988.

72. "Mineral Grains in the 10 and 20 μm spectral features in the Trapezium nebula", F. Hoyle, N.C. Wickramasinghe and Q. Majeed, *Astrophys. Space Sci.*, **141**, 399, 1988.

73. "Archaeopteryx — more evidence of a forgery", F. Hoyle, N.C. Wickramasinghe, L.M. Spetner and M. Magaritz, *British J. Photography*, pp. 14–18 (7 Jan. 1988).

74. "The infrared excess from the White Dwarf star G29–38: a Brown Dwarf or dust?", F. Hoyle, N.C. Wickramasinghe and S. Al-Mufti, *Astrophys. Space Sci.*, **143**, 193, 1988.

75. "Metallic particles in astronomy", F. Hoyle and N.C. Wickramasinghe, *Astrophys. Space Sci.*, **147**, 245–256, 1988.

76. "The organic nature of cometary grains", N.C. Wickramasinghe, F. Hoyle, M.K. Wallis and S. Al-Mufti, *Earth, Moon and Planets*, **40**, 101, 1988.

77. "Mineral and organic particles in Astronomy", N.C. Wickramasinghe, F. Hoyle and Q. Majeed, *Astrophys. Space Sci.*, **158**, 335, 1989.

78. "Modelling the 5–$30\,\mu$m spectrum of Comet Halley", N.C. Wickramasinghe, M.K. Wallis and F. Hoyle, *Earth, Moon and Planets*, **43**, 145, 1988.

79. "Aromatic hydrocarbons in very small interstellar grains", N.C. Wickramasinghe, F. Hoyle and T. Al-Jubory, *Astrophys. Space Sci.*, **158**, 135, 1989.

80. "An integrated 2.5–$12.5\,\mu$m emission spectrum of naturally occurring aromatic molecules", N.C. Wickramasinghe, F. Hoyle and T. Al-Jubory, *Astrophys. Space Sci.*, **166**, 333, 1990.

81. "Extraterrestrial particles and the greenhouse effect", N.C. Wickramasinghe, F. Hoyle and R. Rabilizirov, *Earth, Moon and Planets*, **46**, 297, 1989.

82. "Greenhouse dust", N.C. Wickramasinghe, F. Hoyle and R. Rabilizirov, *Nature*, **341**, 28, 1989.

83. "A unified model for the $3.28\,\mu$m and the 2200 Å interstellar extinction feature", F. Hoyle and N.C. Wickramasinghe, *Astrophys. Space Sci.*, **154**, 143, 1989.

84. "Linear and circular polarization by hollow organic grains", F. Hoyle and N.C. Wickramasinghe, *Astrophys. Space Sci.*, **151**, 285, 1989.

85. "The microwave background in steady-state cosmology", F. Hoyle and N.C. Wickramasinghe, *ESA SP-290*, 489, 1989.

86. "A unified model for the $3.28\,\mu$m and $3.4\,\mu$m spectral feature in the interstellar medium and in comets", F. Hoyle and N.C. Wickramasinghe, *ESA SP-290*, 67, 1989.

87. "Biologic versus abiotic models of cometary dust", M.K. Wallis, N.C. Wickramasinghe, F. Hoyle and R. Rabilizirov, *Mon. not. R. Astron. Soc.*, **238**, 1165–1170, 1989.

88. "The extragalactic Universe: and alternative view", H.C. Arp, G. Burbidge, F. Hoyle, J.V. Narlikar and N.C. Wickramasinghe, *Nature*, **346**, 807–812, 1990.

89. "The case for life as a cosmic phenomenon", F. Hoyle and N.C. Wickramasinghe, *Nature*, **322**, 509, 1986.

90. "Sunspots and influenza", F. Hoyle and N.C. Wickramasinghe, *Nature*, **343**, 304, 1990.

91. "Influenza — evidence against contagion: discussion paper", F. Hoyle and N.C. Wickramasinghe, *J. Roy. Soc. Med.*, **83**, 258, 1990.

92. "The microwave background: its smoothness and frequency distribution as an astrophysical product", F. Hoyle, N.C. Wickramasinghe and G. Burbidge, *29th Liege International Astrophysical Colloquium*, July 2–6, 1990.

93. "Mineral grains in interstellar space", N.C. Wickramasinghe, F. Hoyle, S. Al-Mufti and T. Al-Jabory, in *Dusty Objects in the Universe*, (eds.) E. Bussoletti and A.A. Vittone (Kluwer Academic Press, 1990).

94. "Back-scattering of sunlight by ice grains in the Mesosphere", F. Hoyle and N.C. Wickramasinghe, *Earth, Moon and Planets*, **52**, 161 170, 1991.

95. "The implications of life as a cosmic phenomenon: The anthropic context", F. Hoyle and N.C. Wickramasinghe, *J. British Interplan. Soc.*, **44**, 77–86, 1991.

96. "Cometary habitats for primitive life", M.K. Wallis, N.C. Wickramasinghe and F. Hoyle, *Adv. Space Res.*, **12**(4), 281–285, 1992.

97. "The extinction of starlight revisited", N.C. Wickramasinghe, B. Jazbi and F. Hoyle, *Astrophys. Space Sci.*, **186**, 67–80, 1991.

98. "Extinction properties of infinitely long graphite cylinders", B. Jazbi, F. Hoyle and N.C. Wickramasinghe, *Astrophys. Space Sci.*, **186**, 151–155, 1991.

99. "The case against graphite particles in interstellar space", N.C. Wickramasinghe, A.N. Wickramasinghe and F. Hoyle, *Astrophys. Space Sci.*, **196**, 167–169, 1992.

100. "The absorption of electromagnetic radiation by metal cylinders of finite length", N.C. Wickramasinghe, A.N. Wickramasinghe and F. Hoyle, *Astrophys. Space Sci.*, **193**, 141–144, 1992.

101. "Comets as a source of interplanetary and interstellar grains", F. Hoyle and N.C. Wickramasinghe, in *Origin and Evolution of Interplanetary Dust* (eds.) A.C. Levasseur-Regourd and H. Hasegawa (Kluwer Academic Publishers, 1991), pp. 235–240.

102. "Microdiamonds and the 3.4 micron feature in protostellar sources", F. Hoyle and N.C. Wickramasinghe, *Astrophys. Space Sci.*, **207**, 309–311, 1993.

103. "Absorption properties of astronomical iron whiskers: an accurate crogenic model", N.C. Wickramasinghe and F. Hoyle, *Astrophys. Space Sci.*, **213**, 143–154.

104. "Critique of Fischer-Tropsch type reactions in the solar nebula", S. Ramadurai, F. Hoyle and N.C. Wickramasinghe, *Bull. Astron Soc. India* **21**, 329–334, 1993 .

105. "Biofluorescence and the extended red emission in astrophysical sources", F. Hoyle and N.C. Wickramasinghe, *Astrophys. Space Sci.*, **235**, 343–347, 1996.

106. "Very small dust grains (VSDP's) in Comet C/1996 B2 (Hyakutake)", N.C. Wickramasinghe and F. Hoyle, *Astrophys. Space Sci.*, **239**, 121, 1996.

107. "Eruptions from comet Hale-Bopp at 6.5AU", N.C. Wickramasinghe, F. Hoyle and D. Lloyd, *Astrophys. Space Sci.*, **240**, 161, 1996.

108. "Infrared signatures of prebiology — or biology", N.C. Wickramasinghe, F. Hoyle, S. Al-Mufti and D.H. Wallis, in *Astronomical and Biochemical Origins and the Search for Life in the Universe* (eds.) C.B. Cosmovici, S. Bowyer and D. Werthimer (Editrice Compositori, 1997).

109. "Comet P/Shoemaker-Levy 9 collision with Jupiter: a model of G-site dust composition", D.H. Wallis and N.C. Wickramasinghe, *Astrophys. Space Sci.*, **254**, 25–35, 1997.

110. "Spectroscopic evidence for panspermia", N.C. Wickramasinghe, F. Hoyle and D.H. Wallis, *Proc. SPIE*, **3111**, 282–295, 1997.

111. "The astonishing redness of Kuiper-Belt objects", N.C. Wickramasinghe and F. Hoyle, *Astrophys. Space Sci.*, **259**, 205–208, 1998.

112. "Microdiamonds and the ultraviolet extinction of starlight", *Astrophys. Space Sci.*, **259**, 379–383, 1998.

113. "Infrared evidence for panspermia: an update", *Astrophys. Space Sci.*, **259**, 385–401, 1998.
   Miller-Urey synthesis in the nuclei of galaxies", N.C. Wickramasinghe and F. Hoyle, *Astrophys. Space Sci.*, **259**, 99–103, 1998.

114. "Search for living cells in stratospheric samples", J.V. Narlikar, S. Ramadurai, P. Bhargava, S.V. Damle, N.C. Wickramasinghe, D. Lloyd, F. Hoyle and D.H. Wallis, *Proc. SPIE*, **3441**, 301–305, 1998.

115. "Panspermia in perspective", N.C. Wickramasinghe, F. Hoyle and B. Klyce, *Proc. SPIE*, **3441**, 306–318, 1988.

116. "Cosmological panspermia", N.C. Wickramasinghe and F. Hoyle. *Proc. SPIE*, **3441**, 319–323, 1998.

117. "Towards an understanding of the nature of racial prejudice", F. Hoyle and N.C. Wickramasinghe, *J. Scientific Exploration*, **13**, 681–684, 1999.

118. "Cosmic Life: Evolution and Chance", F. Hoyle and N.C. Wickramasinghe, *The Biochemist*, **21**(6), 1999.

119. "Astronomical Origins of Life: Steps towards Panspermia", F. Hoyle and N.C. Wickramasinghe (Kluwer Academic Publishers, 2000).
120. "Cross-linked Heteroaromatic Polymers in Interstellar dust", N.C. Wickramasinghe, D.T. Wickramasinghe and F. Hoyle, *Astrophys. Space Sci.*, **275**, 181–184, 2001.
121. "A bacterial "singerprint in a Leonid meteor train", N.C. Wickramasinghe and F. Hoyle, *Astrophys. Space Sci.*, **277**, 625–628, 2001.
122. "The detection of living cells in stratospheric samples", Melanie J. Harris, N.C. Wickramasinghe, David Lloyd, M. Turner, F. Hoyle, J.V. Narlikar and P. Rajaratnam, *Proc. SPIE*, **4495**, 192–198, 2002.

Books

Lifecloud: The Origin of Life in the Galaxy: F. Hoyle and N.C. Wickramasinghe (J.M. Dent, Lond., 1978).

Diseases from Space: F. Hoyle and N.C. Wickramasinghe (J.M. Dent, Lond., 1979).

Origin of Life: F. Hoyle and N.C. Wickramasinghe (University College Cardiff Press, 1979).

Space Travellers: The Bringers of Life: F. Hoyle and N.C. Wickramasinghe (University College Cardiff Press, 1981).

Evolution from Space: F. Hoyle and N.C. Wickramasinghe (J.M. Dent, 1981).

Is Life an Astronomical Phenomenon?: F. Hoyle and N.C. Wickramasinghe (Universtiy College, Cardiff Press, 1982).

Why Neo Darwinism does not Work: F. Hoyle and N.C. Wickramasinghe (University College Cardiff Press, 1982).

Proofs that Life is Cosmic: F. Hoyle and N.C. Wickramasinghe (Inst. of Fund. Studies, Sri Lanka, Mem, No. 1, 1982).

From Grains to Bacteria: F. Hoyle and N.C. Wickramasinghe (University College, Cardiff Press, 1984).

Living Comets: F. Hoyle and N.C. Wickramasinghe (University College, Cardiff Press, 1985).

Viruses from Space: F. Hoyle and N.C. Wickramasinghe (University College Cardiff Press, 1986).

Archaeopteryx — The Primordial Bird: A Case of Fossil Forgery: F. Hoyle and N.C. Wickramasinghe (Christopher Davies, Swansea, 1986).

Cosmic Life Force: F. Hoyle and N.C. Wickramasinghe (J.M. Dent, Lond., 1988).

The Theory of Cosmic Grains: F. Hoyle and N.C. Wickramasinghe (Kluwer Academic Publishers, 1990).

Our Place in the Cosmos: F. Hoyle and N.C. Wickramasinghe (Weidenfeld and Nicholson, Lond., 1993).

Life of Mars: The Case for a Cosmic Heritage: F. Hoyle and N.C. Wickramasinghe (Clinical Press, 1997).

Astronomical Origins of Life: Steps Towards Panspermia: F. Hoyle and N.C. Wickramasinghe (Kluwer Academic Press, 2000).

Index

"PATTERSON HAS ALWAYS BEEN AN EXPERT AT CONCEIVING CHILLING VILLAINS OF HIS MANY PIECES, AND WITH SULLIVAN, HE ACHIEVES NEW HEIGHTS (OR WOULD THAT BE DEPTHS?) OF TERROR WITH THE INVISIBLE MAN. AND WHILE THE STORY PROCEEDS AT BREAKNECK SPEED…PATTERSON AND SULLIVAN CREATE PLENTY OF SCENES THAT READERS WILL NOT SOON FORGET…*PRIVATE BERLIN* will make you a fan of this wide-ranging and marvelously conceived series, if you are not one already." —BookReporter.com

"FAST-PACED ACTION AND UNFOR-GETTABLE CHARACTERS WITH PLOT TWISTS AND DECEPTIONS WORTHY OF ANY JAMES PATTERSON NOVEL." —Examiner.com

"ONE OF THE BEST AND MOST ENTHRALLING NOVELS PATTERSON HAS WRITTEN."
—Quick-book-review.blogspot.com

"THIS COMPLEX AND LAYERED EUROPEAN THRILLER GRABS YOU FROM THE START, AND COMPELS YOU TO READ."—TheMysterySite.com

PRIVATE LONDON

"THERE ARE THE USUAL TWISTS AND TURNS AND EVERY TIME YOU THINK YOU HAVE IT FIGURED OUT, ANOTHER WRENCH IS THROWN IN. I WAS CAPTIVATED WITH THE STORY AND READ IT IN ONE SITTING."
—AlwayswithaBook.blogspot.com

"REGARDLESS OF PLACEMENT IN THE PRIVATE CANON, *PRIVATE LONDON* MEETS THE HIGH STANDARD ESTABLISHED BY THE

OTHER VOLUMES, AS IT IS LOADED WITH ACTION, TWISTS, TURNS, AND SURPRISES...CERTAINLY ONE OF PATTERSON'S MORE COMPLEX WORKS OF LATE... If you haven't read anything by Patterson for a while, you should give *PRIVATE LONDON* and the Private series a try and acquire a healthy addiction." —BookReporter.com

"THE STORY CONTINUES ALONG QUITE QUICKLY WITH THE TWO-PAGE CHAPTERS FLYING PAST FASTER THAN YOU CAN IMAGINE. I READ THIS BOOK IN ONLY AN EVENING. If you are a Patterson fan then you will probably enjoy this one as well."
—TheFringeMagazine.blogspot.com

PRIVATE GAMES

"EVEN IF YOU ARE NOT INTERESTED IN THE SUMMER GAMES, YOU WILL WANT TO READ THE BOOK, NOT

ONLY FOR ITS THRILL-A-PAGE
PLOTTING BUT ALSO FOR THE
OPENING DAY SPECTACLE THAT'S
ON DISPLAY…*PRIVATE GAMES*
SHOWS BOTH PATTERSON AND
SULLIVAN TO BE AT THE TOP OF
THEIRS."   —BookReporter.com

"ANOTHER FAST-PACED, ACTION-
PACKED THRILLER THAT WILL HAVE
YOU NOT WANTING TO PUT IT
DOWN UNTIL THE VERY LAST PAGE."
                    —ThePhantomParagrapher
                              .blogspot.com

"[AN] EXCELLENT READ…THE
WRITING IS BETTER, THE CHARAC-
TERS MORE DEVELOPED, AND THE
STORY MORE SUSPENSEFUL THAN
MOST OF THE RECENT JAMES
PATTERSON NOVELS…LOOK FOR
PRIVATE BERLIN."
                    —TheMysteryReader.com

"PATTERSON, HE OF SIX DOZEN NOVELS AND COUNTING, HAS AN UNCANNY KNACK FOR THE TIMELY THRILLER, AND THIS ONE IS NO EXCEPTION…A PLEASANT ROMP."

—*Kirkus Reviews*

"THIS ONE IS SET UP WITH SHORT CHAPTERS THAT YOU CAN'T HELP [BUT] FLY THROUGH…I REALLY HAVE COME TO ENJOY THE PRIVATE SERIES AND HOPE THIS SERIES CONTINUES FOR A LONG TIME."

—AlwayswithaBook.blogspot.com

PRIVATE: #1 SUSPECT

"BETWEEN THE COVERS OF THE FIRST TWO BOOKS (*PRIVATE* AND *#1 SUSPECT*) IS SOME OF JAMES PATTERSON'S BEST WORK…A PLOT THAT CONTAINS SUBSTANCE WITH-OUT SACRIFICING READABILITY AND IS INHABITED BY CHARACTERS WHO ARE SYMPATHETIC AND MULTI-

DIMENSIONAL. [THE] CONCEPT AND EXECUTION ARE FLAWLESS...And if someone doesn't latch on to these novels for a television drama, they should find employment in another industry. Jump on now." —BookReporter.com

"[THEY] MAKE ONE HECK OF A GREAT WRITING TEAM AND PROVE IT ONCE AGAIN WITH [THIS] CLASSY THRILLER, THE LATEST IN A PRIVATE INVESTIGATION SERIES THAT'S SURE TO BLOW THE LID OFF A POPULAR GENRE...If you want to be entertained to the max, you can't go wrong when you pick up a thriller by Patterson and Paetro."
—NightsandWeekends.com

"AN AMAZING STORY FROM A FABULOUS STORYTELLER."
—MaryGramlich.blogspot.com

"I READ ON THE TREADMILL, AND TOWARD THE END I ALMOST FELL FROM ALL THE CLIFFHANGERS AND REVELATIONS…CHOCK-FULL OF INTRIGUE, MYSTERY, AND EDGE-OF-YOUR-SEAT TWISTS AND TURNS! I CANNOT *WAIT* FOR THE SEQUEL!"
—ReadingWritingBreathing.com

"*PRIVATE: #1 SUSPECT* STARTED OUT WITH A BANG…LITERALLY. I WAS CAUGHT UP IN THE ACTION FROM PAGE ONE."
—MeMyBookandtheCouch .blogspot.com

PRIVATE

"A QUICK READ WITH SHORT CHAPTERS AND LOADS OF ACTION…DESTINED TO BECOME ANOTHER SUCCESSFUL SERIES IN THE JAMES PATTERSON ARSENAL… If you are a Patterson fan, you will not be disappointed…Jack Morgan is a great protagonist...Bring on even more PRIVATE."   —TheMysteryReader.com

"A FUN AND ENJOYABLE READ... PATTERSON LETS THE READER FEEL LIKE THEY ARE RIGHT THERE IN THE ACTION...It's such a treat knowing there is so much more to explore about [these] characters...He certainly set things up for some explosive action in future novels."
    —CurlingUpbytheFire.blogspot.com

"*PRIVATE* MIXES ACTION, MYSTERY, AND PERSONAL DRAMA TO CREATE A HIGHLY READABLE AND ENTERTAINING EXPERIENCE. If this first volume is any indication, Patterson and Paetro may well be on their way to rivaling—and possibly surpassing—the popularity of their Women's Murder Club series."   —BookReporter.com

"SLICK AND SUSPENSEFUL IN THE USUAL WELL-RECEIVED PATTERSON

STYLE (SUCCINCT AND SIMPLE CHAPTERS, LARGER-THAN-LIFE CHARACTERS, AND A PLETHORA OF SURPRISES)." —BookLoons.com

"A GREAT READ…FAST-PACED…If you like James Patterson you will be more than happy to add this one to your 'To Be Read' pile."
—LuxuryReading.com

"*PRIVATE* WILL HAVE YOU ON THE EDGE OF YOUR SEAT…WITH TWISTS AND TURNS THAT MAKE YOU WANT TO KEEP TURNING THE PAGES."
—ChickwithBooks.blogspot.com

"EXCITING…A GREAT WAY TO SPEND THE AFTERNOON…COMPLEX AND INTERESTING CHARACTERS."
—BibliophilicBookBlog.com

"*PRIVATE* WILL GRAB YOU FROM PAGE ONE AND FORCE YOU TO SIT

THERE UNTIL YOU TURN THE VERY LAST PAGE…A GREAT START TO A NEW SERIES FROM THE MASTER OF FAST-PACED THRILL RIDES."

—LorisReadingCorner.com

"THE PACING IS FRENETIC…I REALLY LIKED JACK MORGAN…THE SECOND-TIER CHARACTERS WERE GREAT…an impressive array of detectives and specialists."

—BookHound.wordpress.com

"PATTERSON ALWAYS DELIVERS A GOOD READ, BUT IN THIS CASE IT WAS SO MUCH MORE…YOU ARE ENTERTAINED WITH MULTIPLE STORIES. Somehow Patterson blends them all seamlessly together, never letting the novel skip or jump, always keeping the reader flawlessly on track and riveted. I LOVED THIS BOOK. A FANTASTIC READ!"

—DemonLoversBooksandMore.com

Private Down Under

THE PRIVATE NOVELS

Private Down Under (with Michael White)
Private L.A. (with Mark Sullivan)
Private Berlin (with Mark Sullivan)
Private London (with Mark Pearson)
Private Games (with Mark Sullivan)
Private: #1 Suspect (with Maxine Paetro)
Private (with Maxine Paetro)

A complete list of books by James Patterson is at the back of this book. For previews of upcoming books and more information about James Patterson, please visit his website or find him on Facebook or at your app store.

Private
Down Under

James Patterson
AND
Michael White

GRAND CENTRAL
PUBLISHING

LARGE PRINT

Copyright © 2014 by James Patterson

Grand Central Publishing
Hachette Book Group
237 Park Avenue
New York, NY 10017

Printed in the United States of America

Different versions of this book were published as *Private Down Under* by Century, a division of Random House, in the UK in 2013 and *Private Oz* by Random House Australia in 2012.

Grand Central Publishing is a division of Hachette Book Group, Inc.
The Grand Central Publishing name and logo is a trademark of Hachette Book Group, Inc.

PRIVATE is a trademark of JBP Business, LLC.

The publisher is not responsible for websites (or their content) that are not owned by the publisher.

ISBN 978-1-62953-014-7

Private Down Under

PROLOGUE

I HAD GOOGLED plenty of info on Justine Smith before I met her. Funny thing, though: even the most serious, businesslike websites couldn't resist slipping in how great-looking she was.

Justine Smith, the stunning second-in-command to Jack Morgan at Private L.A....

Justine Smith, well known to the L.A. underworld for her unsurpassed police smarts, is also known to the L.A. paparazzi for her unsurpassed figure and face...

I couldn't resist: I pressed the Google Images button and took a tour of some mighty impressive photos. Justine posing with fifteen police officers after a major blow-and-smack bust in Venice Beach; Justine, all businesslike at her desk, with Mayor Garcetti on the other side; Justine in a very snug gray Versace gown at the Oscars.

I was mentally reviewing these images as I waited for her at the Sydney airport Customs arrival. Jack Morgan, perhaps the most important private investigator in the world, was sending her to help me launch Private Sydney, his latest addition to what is probably the most important investigative bureau in the world. Once described by Jack himself as "what Interpol tried to be, what the FBI wants to be, and what the CIA should be."

Private was located in major cities throughout the globe. Now Sydney would house headquarters for the Asia-Pac branch. And Jack Morgan had chosen me, Craig Gisto, to oversee this newest jewel in the Private crown.

Jack Morgan had the resources—both

personal and financial—to do it. He installed scientific police and laboratory equipment that went beyond state of the art. He paid university researchers to bring their findings to him first. And Jack Morgan had something else . . .

He had the brains to hire the best people. Yeah, I know that sounds easy, but it doesn't happen much. A lot of CEOs say they *want* the best people, but what they really mean is "I want someone *almost* as good as me" or "someone who's really good . . . at *taking my shit*." Not Jack. He wanted the best. And in his mind, that meant equal amounts of brainpower and guts.

It was a select group, all right, and I was excited, ecstatic, and frankly terrified that Jack Morgan had put me in it. Having Justine help me was another wise move. She knew Private; she knew Jack Morgan. And, watching her—the first passenger out of Customs—I immediately knew one thing: the Google image search had not done her justice.

I held back and let her family greet her first. There was her sister, Greta, and Greta's husband, Brett Thorogood, my new best bud. Brett was the deputy commissioner of New South Wales Police and was nothing but happy to have Private Sydney opening in his town. Brett and Greta's kids—Nikki, eight, and Serge, ten—ran to their aunt Justine. Hugs and kisses all around. Then I stepped forward. I shook Justine's hand. This was going to be one fine partnership.

I'd parked my Maserati GranCabrio in the pickup zone. The Thorogoods headed off after we'd all synchronized watches for the launch party that evening, and we were off, pulling out of the airport and onto the sun-drenched freeway.

Neither of us knew then that we were in the fast lane—the *very* fast lane—headed toward a great big pile of shit.

CHAPTER 1

THAT SAME NIGHT

HE RUNS LIKE a crazy man, a horribly injured crazy man. The stumbling and falling, long strides of a man overwhelmed with pain and fear.

He runs, gasping, then he hits a hard object—face-first. His nose shatters, sending a cascade of blood and snot down his face, agony through his head and down his spine. The man falls back, slams to the floor. His head cracks as he hits the concrete.

He has been deaf since birth. He can

sense his frantic heartbeat. The rumbling of his stomach. He feels the blood on his stomach and hips. The blood feels more like a thick syrup than a thin liquid.

He is blind. They have tied a leather sack around his head, fastened it at the neck with wiring and rope. The leather sack is tight and painful and totally immovable from his neck.

He vomits. The vomit starts to fill the leather bag that covers his head. The sack begins to fill with the chunky, nauseating upchuck. He is about to drown in his own hideous diving mask.

The man hits a wall. Literally. He clings to it and suddenly feels a searing pain, a slash in his right thigh. Now two deep wounds shoot into his back. He thinks, *Kidneys.* He also thinks he cannot collapse. If there is salvation it will be entirely due to his own willpower.

He feels the vibration of feet, people running after him. A burst of terrible agony in his back. Two thumps propel him to the wall. He smells fresh blood. He smells tire rubber. Another crunch, his thigh exploding. He keeps to the wall; blood drips from his nose, his leg, his back. He feels wet all over.

I'm not really awake. I'm not really asleep. I am afraid that the pain is so great that something will explode. Like in a mad cartoon, the top of my head will shoot right up and off. The pain will burst out of me like the fire spitting out of a volcano. My eyes have disappeared. They have been replaced by swirling pools of agony. How will I see the long white tunnel of death when it comes to take me away?

He keeps trying to feel his way along the rough cement wall. He moves left. He wills his knees to hold him up. And that works. For a moment. A bullet stings his right earlobe. He keeps moving. Sharp flying pieces of the wall bomb his hands and bare arms. Then another bullet. Now the pain is so encompassing that he is not certain where he's been hit. A new source of burning pain erupts. It is the back of his neck. The new blood and the old vomit mingle.

A doorway. He feels an iron handle. The door opens. He can no longer move from the waist down. The man falls through the open door. He falls headfirst onto a concrete floor. The pain eases. Maybe it's a dream. Maybe they've slipped a painkiller into him. *Maybe,* he thinks. *Maybe.*

He lies there.

Blind. Bloody. Deaf. Dead.

CHAPTER 2

PLEASE EXCUSE MY raging ego, but I think most police types would say that I'm a good private investigator. (After all, Jack Morgan chose me to open and operate Private Sydney.)

Now, please excuse my *really* raging ego, but I think quite a few women would say that I'm a pretty good romancer. (After all, I've sported more than a few fine lady types around the hot pubs of Sydney.)

What can Craig Gisto *not* do? Well, high on the list is *throw a party*. The best I'm capable of is bringing in a few cases of Tooheys Extra Dry, setting out a few bowls of crisps, and hoping for the best.

Well, the kind of party—excuse me...*reception*—I had to give for the official opening of Private Sydney was not the kind of party I usually threw.

So I asked Mary Clarke, my second-in-command, how I should handle the situation. Mary looked at me as if I'd been born last Tuesday and said, "Hire a caterer and forget about it." Then she added her favorite phrase: "Now why didn't you think of that?"

As is often the case, Mary had the correct solution. Wild Thyme Catering filled the huge atrium reception area of Private Sydney with tables full of cold prawns and rock lobster, skewers of tenderloin chunks, huge bunches of sunflowers. The bar had good Australian Shiraz and even a few tins of ice-cold Tooheys.

As soon as I felt certain that everything was going smoothly, I was hit by a problem. Suddenly a tremendous crashing sound filled the room. At first I thought one of my fancy caterers had dropped a tray of fancy crystal glasses.

I watched as Mary Clarke spun around on her heels in the direction of the crash. Mary's very tall, muscular—a big-boned woman—but she has the reaction speed of an Olympic gold-medal sprinter.

A nanosecond later I turned to see what had happened. I didn't have to look far. A few feet from where I'd been standing was a human figure, facedown on the floor. Blood was pooling around the body. Mary crouched beside him. I knelt, and as the people in the room gasped and turned dead quiet, Mary and I turned the corpse over to reveal a vision of horror: a male with some sort of hood tied over his face. I had seen a few seconds earlier that he had both stab wounds and bullet wounds in his back. Now that he was lying on his back, his right thigh looked like a piece of ragged, jagged, bloody meat.

I looked over for a moment and saw

Justine Smith and her brother-in-law, Deputy Commissioner Thorogood, standing over me and the dead man.

"Jesus H. Christ," Thorogood said as I tried in vain to undo the wires and ropes that held the leather hood in place.

"Let me," I heard someone say. It was Private's techno and lab genius, Darlene. My usual picture of Darlene was of a skinny woman in a drab gray lab coat, protective goggles, and a thick, white hairnet. Tonight she was dressed in a snug red silk dress, and "skinny" had turned to "curvy."

She slipped on a pair of latex gloves that she happened to have in her pocket-book, and then—I shouldn't have been surprised—she dug into her bag a little deeper and produced a multifaceted knife. One of those facets was a wire cutter. She cut through the wires and rope quickly, and then Thorogood and I eased the sack from the victim's head.

"Holy fuck!" Justine said. She was speaking for all of us.

The victim's eyes had been gouged out. The sockets were red craters. A gray and beige bundle of nerves oozed from the left one. Nerves stuck to the dry skin. The blood around his neck mixed with—I wasn't sure what it was: a disgusting-smelling brownish-yellow-pink liquid.

As if she could read my mind, Darlene said, "It's vomit."

It was tough to guess the guy's age, but what little amount of skin was left intact was smooth and lightly tanned. He was a young kid, maybe in his late teens. At the most he was twenty years old.

I saw Johnito Ishmah, the youngest guy on my team, standing behind Mary.

"Johnny, get everyone out of here. Everyone." Then I looked at Mary and said, "Come with me." We both stood up as I watched Deputy Commissioner Thorogood pull out his phone. I heard him say, "Inspector..." And I knew his boys would be here in a few minutes.

"Well, not your average gate-crasher," I heard Darlene mumble as Mary and I headed toward the bloodstained service door.

CHAPTER 3

WE OPENED THE door and followed the trail of blood down three flights of stairs.

"How'd he manage to get so far when he had to be so weak from blood loss?" I asked.

"Probably had some people helping him along," Mary said. *Yeah,* I thought, *why didn't I think of that?*

As we approached the first floor, Mary said, "Passage ahead leads to the garage."

Holy shit. The concrete garage walls looked like someone had thrown buckets of red paint on them, almost like a macabre kind of modern art. As we picked our way round the puddles, I leaned on the second door, and we were out, onto garage level 1. Plenty of blood still, oval droplets on the rough concrete. The sort of splashes someone makes when they are running and bleeding at the same time.

The poor kid had stopped here; blood had pooled into a patch about two feet wide that was rippling away toward a drain in the floor. The trail led off to the left. Three cars stood there: a Merc, a Prius, and my black Mas. Tire marks close to the bend, more blood.

I bent down and picked up a shell casing, holding it in the tissue still in my hand.

".357 Sig," Mary said. She was ex–military police, knew a thing or two.

"Pros."

Mary surveyed the ceiling. "They've got cameras everywhere," she said.

"The guard at the gate has the security-camera monitors." I led the way to the guard booth, and I found exactly what I was afraid of. The place had been hit.

Glass everywhere, the guard slumped unconscious but not dead, a row of monitors an inch from his head. The cable to a hard drive was dangling. Standard system—record the garage for twelve-hour rotations on a terabyte hard drive. Wipe it, start again.

"They took the hard drive," Mary said, nodding at the lead.

I crouched down beside the guard and lifted his head gently. He stirred, pulled back, and went for his gun. Of course, the gun was gone too.

"Whoa, buddy!" Mary exclaimed, palms up.

The guy recognized me. "Mr. Gisto." He ran a hand over his badly bruised forehead.

"Easy, pal." I placed a hand on his shoulder. "Do you remember anything?"

He sighed. "Couple of guys in hoodies. It happened so bloody fast..."

"All right," I said, turning to Mary.

There was a sudden movement beyond the booth window. I looked up to see a cop in a power stance, finger poised on the trigger.

A moment later Deputy Commissioner Thorogood appeared in the doorway. Thorogood touched the officer's arm. "Put it down, Constable."

It was then that I saw a third guy. He was standing next to Thorogood. Average build, five ten, with a cold, lived-in face. I recognized him immediately and felt a hard jolt of painful memories. I was absolutely sure the guy recognized me also. But he stood motionless, expressionless.

Yep, he was still the same devious son of a bitch.

CHAPTER 4

A COP CAR screeched to the entrance
gate. Right behind it was a van with the
word FORENSICS on its side.

Outside, Thorogood made the intro-
ductions. He was oblivious to the ani-
mosity that was starting to fill the air.
"This is Craig Gisto and Mary Clarke,
Private Sydney—a new investigative
agency started by a friend of mine, Jack
Morgan, in L.A. These guys head up the
Sydney branch. Craig, Mary, this is In-
spector Mark Talbot, Sydney Local Area
Command."

Mary extended her hand. Talbot
didn't shake it.

"And what exactly are they doing here?" Talbot looked straight at me as he spoke. I half smiled back.

"We have an arrangement...," Thorogood responded.

"Arrangement, sir?"

"I sent an e-mail, for God's sake. We help Private, Private helps us. Understand?" Thorogood didn't wait to hear if Talbot understood. Instead he turned to me and said, "So, what do we have here, Craig?"

"Lotta blood. Your forensics guys'll have fun. The hard drive for the security cameras in the booth walked." I motioned with my thumb toward the booth. "And I found this." I pulled the tissue from my jacket pocket and handed the bullet casing to Thorogood.

"That should have been left where you found it," Talbot said.

Thorogood ignored the inspector. He looked hard at the casing and said, ".357 Sig. Okay, so what do you and Mary want from us?"

"Give Darlene access to the crime scene and ten minutes with the body before it's taken to the morgue," Mary said.

Thorogood nodded. "Fine."

"That's bullshit!" Talbot exclaimed, and glared at us. Then he saw Thorogood's expression—icy and pissed off.

Talbot shut up. But I knew it wouldn't be for long.

CHAPTER 5

DARLENE'S LAB STOOD just off the atrium corridor where Private Sydney's launch party had been. It was Darlene's own little kingdom. In spite of the grim work she often dealt with, in here she felt relaxed, isolated from the troubles of the outside world.

She had designed the lab herself and been given carte blanche to install the best equipment available. Better still, through her contacts, she had some technology no one beyond Private would see for years to come. Frankly, there was no one in Australia — perhaps even in the States—who was as superb at this job as Darlene.

She was completely revved up to deal with the grotesque, uninvited corpse from the party. Police Forensics had worked through the night and cataloged everything before passing on the samples to Darlene an hour ago. A courier had delivered a case of test tubes and a USB at six a.m. She'd already been at Private for an hour.

She opened the clasps of the sample box and looked inside. Each test tube was labeled and itemized by date, location, and type. They contained samples of the corpse's blood, scrapings from beneath his fingernails, individual hairs from his jacket. She had put together a collection of her own photographs and a file from the police photographer.

There was no ID on the body. The victim was male, Asian, between eighteen and twenty-one years old. Both eyes had been hacked out with a sharp instrument. It was clearly not a professional job. By the condition of the wound, it was done at least thirty-six hours before death. The sockets were infected. His clothes were thick with filth. They stank of sweat, urine, and excrement. He'd probably been wearing them for days, held captive someplace. But the jacket he'd worn was expensive—Prada. His hair had been well cut, maybe two weeks ago. This was a rich kid.

So it seemed likely they were looking at kidnapping, Darlene thought. Maybe the kid had escaped his captors. Maybe he'd stopped being useful. Maybe the family had refused…No way of knowing—yet.

She removed seven test tubes from the case and walked over to a row of machines on an adjacent bench, each device glistening, new. She slotted the test tubes into a metal rack, pulled up a stool, switched on the machines, and listened to the whir of computers booting up and electron microscopes coming on line.

The first test tube was labeled "Nail Scraping. Left *digitus secundus manus*." With the tweezers, she slid out the piece of material. It was a couple of millimeters square, a blob of blue and pink. She placed it on a slide, lowered a second rectangular piece of glass over it, and positioned it in the crosshairs of the microscope.

The image was a yellowish white. Set to a magnification level of x1,000, human flesh looked like a blanched moonscape. She tracked the microscope to the right and refocused. It looked almost the same; only the details were different. She set the tracking going again, back left, past the starting position. Refocused. Paused. Sat back for a second, then peered into the eyepiece once more. "Now, that's just weird," she said.

CHAPTER 6

AS I PULLED into the parking lot of Private, I wiped away a trickle of sweat running down my cheek. My car's thermometer read ninety-two degrees. Then, as I eased into my parking spot, my cell phone rang. I guessed that it was Darlene. I guessed right.

"Hey, Darlene. What did you find out?"

"The police have ID'd the victim. His name's Ho Chang, nineteen, left Shore School last year. His father is Ho Meng, a very well-known and very wealthy importer-exporter. The boy was reported missing more than two days ago."

"Well, that's something."

"I found out some other stuff too."

"Great...what?"

"I'd rather show you—in the lab."

"See you in a minute."

I got to Reception and was surprised to find Mary and Johnny waiting for me. It was only eight a.m. I was even more surprised to see a tall man in a finely tailored silk suit standing with them. Beside him stood a guy in a gray suit. A bodyguard, I guessed. He had that certain boneheaded look about him. And his suit was neither silk nor finely tailored.

Johnny nodded to me and exited. Then Mary said, "Craig, this is Mr. Ho Meng. Mr. Ho, my boss, Craig Gisto."

We shook hands.

"I just heard the sad news," I said. "Please accept my..."

He raised a hand, shaking his head slowly.

I stopped talking. Then I said that we should go to my office.

Mary and Ho sat on opposite ends of my sofa, and I pulled round a chair. Meanwhile the bonehead stood by the door, arms folded.

"Mr. Ho and I have met before," Mary began. She was wearing cargo pants and a tight, short-sleeved black T-shirt that accentuated the girth of her arms. "Mr. Ho was a commissioner in the Hong Kong Police Force. I met him when he delivered a special lecture at the Military Police College a few years back."

"I would like you to find my son's killer," Ho responded. His voice was remarkably refined. I guessed Oxford or Cambridge.

"I assume the police are—"

"I do not trust the Australian police, Mr. Gisto."

I watched him. He'd drifted off into grief for a second, but then his expression hardened, and he spoke again.

"My son was reported missing more than two days ago. His death was preventable. The police did nothing."

"I'm sure they tried."

"Please do not make excuses for them, Mr. Gisto." He held his hand up again. "They're either incompetent, lazy, or lacking resources. Whatever it is, I won't work with them."

Mary, of course, realized that I didn't want to get into a debate on this subject.

"Mr. Ho, what can you tell us about your son? Any clues how he got into this trouble?" Mary asked.

He sighed. "Chang was a wonderful boy. Headstrong, for sure. Like his father. He was profoundly deaf but struggled for independence. He was a highly accomplished reader of lips. Insisted he have his own apartment as soon as he left school."

"He was deaf?" I said, surprised.

Ho nodded. "From birth." He glanced at Mary. "Listen. I would be the first to admit that I've not always been a model father. Chang's mother died twelve years ago. I've been obsessed with my business. I could never find the time. I shouldn't have let him leave home so young."

"When did you last see your son?" I asked.

"Thursday night. A family dinner—rare—at Rockpool." I knew Rockpool, the most expensive restaurant in Sydney.

"So that would be three days ago?"

"Yes. I went to his apartment on Friday morning. He wasn't there. I tried to SMS him, e-mailed him. Nothing. I reported him missing by late afternoon."

Another glance toward the window.

"The police called me just after midnight, when they'd identified Chang's body. I went to the morgue at six this morning." His voice was brittle. "I saw what they did to him." He looked at Mary, then at me, his face like a mannequin's. "You have to find the killer, Mr. Gisto. I am a very wealthy man. It matters little to me what it costs."

CHAPTER 7

DEPUTY COMMISSIONER THORO-GOOD was coming into Reception just as I was escorting Ho Meng and his bodyguard to the elevators.

Thorogood and I then walked silently back to my office.

"That was the father of the murdered kid," I said as we sat down. "He's mighty pissed with your people."

Thorogood's face creased into a frown.

"He can't understand why you didn't save his boy."

"So he's come to you?"

I nodded. I could tell that Thorogood was trying to remain calm.

"Well, you know our agreement, Craig. We share intel."

"He doesn't want to talk to the police."

The deputy commissioner blanched, anger in his eyes. "Well, it's not up to him, is it?" he snapped. "If he's withholding evidence…"

I let it go, went to change the subject. There was a knock on the door. Darlene poked her head round. "Bad time? You said you'd—"

"Sorry, Darlene," I said quickly. "Come in."

"Deputy Commissioner, you've met Darlene Cooper, haven't you?"

He stood up, extended a hand. "We…ah…met last night at the…"

Darlene gave the man a brief smile. The girl was a cool paradox, beautiful *and* brilliant—a nerd who would also look great in a *Sports Illustrated* swimsuit edition. She'd done the whole modeling shtick for a year after finishing her degree in forensics at Monash, became a disciple of Sci, Jack Morgan's resident lab genius at Private L.A. Then she'd come back to Oz and helped us establish Private Sydney.

"You wanted to know the latest," she began, before flashing her baby blues at the deputy commissioner.

"Absolutely," I said.

She handed me a couple of sheets of paper. They were covered with graphs and numbers. I turned them sideways, then back again.

"Analysis of skin samples and DNA," she explained.

"Oh, great."

"That was bloody quick!" Thorogood said.

"So, what're your conclusions?" I asked.

"I took a range of samples from the body. Unfortunately I haven't been able to get any prints, but I found three distinct DNA profiles. One of these is certainly the victim's."

"No luck finding a match for the other two?" Thorogood asked.

Darlene shook her head. "Nothing close on any database."

"Anything else?" I asked.

"Well, yeah, actually. I took a sample of material from under Ho Chang's fingernails." She handed me a photograph. I stared at it for several moments, passed it to Thorogood. He sat back, held the photo up to the light.

"It's human skin. I suspect there was a serious struggle. Ho must have taken a chunk out of the other guy."

"But what's the blue?" Thorogood asked, studying the image. It showed a highly magnified, ragged rectangle of skin. One corner was dark blue.

"Stumped me," Darlene replied, "for a few seconds. Then I realized it was probably a bit of a tattoo."

Thorogood looked at Darlene, back at the picture.

"Very clever," I said.

"Oh, I'm even cleverer than that."

I shot a glance at Thorogood, who was now giving Darlene a skeptical look.

"I took a sample and ran it through a gas chromatograph that separates out the constituents of a blend. Tattoo ink is a cocktail of many different ingredients. The gas chromatograph pulls these away from each other and gives a readout to show everything that makes up the blend. This is what I got."

I took another sheet of paper from my science whiz. It showed a graph with different-colored bars lined up across the paper.

"There were forty-seven different elements in the ink—vegetable dyes, traces of solvent, zinc, copper. But one thing stood out."

I handed the sheet to Thorogood.

"An unusual level of antimony," she said.

We both looked at Darlene blankly.

"Only Chinese tattooists use that type of ink."

She paused, then said: "It's commonly found in the tattoos of members of the..."

Darlene paused again, perhaps for dramatic effect. But I always predict the worst. So I finished the sentence for her.

"The Triad gang," I said. It was a good guess. Unfortunately.

Darlene nodded. Thorogood said, "Oh, shit." And suddenly my mind was racing as if it were on fire.

I was suddenly dealing with a ritual murder *and* the most notorious gang on the continent of Australia.

As if I didn't have enough chaos on my hands, Thorogood unknowingly brought up another problem.

"Darlene, please text this info ASAP to Inspector Talbot," he said.

Inspector Talbot. Inspector Mark Talbot. I had forgotten for a while that he was my assigned municipal police contact on the case.

He was also my cousin. He was also in love with the woman I married.

CHAPTER 8

THREE YEARS AGO

IT WAS A perfect Sydney morning. Pristine blue sky, not a cloud in sight, a crispness to the air that made you kid yourself that everything was right with the world. Even the traffic was light, for seven a.m. I had the roof down on the old Porsche convertible I'd bought fifth-hand ten years before.

We were en route to the airport. Becky, my wife of nine years, our three-year-old son, Cal, and I. Becky looked amazing. She was wearing a diaphanous dress and a thick rope of fake pearls. She was tanned from the spring sunshine. When she moved her hands, the collection of bangles on her wrists jangled. She'd put on a bit of weight and looked better for it. We'd made love that morning while Cal was asleep, and I could still visualize her.

I was ever grateful that I'd ended up with Becky as my wife. It was a tale that started with tragedy and turned into joy.

You see, when I was an eight-year-old lad in London, my crackhead mum died on the streets of Hackney. The grown-ups bundled me off to my uncle Ben Talbot in Australia. From the dreariest place in the world to the sweetest land on earth. It might have been heaven except for Uncle Ben's devil of a ten-year-old son, Mark. He hated me from the start; I was an unwelcome intruder, and Mark showed it by using my head as a human speed bag.

I was set free at eighteen, when I went to college to study law. In my second year I joined an exchange program with

UCLA, spent a year in the States. It was the best year of my life.

When I returned home to Oz at Easter, Mark and a sweet chick named Becky were having an engagement party. Mark seemed a bit nicer, a bit friendlier. And his wife-to-be seemed—well, she seemed too good for him. When Becky and I began to talk, the old Mark came out of hiding. He accused me of trying to seduce his fiancée, and Becky accused him of being crazy, and so we ended up tearing into each other. Becky, who now saw Mark in the fullness of his madness, ended up calling off the engagement.

I never saw Becky again. Well, not *never*. Nine years after Mark's engagement party Becky, Cal, and I were on our way to holiday. Sweet Jesus, I was a lucky man.

I glanced around and saw her long auburn hair blown back by the warm breeze. She was excited about our trip to Bali. We all were—our first holiday in two years. I'd been working hard to build up my PI agency, Solutions, Inc., and I was only now able to take a week off, splash some cash on a fancy resort.

I'd woken up that morning feeling more relaxed than I had in years. I'd had nice dreams too. I was back on our wedding day. Nine years before. It was a bittersweet occasion. I'd bumped into Becky by chance one morning at Darling Harbour. The old spark was there; we were both single. It just happened. We were meant for each other. Within a year we were married.

Mark must've heard I was with Becky, but, seeing as I hadn't spoken to him since my second year in college, I had no idea what he'd thought about it. Though it was his craziness that brought it on, he would never forgive me for what happened at his party. But, hell, it was such a long time ago.

Cal was strapped in the back, a suitcase next to him. On top of that was the brightly colored Kung Fu Panda carryon bag he planned to wheel to the plane and put in the overhead locker. He'd not flown before, but I'd told him all about it the previous night in lieu of a bed-time story. Cal had the same auburn hair as his mother, the same eyes. In fact, there wasn't much immediately obvious about his looks that confirmed he was mine. But he definitely had my temper-ament—patient and calm but vicious when riled.

"So, you looking forward to the trip, little man?" I called to Cal over the noise of the road and the wind and the power-ful engine. "I know I am."

He nodded. I saw him in the rearview mirror, a big smile across his face, baby teeth gleaming.

"What you looking forward to most, Cal?"

He thought for a moment, forehead wrinkled. Then hollered: "Catching fish!"

I glanced over at Becky, and we both laughed. I turned back and saw the pickup truck on the wrong side of the road, coming straight for us. And I knew immediately that this was the end. I could feel Becky freeze beside me, watched as the ugly, great vehicle covered the distance between us. With each vanishing yard, I felt my life—our lives together—drain away.

CHAPTER 9

I DON'T REMEMBER the impact—no one ever does, do they? The horror began when I started to open my eyes. But at first, everything was blurred, and I was stone-deaf. I just saw colored shapes. Then my hearing came back—but I couldn't make out a single human sound. Instead, a loud, shrill whine, the engine freewheeling in neutral.

I felt a *drip, drip, drip* on my face.

My car had rolled and ended up driver's side to the tarmac. I could see a shape close to—almost on top of—me. Gradually my vision cleared enough to make it out. Becky's face. Her dead eyes open, staring at me…droplets of her blood falling onto my cheek.

I tried to scream, but nothing came out. I couldn't speak, just produced animal noises in my throat. Tried to pull away, horrified. I turned my head slowly. A pain shot down my spine. I could just see Cal in the back. He'd slumped to the side, body contorted.

I managed to twist in the seat and had the presence of mind to feel for Becky's pulse. Then I saw the cut in her neck. She was almost decapitated.

I felt vomit rise up, and I spewed down my front. I thought I'd choke, and a part of me wished I would. I could visualize the new life if I were to survive. A life alone, my family gone…*just like that.*

I turned back to Cal, unbuckled my seat belt, gained enough leverage to slither into the rear of the car.

"Cal? Cal?" My voice broke. "Aggghhh!" I screamed again. Another stream of vomit welled up and out. I started to cry.

"Cal?" I pulled him up. His head lolled, blood trickling from the side of his mouth.

I thought I saw his eyelids flicker. "Cal?" I shouted again. I got his wrist, pulled it up, tried to find a pulse. His arm wet with blood. My fingers wet with blood. No pulse.

"CAL…CAL." I shook him.

I reached for my cell, pulled it from my jacket, but it fell to pieces in my hand.

There's a gap in my memory after that. Next thing I knew I was clambering through the passenger window. The buckled window frame and remnants of glass were cutting me open, but I didn't care. I landed on the road, guts churning, blood in my eyes diluted by tears flowing down my cheeks. I groaned—a primordial sound.

There was a revolting smell—gas, rubber…I managed to get to my knees, leaned on the car and pulled myself into a hunched, twisted figure, feeling like an octogenarian suddenly. The front of the pickup truck stood ten feet away, hood crumpled, windshield smashed. I could see the top of the driver's head above the steering wheel.

I shuffled over. From far off came the sound of sirens.

The door of the truck fell away as I yanked on the handle, and I just managed to step back before it landed at my feet. It was an old, screwed-up wagon. The driver hadn't been wearing a belt. His face was smashed in, spine snapped. A vertebra protruded from his shirt back.

I leaned in, caught the smell of alcohol. Then I saw the can of beer on the floor of the passenger side. It lay in a puddle of foaming liquid.

Fury hit me in a way I'd never experienced before or since. It was pure, all consuming. I grabbed the guy's hair, yanked his head back. His features were just recognizable. He was maybe twenty-five, blond, little goatee.

I felt the vomit rise again, but this time I held it down, lifted my fist, smashed it into the dead driver's face. I hit him again and again. "BASTARD! AAGGGHH! MOTHERFUCKER!"

I kept hitting him and hitting him, the dead man's shattered head lifelessly falling back and forth.

Then I felt a hand on my shoulder.

"Stop it, Craig," a man said.

I was in so much pain. I was so exhausted. I stopped on command. My vision was blurred. I could tell that it was a cop talking, but I couldn't make out the features.

"You're lucky to be alive, mate," the cop said.

I heard the words, but they didn't really register. I managed to turn my head a little to the left, then the right.

I looked at the face, focused. *Shit, man!* It was my cousin Mark. Mark fucking Talbot.

Strangely, I didn't care. Mark didn't matter. Nothing mattered. Becky and Cal were dead. Mark shouted toward the ambulance.

"Let's get a gurney over here. One victim's still alive!" Mark yelled. Then, with a smirk, he looked down at me and spoke very softly.

"You smashed that driver's face to a pulp," Talbot went on.

I didn't care what he said. It was not worth the pain to try to respond.

He spoke again. "But you've always been lucky, Craig. A lucky man you are. Sure, you smashed the lorry driver's face up. But..." He lifted a thin, beige folder into view. "Forensics preliminary report. The driver died on impact."

I didn't care. I was alive, but I wanted to be dead. Three emergency medics eased me onto the rolling stretcher. They carefully wheeled me toward the ambulance.

Mark started to walk away. Suddenly

he stopped. He walked back to me and leaned in close to my ear.

"You got what you deserved, you fucker. And you'll go to hell."

Then he was gone.

And I didn't care.

CHAPTER 10

THE VIEW OF Darling Harbour from the top floor of the Citadel Hotel was breathtaking. When Justine Smith walked into the room, her beauty totally dwarfed the beauty of the view. Luxurious room, shimmering evening sun. A fabulous woman in a fabulous room. Sliding doors opened onto a walled deck, a Jacuzzi sunk into the balcony.

She'd naively hoped that the opening of the Sydney branch of Private would offer some welcome relief from the usual death and destruction back home in L.A. Fat chance!

She kicked off her shoes and walked into the bedroom. The air-con was set to the perfect temperature, the bedding turned back, an Adora chocolate truffle placed on the pillow. The room smelled faintly of orange.

Unbuttoning her blouse, she turned and caught her reflection in a wall of mirrors. Slipping off her skirt, bra, and panties, she stood naked and considered her body.

"Not bad, baby," she said. Did a half turn to her left. She had a narrow waist, a flat tummy, firm breasts. "Gotta be some benefits from eating nothing and having no bambini, I guess." She did a pirouette and headed for the bathroom.

Then she changed her mind. Pulling on a robe, she went back into the main room, slid open the doors, and felt the crisp heat. A refreshing breeze came in over the harbor. She strode to the chest-high wall, admired the view.

Two minutes later, Justine was naked and immersed in bubbles, a glass of Krug on the side of the Jacuzzi. "God! This is the life!" she said aloud, and rested her head against the soft cushioning behind her neck. With her eyes closed, she reached for the champagne flute, brought it over, and let the bubbles tickle the inside of her mouth.

Her cell rang.

She groaned, and a voice in her head said *Ignore it.* But that wasn't in her nature. She lifted herself from the Jacuzzi, padded over to the phone wet and naked.

She saw the name on the screen—GRETA. Stabbed the green button.

The first thing she heard was sobs.

"Greta! What is it?"

Something unintelligible.

"Hey, sis—slow down."

More sobs. Finally, a sentence. "Oh, Justine. One of my friends has just been murdered."

CHAPTER 11

JOHNNY ISHMAH AND I were in my office, reviewing the police report on the Ho kid.

Johnny was only twenty-three, not much older than the victim. He was born in Lebanon and came over here with poor immigrant parents when he was three. Johnny could have ended up a criminal or dead, but he was smart and got out of the ghettos of Sydney's western suburbs. Fridays, Saturdays, and Sundays, he worked as a bungee-jump instructor. At night he was working on a psychology degree. And, in case the bungee-jumping job wasn't dangerous enough, all his other time was spent at Private Sydney. I trusted Johnny, and trust is always top of my list when it comes to the job.

"There are two Ho boys, right. Dai and Chang. Chang's the younger by three years," Johnny said. "The mum died when he was seven. Rich business-man father, probably never home."

I nodded. "Severely disturbed by his mother's death?"

"Definitely. His deafness made him determined to prove to his father he's every bit as good as the older brother."

The phone rang.

"Justine—" I began, and she cut over me. Johnny could see my expression darken. He raised a questioning eyebrow.

"What!" I exclaimed. "How long ago? All right, we'll go to the Thorogoods' place together. I'll pick you up in five. The Citadel Hotel, right?"

"What's up?" Johnny asked as soon as I clicked off.

I was already out of my chair. "A woman named Stacy. She was a really good friend of Justine's sister, Greta."

"Christ!"

"The cops are all over the street."

"What's the neighborhood?"

"Bellevue Hill. A very convenient location."

"Whaddaya mean?"

"The slaughter took place on a fancy, quiet street, just across from Deputy Commissioner Thorogood's house."

CHAPTER 12

I SPED OUT of the garage and pulled onto George Street. I checked my watch: 6:57 p.m. The city was packed with shoppers, bargain hunting in the January sales.

The heavy traffic threw me slightly off schedule. It took me eight minutes, not five, to pick up Justine, who was waiting in the drive-through outside the hotel. It was impossible not to register how amazing she looked: white linen pants, a tight beige top, her slightly damp hair flowing over her shoulders.

We merged with the highway traffic.

"Did your sister offer any details?" I asked, and I tried to put out of my mind the intoxicating smell of Givenchy Dahlia Noir perfume that wafted toward me.

"She was a mess. She could barely speak."

I drove east down Park Street and onto William Street, and we fell silent. I could hear a siren far off.

Bellevue Hill is mostly old money, with a sprinkling of nouveau business gurus and gangsters. The Thorogoods' house was an ultramodern place that backed onto the Royal Sydney Golf Club. Its wide, glass-balustraded balconies offered views east, toward the ocean.

I followed Justine up the granite path.

Greta, eyes moist, mascara totally messed up, opened the door before we reached it and beckoned us in.

"Tell us what happened," Justine said as her sister fell into her arms. We walked into a vast living room and sat in a horseshoe of low-slung white leather sofas.

"It was about six o'clock. Brett telephoned me. He was on his way home. He said there was a 'problem' in the neighborhood and I should stay in the house. I had heard sirens, and then I saw the blue and red police lights. Then squad cars pulled up...just over there." Greta pointed through the window. "Brett asked me if I knew what was happening, and, of course, I didn't. And then he said again that it was probably better if I stayed inside. I agreed, but then...I thought... *The kids are both on sleepovers. What the hell?* I snuck out."

Her face froze for a second. She looked at us, her eyes watering. "I wish I hadn't." She swallowed hard. "Stacy's got three kids...There was blood everywhere." She broke down, and Justine encircled her in her arms, letting her younger sister sob into her shoulder.

"Greta," I said, as sympathetically as I could, "is there anything at all unusual about Stacy? Anything that could suggest she would be targeted?"

She looked lost. "No. Stace was just a regular mum. We got to know each other through the school. Her eldest son's the same age as Serge."

"Okay, Greta, I know this might sound insensitive, but were Stacy and her husband happy?"

She shook her head. "Craig, please! I'm upset, but I'm not stupid! My husband *is* the deputy commissioner!"

"Yeah... sorry."

"As far as I know, Stacy and David are—*were* happy. You never can tell, though, right?"

I glanced at Justine. "I'm going to..." Flicked my head toward the street. Justine nodded and turned back to her sister.

As I walked down the front path, I kept thinking of Greta's question: "You never can tell, though, right?"

CHAPTER 13

IT WAS EARLY evening, but the summer sun, along with the glow of headlights and police floodlights, made it look like midday. I jogged down a side alleyway.

The end of the lane was cordoned off with crime-scene tape. A cop was standing just to my side of it. I showed him my ID. He glanced at it, then asked me to wait a moment. Two minutes later, he was back with a young guy I'd seen with Thorogood last night.

"Is the deputy commissioner around?" I asked.

"He just left for HQ, Mr. Gisto. Inspector Talbot's given you the green light, though." He offered little more than a nod, lifting the tape to indicate I should follow him.

I could see the back of a car in the alley. It was a new Lexus SUV, an LX 570. All the car doors were opened. The intense white of the floodlights lit up the number plate: STACE. Forensics was already there—blue-suited figures picking and poking around.

I walked toward the driver's side. The dead woman was strapped into the front seat. The seat had been lowered back, almost to a completely horizontal position.

Mark Talbot saw me and came over. "I'm only agreeing to you being here because Thorogood insisted," he said woodenly, and lifted his cell to indicate that he'd just spoken to his boss.

I wanted to say, "Who gives a fuck about your opinion?" but I held my tongue, ignored him, and walked over

to the body. The woman's face was disfigured with fifty or more cigarette
burns, all over her cheeks and down her
neck.

She was, I guessed, early forties, with
a blondish bob, a well-preserved figure,
smooth skin. All that moisturizing and
sunscreen had helped, I guessed. She
wore an expensive watch. There was a
big diamond—my guess? four carats—
next to a simple gold wedding band.
She was dressed in a flimsy pink cotton
dress. Someone had placed a green sheet
over her from the abdomen down. I still
wasn't sure of the cause of death.

"Tortured and then stabbed repeatedly in the back," Mark Talbot said, and
pulled the woman forward. A mess of
congealed blood, three—no, four long,
black gashes.

"What's with the sheet?" I asked.

"Look for yourself."

I pulled aside the fabric—and took a
step back.

CHAPTER 14

"WELL, YOU ALL know the basic story," I said as I walked into the conference room. "A close friend of Greta Thorogood was tortured and killed a few yards from Brett and Greta's front door. Bizarre MO."

I looked around the table. I'd called in the entire team, plus Justine.

Yeah, they already knew the basics of the homicide. Bad news travels fast.

"Let's see if anything I have brings something to the party—you should excuse my choice of words."

I flicked a remote, and the blinds

closed. A second touch on the rubber pad, and a flat-screen lit up at the far end of the room. "I shot this on my phone."

My homemade video was a jumbled mess at first, but then it settled down as I steadied my hand and set the phone to Stabilize Video.

The inside of the victim's car.

"Stacy Fleetwood," I said flatly, as the horrific image of the dead woman appeared. "She was murdered sometime around five thirty yesterday evening in the car alley next to her house in Bellevue Hill. Facially disfigured and stabbed four times in the back as she got out of her vehicle. She was then returned to the car...postmortem." The camera moved to show the dead woman straight on. I had panned down, zoomed in.

"Get ready for some terrible shit," I said.

The camera revealed that the victim's lower garments had been removed, her legs spread wide. A bunch of money had been inserted into her vagina. You could see the golden yellow of Australian fifty-dollar bills.

The video ended. The blinds came up. No one spoke.

I looked round the room. Darlene was staring straight at me. Justine studied the table. Mary was still glaring at where the image had been a few seconds ago. Johnny was counting his shoes.

"Not nice, I know, but there you have it."

"Pretty fucking sick, actually," Mary said with a steely look.

"Yep. Certainly is," I said. "Pretty fucking sick."

"What've the police found out?" Darlene asked.

"Not a lot. Their forensics people have promised to get a complete set of crime-scene samples over to you by midmorning. Thorogood is being very cooperative. I guess Greta is putting pressure on him to keep us fully involved."

"So am I, Craig," Justine remarked.

"Brett's subscribing to the idea that two heads are better than one. He knew Stacy too. He's genuinely upset."

"So what now?" said Mary.

"Darlene, you go to work on the samples soon as they arrive," I said.

She nodded.

"Justine, you and I should take a trip to the police morgue. Find out anything we can."

"Can I say something?" Johnny asked.

"Of course," I answered.

"I've got a very nasty feeling that the unfortunate Stacy Fleetwood is only the first victim."

"What makes you think that?" I asked, swiveling my chair.

"Because—and Justine will verify this," Johnny began, glancing over to where she sat, "the murder was ritualistic."

Justine nodded solemnly.

"So?" I persisted. And Justine elaborated.

"One-off murders are a type—the

most common sort. Someone dies in a violent crime—a bank raid, a gang killing, domestic violence: collateral damage. Or people are slaughtered clinically: revenge, jealousy. But this one, a woman who's tortured, killed, dumped in her car, and has her vagina stuffed with banknotes, is not the victim of a spontaneous act. It was planned, and everything about it has meaning. I hope it's not the case, but I think Johnny's right—Stacy Fleetwood is just the first."

CHAPTER 15

I CALLED MARY over as the rest of the team filed out.

"What's up?"

"The Ho murder. You gotta stay on that too. Darlene's found some interesting stuff."

"Yeah, I heard...the Triad gang. You're thinking drugs?"

"Possibly, but from what Ho Meng said, his son was hardly the sort."

"And Darlene found no evidence he was a user—totally clean."

"Maybe the kid was dealing."

"Well, yeah. Maybe, but it might not

be drugs; the Triads are involved in all sorts of shit."

"Maybe it wasn't the son," Mary replied. "What about the father? Could've been something to do with him. I'd be surprised, but we have to consider it."

"It crossed my mind. I don't think Mr. Ho gave us everything he had yesterday."

"I agree."

I looked at Mary. I'd known her for years, and I knew she had a soft side, but I think only a handful of people had ever seen it.

"You know the guy a little. Reach out to him," I suggested. "Find out if he has connections with the Triads."

"He must. But he won't like us asking about it."

"No, he won't," I replied. "But that's too goddamn bad. He needs reminding that if he wants us to find his son's killer, we have to have every single thing he can give us—not just about Chang, but about himself too."

She nodded and looked straight into my eyes.

"You okay with that, Mary? The Triads are not nice."

"Oh, please! I'm a big girl, and I thrive on not nice."

CHAPTER 16

I KNOW THIS from way too much experience: all morgues smell the same.

The scent of the New South Wales police morgue in Surry Hills, a couple of miles from the central business district, was no different. A mixture of chemicals, blood, and heartbreak.

A tall, fiftyish-looking man with a gray beard and round tortoiseshell eyeglasses met us in the small, overlit anteroom. A pass was clipped to his lapel—photo and name, Dr. Hugh Gravely. Yeah, I know. Very appropriate name.

Dr. Gravely was friendly enough and

showed Justine and me into the main part of the morgue. The ceiling and the room were lit brightly by fluorescent strips. The stink was much more intense here.

Stacy Fleetwood lay on the slab. Gray skin, wet hair pulled back, a red, crudely restitched Y-shaped incision on her upper torso. She would have been a very handsome woman yesterday.

I suddenly felt like I needed to sneeze. A horrible pain exploded in my chest. I almost let it show, but I reined it in. I knew exactly what this was. I had been to a very similar morgue…after the crash. I'd had to identify Becky's and Cal's bodies.

"The victim was thirty-nine," Dr. Gravely said, his voice emotionless. "Died from multiple stab wounds. Two distinct thrusts to the thoracic spine, two more to the lumbar. Each one approximately two inches deep. The knife had a serrated blade approximately eight inches in length. It punctured her liver and right kidney. The lumbar penetrations perforated the large intestine. The victim almost certainly died from heart failure precipitated by shock."

Justine stepped forward and inspected Stacy's lower half. "You've removed the banknotes."

"They've gone to Police Forensics, along with the woman's clothing, jewelry—everything on her person."

Justine nodded.

"I did examine them first, of course. But you'll know about them from the police, right?"

"No," Justine and I said in unison. "What about them?" I added slowly.

"Well, the money…They're fake notes—photocopies."

CHAPTER 17

EVERYTHING ABOUT THE bar of the Blue Sydney hotel in Woolloomooloo was ultramodern and beyond—over-sized concrete buffet counters, exposed brushed-chrome pipes, metal grills.

Mary sat at the huge steel bar, drinking her third espresso of the morning.

"Good morning," Ho said as he walked toward her and extended his hand.

"Thanks for agreeing to meet with me," Mary said.

Ho ordered a glass of Evian "with no ice, no lime, no lemon. And please open the bottle in front of me."

After he got that all straightened out, Ho turned to me.

"I wasn't being entirely forthcoming with you and Mr. Gisto yesterday. I don't know Mr. Gisto. But I've done some checking, and he seems like a worthy man. And besides," he added with a small smile, "you obviously trust him, and that is good enough for me."

Mary stayed stony faced. She didn't care about Ho's opinion of Gisto.

"The fact is: I believe my son was kidnapped and killed by the Triads."

"I know."

"How do *you* know?"

"Our forensics experts have found compelling evidence to support that idea."

"I see. Well, I have a lot of experience with the gangs, going back years. I know how these animals operate."

"From your time in the Hong Kong Police Force?"

"Before I met you at the Police Training Academy, I had already been one of the senior officers in Hong Kong involved with breaking up the Huang gang in 'ninety-four. Then I headed up the task force who smashed two other big Triad teams, in Kowloon and Macau."

"You told me you immigrated to Australia because it was so beautiful here."

He smiled slightly and said, "It was beautiful...beautiful to get away from the Triads."

"And you think this attack on your family was some sort of revenge?"

"I'm convinced of it."

"Why?"

Ho was silent for a few moments, gazing around the almost-empty bar. "I received a ransom note."

Mary raised an eyebrow. "Maybe we should start at the beginning, Meng."

"Yesterday I told you that the last time I saw my son was on Thursday. I reported him missing the following day. That night, Friday, I received a package. The note demanded that I cooperate with a gang that's planning to smuggle heroin from Hong Kong. The note was delivered in a box…along with one of my son's eyes."

"And you didn't go to the police when this happened?"

Ho shook his head. "No, as I've told you—"

"You don't trust the cops. Why?"

"I'd rather not say."

Mary rested an elbow on the table and rubbed her forehead. "Okay," she said. "What happened next?"

"Saturday night I received a call from the man who called himself Big Gang Leader. He said I had twenty-four hours to agree to their 'request,' or my son would be killed."

"That would give you until Sunday night. And they *did* murder him." Mary shook her head slowly.

"I've concluded that they were going to kill Chang and dump his body in a public space—a building lobby, a church, a parking lot."

"But why?" Mary asked. "Surely they would have been more discreet."

"Quite the opposite. They would have wanted to advertise it. I'm not the only Asian businessman in this city. If I kept refusing, they could go elsewhere. They wanted to broadcast the murder, as a warning to others—that's how they operate: fear and arrogance."

"But you did refuse them," Mary said.

"I *could not* agree to their demands. They are targeting me because of my past. Helping them smuggle heroin would go against everything I believe in."

Mary looked away from him.

"You may seem outraged, Mary. But I had to do it for the greater good of the public."

His eyes filled with tears. Ho Meng banged down on the bar with his fist and growled, "And, yes, my heart will break every day for the rest of my wretched life!"

CHAPTER 18

I'D JUST WALKED into the lab at Private. Darlene was tapping at a computer. She used only her thumbs to tap, like a teenager urgently text-messaging on a cell phone. I had made certain that the police had sent over absolutely everything from the Stacy Fleetwood murder scene for her to study.

"Find anything?" I asked.

"Not a lot more than the Police Forensics guys have found, I'm afraid. The banknotes are photocopies…high quality—about the grade of a top-end domestic printer."

"Fingerprints?"

"I wish! No . . . zip. Actually, to be honest, I didn't expect anything. The killer wore latex gloves. I found traces of the cornstarch powder that coats standard gloves."

"And nothing special about that?"

"Nope. These gloves could have come from any one of a hundred outlets, a thousand—Coles, Woolworths, any drugstore."

"Okay. Anything else?"

"Biological matter from the woman's vagina. I could tell you where she was in her menstrual cycle and whether or not she'd had sex during the past twenty-four hours. But I can't give you anything practical about what was put into her."

"She wasn't raped?"

"Definitely not."

I looked round the lab. Benches on both sides. On top of these benches stood impressive-looking machines with elaborate control panels and flashing lights. I recognized a powerful microscope and a centrifuge, but that was about it. The rest might as well have been Venusian technology.

"The cops gave you all the material you need?"

"Yeah, personal effects, plus a file containing several hundred photographs of the crime scene. I've analyzed Stacy Fleetwood's jacket. I can confirm the police pathologist's assessment of the attack—the number of stab wounds, the angle of entry, the type of knife. Although, of course, the weapon hasn't been found. I wish I could have been at the crime scene. It's hard, working secondhand like this. I might have caught something the cops missed."

"I understand," I replied. "And you found nothing unusual with anything Police Forensics handed over?"

"No, Craig. I'm sorry. Hate to admit it—but right now I'm drawing a complete blank."

CHAPTER 19

LOVE WAS IN the air-conditioned air. At least, love was in the small space of air-conditioned air that existed between our ridiculously hip receptionist, Cookie, and our ridiculously love-struck assistant, Johnny. Johnny was leaning over Cookie's desk so far that he looked like he might have been bodysurfing.

"Don't fall over," I said. I was on my way back from Darlene's lab.

"I just stopped to say good morning to Cookie," he said as he brought his feet back to the floor and headed toward his cubicle.

Then another voice, Justine's: "Am I just stupid? I never knew that the city of Sydney is so hot. This heat'll take the hop out of your kangaroos."

Justine was dressed for the weather, her hair pulled back in a yellow scrunchie. A good, old-fashioned ponytail. The white linen sundress she wore would fit in just fine on a tennis court.

"Well," I said, "the Internet said it was thirty-nine degrees at six o'clock this morning."

"Jesus! That's almost a hundred and three Fahrenheit," Justine said.

"Now, that's impressive. An American who doesn't need a thermometer translation."

We both laughed.

"If you're willing to go back outside and brave the heat for a few more minutes, I can show you something that you'll really like," I said.

Justine pulled her iPad from her big, white canvas bag, tapped quickly at the screen, and declared that she was "yours for a half hour."

Moments later we were walking down Macquarie Street, looking out at Circular Quay. If it weren't for the fact that she was the sharpest woman investigator in the world, if it weren't for the fact that two hideous murders were nowhere near a solution, if it weren't for the fact that I was launching the most prestigious investigative firm in Asia-Pac, and if it weren't for the fact that Justine was intimately involved with my boss...well, then we were just an ordinary couple strolling along Sydney Cove.

Straight ahead of us stood the opera house, the tiers of wide steps leading to its massive windows just a couple dozen yards away. People were sitting on the steps, drinking Slurpees and Cokes and bottled water.

"Bottled water," she said. "I never got it. Why not just take it from the tap?"

"I think we agree on that," I said.

We turned onto the quay and walked in the shade, an arcade of shops to our left. An Aboriginal man was playing a didgeridoo over a hip-hop beat spilling from an iPod plugged into a big speaker.

"Very postmodern!" Justine observed. "So, where exactly are you taking me?"

"I've decided it's going to be a surprise."

We came to a bar, with tables and umbrellas outside, a few people eating a late breakfast. A big flat-screen TV on the wall inside was showing a soccer game from the English Premier League, Chelsea versus Tottenham. I led the way through the bar and up a flight of stairs. On the wall was a sign that read: ICE BAR.

"What's this?" Justine asked.

"In a minute," I said.

I stepped up to the counter. A few other customers milled about. Sixty seconds later, I had two tickets in my hand and guided Justine around a corner. A perfectly tanned blond woman was waiting for us by a rack of fur coats.

Justine turned to me again.

"Okay, this is the deal," I said. "You want to cool down? The ice bar is set to minus twenty—Fahrenheit. Everything is made from ice, including the cocktail glasses. We stay in for a drink—twenty minutes. You'll feel a lot cooler by the end of it."

"I'm ready," she said, and the blond babe helped her into a totally unfashionable brown, fur-lined anorak and mittens. She seemed to be loving it all.

We went into the antechamber to acclimatize. Here I told her that it was just eighteen degrees Fahrenheit. From there we went into the prep room.

"Here we're down to five degrees Fahrenheit."

"I get it, Craig. You're talking Fahrenheit English," she said.

Then the door to the bar swished open, and we were inside. The digital thermometer on the wall told us it was minus twenty degrees Celsius—and it felt it, even through the fur-lined boots, the fur-lined anorak, and the fur-lined mittens.

The floor was covered with ice. The chairs around the walls were made of ice, the bar was ice. Everything was backlit electric blue.

"This is fantastic, Craig!" Justine beamed, her breath steamy and fragrant. I told her to drink her sour-cherry daiquiri quickly... before it froze.

"Since we're here, and since we're cool, do you mind if I take this opportunity to have a small meeting?" I asked.

"It's never the wrong time for a meeting," she said.

I brought her up to date on where we were on the Chang murder. I also brought her up to date on the obstacles ahead of us and the limitations that Darlene had come up against.

"If it can be solved, Darlene can solve it," Justine said.

We discussed Stacy Fleetwood's gruesome murder.

"Do you think there are any connections in these things?" she asked.

"If there are, I haven't found them."

There was a pause, and then Justine said, "I know what you're thinking."

"You do?"

"You're thinking Jack is going to be disappointed in you. Let me tell you something, Craig. Jack doesn't expect miracles right away. He expects hard work, out-of-the-box thinking. *Then* he expects miracles."

"I'll try not to disappoint him."

"And I'll help you. But Craig, listen. Before my butt freezes permanently, I need to tell you something."

Uh-oh. Watch out. Some load of manure is about to fall on me.

"I know that Mark Talbot is the cousin you lived with when you came over here as a kid from England."

"You did a background on him?"

"I do a background on every person I come in contact with."

"This is the one time I skipped

it. I figured they were Thorogood's boys."

"I'd have assumed that too. And nothing serious showed up, just the info that you and Talbot have some bad blood between you."

"I'll tell you about it someday," I said.

"You don't have to," she said. She smiled.

Then, with deadly seriousness, she said, "There's one more thing I've got to tell you, Craig."

What the hell was coming now?

"My goddamn daiquiri froze."

CHAPTER 20

THE HO MANSION was in Mosman, a few hundred yards from Taronga Zoo. It was an easy house to spot. Most houses in the area were elegant Victorian mansions, meticulously preserved. The Ho house was a total McMansion—an ornamental mess of marble, flagstone, and vulgarity.

Buzzed in through an electrically operated gate, Mary and I strode up a gravel path that passed over a pond filled with koi. A Malaysian maid met us at the front door and showed us in to a grandiose, circular hall. A young Chinese guy in a perfectly tailored blue suit appeared in an archway to the right of the hall. He had an earpiece in place, a wire disappearing into his shirt collar. It was hard to miss the bulge of a firearm under his jacket.

I was about to show him my ID when he said, "That's unnecessary. You've been ID'd and cleared, and you're four minutes early."

Mary and I followed him down a corridor. We hung a right, then a left. We passed huge rooms—a gym, a home theater, a couple of living areas, each with the floor space of your average basketball court.

We reached a door on the right. Another guard—identical blue suit, identical earpiece, identical jacket bulge. He stiffened as we came around the corner.

The first guy walked off without a word. The second guard opened the door and nodded us in.

"Wait here," he said. Then he left the room.

I could have, by now, predicted the look of the room: a four-meter-high ceiling, gaudy red-silk sofas, a mahogany carved desk that could easily have served as a banquet table. Ancient framed prints, potted palms, thronelike chairs. Everything but Ho himself.

I heard a faint sound from the far corner. There was a door into another room. I noticed a flickering light coming from beyond the doorway but couldn't make out the sound.

I turned to Mary and put a finger to my lips. Stopping a yard from the door, I pulled up close to the wall, peered in. Mary was right next to me.

There was a wide flat-screen on the far wall.

I looked at the screen. The film was beautifully shot: a small Chinese boy, perhaps five years old, playing with a toy train. The boy lifted his head and looked directly at the camera, a beatific smile on his face. Then the film cut to another scene. The boy was a little older here. He was flying a kite on the beach. The camera shot widened, and I saw the Bathers' Pavilion, the landmark café on Balmoral Beach, a mile from here.

Ho Meng sat in half profile on a sofa. He stared at the screen. A line of tears ran down his cheek. His body shook.

I felt a hand on my shoulder. Now both hands were gripping my shoulders. Whoever was touching me now turned me gently from the sight of the weeping Ho Meng.

When I looked around at the person who was touching me I couldn't help but blurt out: "Holy fucking shit!"

CHAPTER 21

"NO, MR. GISTO. I am not a ghost. I am Dai, Chang's brother," the young man said. Dai then led us on the long journey halfway across the room and indicated that we should sit on a sofa. He pulled up a chair and leaned forward.

"I'm sorry you had to see that."

I started to reply, but he lifted a hand.

"Please. I'm sorry because my father would have been so ashamed if he knew you were there. I'm sorry for him, for me."

I nodded.

"We didn't mean to intrude," Mary said.

"What is it you want?"

"We had a meeting set up to talk to your father about your brother's murder."

Dai shook his head.

"You don't have to be discreet with me. My father has told me all about the Triads. I grew up with them as a dark presence in our lives."

There was a sound from the doorway. Ho Meng was standing there. He quickly walked toward us as Mary and I stood up. He gripped my hand and then kissed Mary on both cheeks. He had transformed from the grief-stricken father we had just seen in the home theater and was once again the upright businessman.

"Please, everyone, sit," he said. "I heard what my son told you, and it is absolutely true. The Triads have hung over our lives like a black shadow, and they still do. In fact, their shadow has grown even darker."

I had to say what I had to say.

"Meng, this morning we knew that you were holding back."

Mary spoke. "If you want us to work with you in hunting down your son's killers, you have to tell us everything."

"You are right. Let me tell you what I know *and* what I think. First of all, I am certain that my wife, Jiao, was murdered by the Triads twelve years ago, soon after we came to Australia. She was last seen in Chinatown, in the middle of the day. Next morning her headless body was discovered in Roseville. The police were convinced it was the work of a psycho killer. They connected it with two similar unsolved murders from three years before. But they never found the killer."

"And *that* is why you don't trust the cops," I said.

Ho merely nodded. "They have consistently let me down. First Jiao, then Chang. I reported him missing. They did nothing. Then he died."

I felt like saying that the police could

not be everywhere all the time but thought better of it.

Then Mary said, "But, Meng, I know you and I have been over this, but . . . what I don't understand is this: if you are convinced the Triads killed your wife, then surely when Chang was kidnapped, you knew they would be serious about killing him if you didn't agree to work with them?"

Dai was about to speak, but his father silenced him with a look. "You're missing the point, Mary. The members of the Triads are not honorable men. They would have killed Chang either way. They would have kept him until I fulfilled my side of the bargain, then they would have slit his throat—he knew too much about them to live. Now perhaps you begin to understand why I don't trust the police."

His lip curled just a bit. A sneer enveloped the rest of his mouth. His eyes filled with tears.

"Do you have a son, Mr. Gisto?"

Only Mary understood the piercing pain of Ho Meng's question.

"No, I do not."

"Then you cannot understand how awful the loss can be."

CHAPTER 22

JOHNNY AND I were in my office. We huddled over a laptop that mapped out various undercover positions in the Bellevue Hill neighborhood, the area where Greta's friend Stacy had been murdered.

"We're placing too many fake gardeners," Johnny said. He was right. I pressed a few buttons, tapped a few arrows, and replaced one gardener with an au pair girl and another with a pool-maintenance worker.

"Much better," Johnny said. He

pointed to the screen and continued. "This way you can position one gardener near this patio and—"

Suddenly Johnny froze. I mean really froze. His right arm in the air. His left hand near the screen. Then he spoke.

"That voice. I'd know it anywhere," he said.

I listened. "I don't hear anything."

"Over here," Johnny said softly, and we both walked to my open office door.

"There. Listen," he said.

I listened and could hear the faint sound of a rough Australian voice coming from the reception area.

"Don't fret me, sweetheart. Mr. G. will be happy to see me. It'll be a bloody fine surprise."

"All right," Cookie said. There was an unusual dreaminess to her voice. "Mr. Gisto's the first office on the right."

And within seconds Johnny and I were face-to-face with Mickey Spencer.

If Mickey Spencer was not the most famous rock star in the world, he was certainly the most famous *Australian* rock star in the world. This guy spent his nights singing in eight-thousand-seat arenas, and he spent his days fighting with the paparazzi.

"Sorry to barge in," he said. He even sounded as if he meant what he said.

"Oh, no. That's perfectly fine," said Johnny, clearly stunned that he was talking to the most important entertainer on the continent.

"No, it's not perfectly fine," I said. "But since you've already barged in, come on inside, and bring your...um...friend with you."

His "friend" was apparently his bodyguard, a massive, bald Maori in a tight-fitting suit. He easily weighed 140 kilos or more.

As for Mickey Spencer, he was quite a bit shorter than I'd have guessed. Fame seems to make celebs *look* taller. He wore black leather pants, a brown T-shirt, and a black suit jacket. His hair was carefully gelled and carefully messy.

A few days' growth and far more wrinkles on his face than any guy in his twenties should have.

"You must be Craig Gisto," Mickey Spencer said as he took a step into my office. He had a light, jaunty voice, and I could hear one of his songs in my head as he spoke.

"How did you work that out?"

"Got the biggest office," he said, and he glanced around. "You're obviously top dog here."

Johnny was now staring at the bodyguard.

"Oh, I'd best introduce you. This little guy here is Hemi," Mickey said. "He watches out for me. Looks really mean, yeah? But only with the enemy—otherwise, he's a pussycat. Aren't you, Hemi?" No response.

"What can we do for you?" I asked.

He spun on his heel, lowered his voice. "Can we go...somewhere?"

We walked into Reception. The pop star gave Cookie a brief, professionally flirtatious smile. She'd been chewing the end of a pen and staring at him with a lost expression.

I took Mickey and Hemi along the hall and indicated to Johnny that he should come with us. "We've a comfortable lounge through here," I said.

"Coffee?" I asked as I closed the door to the room.

"Hemi'll have water—sparkling, if you have it. I'll have something a wee bit stronger—bourbon, if you have it."

I had a bottle of Blanton's Private Reserve bourbon in the small bar against the far wall.

"Great choice, man!" Mickey said as I poured him a very generous measure.

I waited for him to take a sip, but he downed it in one gulp. Meanwhile, Johnny had found a bottle of San Pellegrino and a glass. He handed them to Hemi.

"Now I'm feeling better," Mickey said.

I decided to wait for him to start talking, but he seemed a bit confused.

"Not used to this sorta thing," he began. "Feels like we're in a Raymond Chandler novel!"

I was a bit surprised by that and must have shown it.

"I'm a big reader. Hated it at school, of course, but on tour there's only so much drinking, snorting, and screwing you can take…gets boring." He produced a megawatt smile. "Anyway." His face straightened, and he looked quickly at Hemi, then back at me. Then he continued.

"I'm here about Ricky Holt."

Both Johnny and I looked at him blankly.

"Ricky Holt is my manager. He's quite well known, dudes!"

"Sorry," I said. "You may have noticed—I'm not a teenage girl. So I haven't really followed your career."

My sarcasm obviously whizzed right past Mr. Spencer's enormous ego.

"No prob," Mickey said as he held up his hands. "You got another little taste for me?" He flicked a nod at the bourbon.

"Sure." I refilled his glass. "So what is it about Mr. Holt?"

Mickey knocked back his second big glassful, wiped his mouth, and said, "Well, you see, it's like this. It's sorta like…Well…How can I put it? Oh, fuck it all. Why don't I just say it outright: *Ricky Holt is trying to kill me.*"

CHAPTER 23

WHATEVER ELSE MICKEY Spencer had already drunk, smoked, or inhaled before he showed up at Private and drank a few pints of bourbon, only God—and possibly Hemi—knew. Miraculously, besides a slight shaking in his hands and some seriously bloodshot eyes, he seemed completely compos mentis.

"Okay, Mickey. What makes you think your manager is trying to kill you?" I asked.

"Simple equation. I'm worth more to him dead than I am alive."

"That doesn't mean—"

"Listen, Mr. G., the bastard's bent. I've

been with him for three years. He picked me up when I was at my lowest point, right after I left my old band. He's a ruthless mother. You need that in a manager, but I know he wants me snuffed out." Mickey clicked his fingers in front of his face.

"If you really think that, why don't you just leave him?" Johnny asked, and glanced at me for affirmation.

Mickey laughed. "Wish I could! How I wish I could. But I'm bound by a watertight contract. I've spoken to every lawyer from here to New York City. Holt has me by the balls."

"There must be—" I began.

"Listen, Craig, you've got to understand. Forget it—there's no way out of the contract." He drew a deep breath. "Look, man, it's all about Club Twenty-Seven."

I flicked a glance at Johnny. He stared back, shrugged.

"What is Club Twenty-Seven?" I asked.

"Christ! You don't know?"

"Sorry. The teenage-girl thing again."

"Almost every dead pop star checked out when they were twenty-seven."

"Really?" I turned to Johnny, who seemed suddenly animated.

"Actually, yeah, that's right," Johnny said.

"Kurt, Hendrix, Janis, Morrison, Amy Winehouse...It's a mighty long list, man," Mickey added.

"So?" I said.

"Dude...I'm twenty-six."

CHAPTER 24

"WELL, WHAT DO you make of that?" I asked Johnny as the doors of the elevator closed on Mickey Spencer and the ever-talkative Hemi.

"Seems genuinely scared, boss."

We walked back into Reception and saw Cookie on the phone. She did a good job trying to disguise the fact that she was telling a friend about her celebrity encounter.

"Tell her you'll be selling autographs tonight," I said. Cookie rolled her eyes. But she did not look flustered.

Back in the meeting room, Johnny settled himself into the chair that only moments ago had held the butt of rock-and-roll royalty. Highly paranoiac royalty *or* highly endangered royalty.

"Refresh my memory," I said. "I was never a big fan. He was in Fun Park, right? Before he went solo and became a massive star?"

"Yeah, Granddad," Johnny replied with a grin.

"I'm more a Nirvana and Chili Peppers kinda guy."

"Yeah," Johnny said, "during the Sinatra era."

"Back to the question, sonny," I said.

Johnny turned serious. "Fun Park was big. Three number-one singles, a hit album. They just got together again—without Mickey, of course."

"But when Mickey left them, his solo career eclipsed his old band, right?"

"Definitely. He is—*was*—huge."

"Was?"

"Gone off a bit recently. Last hit was well over a year ago."

"Which is an eternity when most of your fans are seven-year-olds."

Johnny laughed. "A bit of an exaggeration. But not *that* big an exaggeration."

"Okay," I said. "Could he just be delusional? He obviously has issues."

"I guess we have to take him seriously," Johnny offered.

"We do? Why?" I paused a beat. "Look, okay. I get it. He's Mickey Spencer—megastar—and, I dunno, he seems like a pretty nice guy, actually. But do we believe him?"

"We obviously need to know a lot more about his manager."

"Okay. Let's take Mickey seriously—at least until we know otherwise."

Johnny seemed to be lost in thought.

"I reckon this one's for you, Johnny."

"Me? On my own?"

"Most definitely. Right up your alley."

"Me? Alone?"

"That's right. But I have one piece of advice. Only one." I paused.

"And that is?"

"Be sure to signal if you think you're drowning."

CHAPTER 25

SYMPATHY AND GRIEF can really get trampled on when a murder investigation has to move forward. Urgency and heartbreak are usually not good companions.

I had wanted to see David Fleetwood, husband of the late murder victim Stacy Fleetwood, as soon as possible, and Greta Thorogood had eased the way for me. So at six o'clock on the day after Stacy's funeral, I was shaking hands with the very recent widower in his smart office on the thirty-fifth floor of Citigroup Centre.

David Fleetwood was very tall, very

handsome, and clearly very sleep deprived.

"You haven't taken compassionate leave, Mr. Fleetwood?"

"I was offered it, of course," he said, his voice a smooth baritone. "But I didn't see the point. Why would I want to kick around the house? My mum is there for a while to tend to the kids. If I'm working I can focus on something other than..." He stopped.

"Makes sense."

He gestured to a seat on the gray leather sofa and said, "I've given a full report to the police. Not sure what more I can..." He trailed off again.

"Look, Mr. Fleetwood, I know this is tough, but I have to ask some fairly personal questions. I need to get some background. I appreciate it's a raw time. I've been through it myself."

"You have?"

I looked around at the white walls, a Balinese wall hanging softening things a little. "I lost my wife and son three years ago."

He stared into my eyes; his expression was absolutely vacant.

"An accident," I added.

It felt so odd, speaking about it to a complete stranger, as if I were just meeting a brand-new grief counselor.

He shrugged. "Ask away."

I paused for a second. "Were you happily married, Mr. Fleetwood?"

"As far as I'm concerned, I was. I think Stace was also. And I'll save you asking, Mr. Gisto. I wasn't having an affair, and I'm pretty sure my wife wasn't either. I do realize this is your first port of call. It would make life easier if she had been ... or if *I* was, I guess."

"Okay, sensitive question number two. Money. Everything all right?"

He waved a hand around the big modern corner office. "I'm third in line to the throne. Sort of the baby Prince Georgie. There's the boss, Max Llewellyn, then Max's son, then me. I pull down a seven-figure salary."

I knew that this didn't necessarily

mean that everything was cool, but I moved on. "It may sound ridiculous, but can you think of anyone at all who may have hated your wife?"

"Stace was a normal wife, a normal mum, Mr. Gisto. She cared for the kids, had her book club, her Pilates class. Who would hate her enough to murder her...? It's nuts."

"You're absolutely sure? Within your social circle? Any grudges? Any big bust-ups recently? Or for that matter...ever?"

He was shaking his head. "No. We are—we *were* part of a big social circle: golf club, yacht club, neighbors, work colleagues." He stared straight at me. "But nothing...We were...rather boring, actually."

"What about you, Mr. Fleetwood? Do you have any enemies?"

His expression changed for the first time. A bleak smile. "Me? Mr. Gisto, in my business I've acquired so many enemies, if I lined them up, they'd stretch from here to the Harbour Bridge."

CHAPTER 26

"WELL, IT COULD be a lead," Justine said.

She was at my apartment in Balmoral, sitting on one of my sofas. She was drinking a cup of coffee. By the way, I thought that she looked exquisite in a bright orange dress with a simple, rope-like belt around the waist.

Don't go getting any ideas. This was not a seduction setup on my part. I had called her while driving home after my meeting with David Fleetwood. I wanted her to know everything. I got her in her car. She was driving out to Bonnie Doon Golf Club to get in nine holes. My place seemed a logical meeting locale.

"I guess these money guys live close to the edge... Plenty of wars," I said.

"And there's also the symbolism of the money—the fake money."

"Of course. All a bit vague, though, right?" I said.

"Oh, totally. But we have to start somewhere, don't we?"

"You've talked to Greta. Did she have any insight? Anything?"

"Just confirmation of what we already know. My sister is part of the same social scene. There are always silly feuds between the moms, she says—the usual thing, rich women, bored, overindulged; husbands never there. The ladies crave excitement. So they invent problems between themselves. Same in L.A., London, anywhere."

"Yeah, but I can't get past the relationship angle. You said it—bored women, husbands never there. Perfect recipe."

"Sure. Look, Craig, Greta told me stuff.

Half the women she knows are having affairs with their personal trainer or tennis coach or even the interior designer, if he's so inclined. But Greta reckons that Stacy and David were simply not like that."

"She's sure?"

For the first time since we'd met, Justine sounded impatient with me. "Would you like to give Greta a shot of amobarbital to be certain?"

I assured her that "truth serum" wouldn't be necessary, and I said I wasn't doubting the veracity of Greta's opinion.

"So, let's check out David Fleetwood's associates. See if any of his enemies hate him enough to kill his wife," I said.

"Find out if he's been a naughty boy, you mean?"

"Oh, don't even question that!" I said. "The guy lives in a five-million-dollar mansion and earns a seven-figure salary. As he more or less told me himself, he's definitely been a naughty boy, if not with women, certainly with money."

Justine gazed out at the view across Middle Harbour, checked her watch. "I'd better go if I want to catch nine holes before dark."

As I was leading her to the door she turned suddenly. "Nearly forgot — would you like to come to my sister's fortieth birthday party?"

I was startled for a second. "Well... er... yeah."

"It's at a restaurant called Hurricane's Grill, at Bondi. Greta raves about it." She took a breath. "She almost called the whole thing off, but Brett and I talked her round. I told her that she couldn't let the bastard who murdered Stacy rule her life. That got her blood up. She can be quite fierce when she's riled!"

"When's the party?"

"Tomorrow night."

"Well, I'm honored."

Justine held my eyes and grinned mischievously. "Don't be. You and young Johnny are the only two men I know in Sydney."

Then she pecked me on the cheek and left.

CHAPTER 27

THE ONLY SPACE at Private Sydney smaller than Johnny's cubicle was the room that housed the brooms and mops and slop buckets. But Johnny's location did have two distinct advantages: number one, he could never fall asleep, because he had the constant, mind-numbing sound of the whirring, clanging photocopier machine next to his cubicle wall, and, number two, if he swung around in his chair, he had a direct view through the glass entry doors of the luscious-looking Cookie, the receptionist.

At the moment he had taken his eyes

from Cookie and was staring intently at his computer screen. He'd been following a paper trail—well, a cyber trail—to find anything helpful about Mickey Spencer's manager, Ricky Holt. But the facts were scant.

Holt was fifty-six, American, born in Utah. Went to Brigham Young University, studied economics. He dropped out after two years and became a minor pop star himself. Played on the New York, CBGB scene in the late seventies, fronting a band called Venison. Then he became a manager for Toys and later Rough Cut, who were pretty successful. He left for Australia in 2010 (for no discernible reason), hooked up with Mickey Spencer as the singer was leaving his own boy band, Fun Park. Six months later Holt had turned Spencer into a big-deal solo star. (The fact that Spencer had a huge voice, played terrific guitar, and exuded sexiness helped Holt work that magic.)

Johnny tapped a few more keys. The

screen showed sales figures for Mickey Spencer's three solo albums. Mickey had peaked with his first album, *Love Box*, which had made the U.S. Billboard top 10. But since then, his career had begun to falter. His latest CD, *Much 2 Much*, had bombed everywhere except in Australia. Yeah, Johnny realized that MP3s and pirated music and iTunes were changing everything, but he also knew that other artists were in the top 10 and Mickey wasn't even close anymore.

So, there's your motive, Johnny thought. *If Mickey is right and the manager is trying to have him snuffed out, it's because his career is on the ropes. Holt's going for the dead-pop-star revenue.* Johnny began typing as fast as he could.

The next ten minutes were a waste. He went through all the official sites linked to Spencer—everything from *Metal Hammer* to *Rolling Stone* to the hundreds of personal teeny-bop blogs where young girls imagined some pretty randy exercises they could personally offer Mickey

Spencer. Johnny kept searching the old material on the bands Holt had managed in the eighties. Nothing.

Well, whoa, wait one goddamn minute. Here was one of those websites where you could learn about the criminal record of anyone in the world. It was worth five Australian dollars on his Visa account to learn that...Holt had been a junkie, had served six months in L.A. Men's Central for possession (it didn't say possession of what) in 1979, spent time in rehab. Okay. That was something.

Johnny went back to some of the personal blogs about Mickey. Okay, this one looked, well, at least interesting—a blog called *Spencer Hate-On,* apparently a collaboration of people (mostly young men) who, as the title implied, detested Mickey so much that they wished him dead or, at the very least, castrated, with the severed organs then fed to Ricky Holt.

Most of it was inane garbage, and Johnny began to scroll down faster and faster, until a sentence jumped out: "Holt's bad times in the States were the best thing that ever happened to Mickey Spencer. What would the useless douche bag have done after Fun Park if Holt hadn't come to Australia to start over again?"

Johnny stopped scrolling and reread the two sentences. Then he checked the responses. There were no more comments.

Johnny leaned back in his chair. A tingle of excitement zipped right through him. It wasn't much, but it was interesting. And then again, Craig Gisto always said, "You've got to start somewhere."

CHAPTER 28

ELSPETH LOMBARD HAS put the kids to bed and is walking down the stairs when she realizes just how very much she needs a glass of Jamsheed Cabernet, her favorite. "Chewy." That's the word her husband, Alex, uses to describe the wine. "Red." That's the word Elspeth uses.

Alex is away on business in Frankfurt. He won't be back until next week. Elspeth is not exactly sad, certainly not depressed. She's just...lonely. Yes, that's the word—lonely.

She opens the temp-controlled wine-storage closet. Damn. The least expensive wine there is a Penfolds Shiraz. The Shiraz is "chewy" also, but it's also about three hundred dollars. Alex would not be a happy guy if he found one of his treasured wines had gone missing. She'll just take a quick walk to the liquor store two streets away.

Five minutes later, she's thirty yards from her house with her Jamsheed Cabernet. Two bottles, actually. After all, there's tomorrow evening to plan for.

It's quiet, sticky, hot. Most of Elspeth's neighbors are indoors watching TV or lounging by their blue-lit pools with cocktails in hand.

She hears a clicking from behind. Some animal. Some bug. She ignores it. Then a shuffling sound. She turns. Nothing. Sidewalk clear. Elspeth spins back again.

The blow comes from behind.

She falls to her knees, as confused as she is hurt.

There's a blur of houses, concrete, darkening sky. She falls forward and hits the sidewalk hard. The wine bottles smash—red liquid everywhere. Pain shoots up her neck, then streaks across the left side of her face. She tries to turn, makes it halfway, and sees a figure in a cheap winter coat leaning over her. Elspeth can actually smell the pizza on her assailant's breath.

She has no chance, no time, no strength to get up. Her attacker is bigger, stronger. She feels herself being dragged into a narrow alleyway between two gardens—her own and the Pressmans', next door. She can see a huge cloud of yellow. She realizes she is being pulled past the yellow tea roses, the little bushes she so carefully nursed through the heat. She tries to scream, but as soon as she opens her mouth, a gloved hand comes over it, grips her lips, and pulls hard. The hand rips and crushes the flesh around her mouth. Elspeth feels a tooth snap inward. More pain. Terrible pain. It spreads out across her face and around her skull.

She's pushed up against the high fence that surrounds the swimming pool. A vile-smelling rag pushes up against her insanely bloody mouth. The attacker is leaning over her. Then the attacker is knotting a part of the rag behind her neck. She struggles, but she's drained, and the assailant is too strong. Elspeth feels a wire being wrapped about her wrists. Then her arms are pinned behind her back.

She can't resist anymore. Her vision is blurred. She sees a head appear in front of her. No details. The face is in shadow, hooded. She sees a match light, a cigarette lit. The flame illuminates part of the hooded face, but only the mouth—pale, thin lips.

Elspeth screams as the cigarette burns her face, but the sound is soaked up in the gag. She can smell her own burned flesh, and now she screams a muffled scream, helpless, as the cigarette is pushed into her again, just beneath her left eye. She starts to cry. The pain sears her insides. It feels as though her head is going to explode. She vomits into the cloth in her mouth and starts to choke on it.

The attacker grabs her, spins her over onto her front. Elspeth's disfigured face hits the sandy ground of the alleyway.

Next comes the knife. She just knows something has pierced her back. She feels a strange dislocation in her spine. In her confused state, submerged in agony, she imagines she's a puppet and her strings have been cut.

The knife goes in again, and Elspeth convulses and gasps. But now the pain has gone. She is now far beyond pain, beyond life.

CHAPTER 29

TONY MACKENZIE RAN the same route at the same time every day of the week. This morning he was coming to the end of his five-mile run. He always felt that runner's sense of euphoria at this point in his circuit. He had just turned onto Wentworth Avenue, the final stretch before the wind-down.

This morning he felt extra energized. The sun was coming up, but the oppressive heat hadn't shown up yet. He passed the end of an alleyway leading off the sidewalk. Tony just kept running. But then his energy began to flag. Something was wrong. He couldn't figure out what it was, but it nagged him. He tried to brush it aside, but the nagging would not let up.

He'd had this feeling before. It was an almost extrasensory talent. Tony would be running and feel that something grim was waiting around the block. Once that happened and he discovered dead koala roadkill; another time there was a woman standing naked at her front door, wildly throwing clothing onto the front lawn (her husband's clothing, presumably).

It was that small feeling of dread that ran through him now. The day was sunny. Bright. A coolish sort of breeze. Sprinkler systems in full swing, an occasional fellow runner.

"G'day."

"G'day."

And yet . . .

A few yards beyond the alley between the Lombards' property and the Pressmans' property, Tony Mackenzie decided to stop. He'd seen something. Something wasn't quite right.

He turned and jogged back toward the entrance to the alley. Looking down the narrow lane, hands on hips, he steadied his breathing. Ten yards ahead, to the side of the alley, lay a dark object. He might have guessed it for a pile of leaves, a delivery of gravel. It could have been a bundle of rags.

He walked toward the object. Sweat dripped into his eyes, leaving a salty kind of burning sensation.

As he drew closer he considered the "pile of rags" theory. Now he thought it might be a homeless person. He stepped forward cautiously, walking not too close to the pile of—

"Holy shit," he whispered. "It's a person." He half expected it to jump up and attack him. He moved back. Then forward again. Closer. He bent over. A few beads of sweat dropped from his own face onto the bloody, ragged figure on the ground. A woman. A blond woman. A woman he knew.

Tony Mackenzie gagged and clutched himself as if a very strong, cold wind had suddenly appeared. He felt a surge of terror in the pit of his stomach. The nerves all over his body seemed to fire simultaneously.

"Oh, fuck, man," he whispered. Then he stumbled back and leaned against the fence. Then he began to cry.

CHAPTER 30

I WAS JUST pulling onto the Harbour Bridge. Glanced at the dashboard clock. It was 6:59 a.m. I felt like shit—I'd hardly slept at all last night. In my nightmares and half sleep, I kept going over Stacy Fleetwood's murder. And the worst of the groggy nightmares was that she sometimes looked like my dead wife, Becky.

I'd gulped two strong coffees before leaving the house, and then I stopped for a Red Bull at my regular gas station in Mosman. Any more caffeine and I could make it to the office faster on foot than in the Maserati.

I moved my thumb to the remote on the steering wheel and switched on the ABC news. My cell rang. I pushed the Receive button and heard Justine's voice. "Craig?"

"That's me! Hi, Justine."

She got right to the point.

"We've got a second murder."

I glanced in the mirror, sped into a gap to my left. "Any details?"

"No. Brett's there now. It's a street away from his and Greta's house."

"No way!" I changed lanes and accelerated along the Cahill Expressway. The traffic was building. "Where's the body, exactly?"

"Wentworth Avenue. Runs parallel to Brett and Greta's street."

"Know it. How did you learn of the murder?"

"I'm *at* Greta and Brett's. Stayed over last night. Brett got the call just as he was leaving for HQ."

"Okay. I'll be there in fifteen...hopefully."

It was pretty much a straight run, and I was there in twelve. I stopped just beyond the police cordon and walked briskly toward the tape. A very self-important-looking constable was guarding the sidewalk. I showed him my ID, and I was relieved when he let me through without any arguments. *Maybe this liaison with the cops could actually work after all,* I thought as I ducked under the yellow tape and walked quickly to where the forensics team was poking around.

Brett Thorogood spotted me and waved me over. I saw Mark a few yards away, his back to me. He was talking to a man in Lycra running shorts.

"Runner found the body," Thorogood explained, his expression grim.

I followed the DC over to where the victim lay—another Stacy, really: a woman, about forty, shoulder-length blond hair. She was dressed in a blood-soaked dress that had been severely torn. The exposed label said DOLCE &

GABBANA. The soil under her and around her was discolored—an ugly blackish purple. Blood. Her face had been mutilated—cigarette burns.

Her dress had been hitched up over her hips, legs splayed. The end of a roll of fifty-dollar bills could just be seen protruding from between her legs. Blood had dried on the insides of her thighs.

"Same MO," I said unnecessarily. Thorogood just stared at the dead woman.

I turned to see Justine at the tape. The cop who'd let me through was questioning her. I strode over, and just as I reached them, he let her under the barrier.

"Same thing as before," I told her as we walked along the alley. Thorogood had moved to one of the police cars on the street. Justine put a hand to her mouth, but as I went to turn her away, she shook me off. "It's okay, Craig!" she said. "Not very much shocks me anymore."

I saw Talbot finish up questioning the jogger and decided to leave Justine to it. I walked over to Mark just as another cop escorted the runner toward Wentworth Avenue.

"Oh...how nice!" he said.

"History repeating itself."

He nodded toward the dead woman. "Doesn't help that poor thing."

"Might help us, though. What do you have?" We weren't talking as if we hated each other; we were talking like two guys with a job to do.

He let out a heavy sigh. "Jogger found her about five forty-five. Jogger's name is Tony Mackenzie, some big-shot trader. The woman had been stabbed repeatedly in the back."

"Do we know who she is?"

"Name's Elspeth Lombard. Address: Forty-Four Wentworth Avenue."

"That's just two houses behind the deputy commissioner's house." I gestured back toward the main road. "Any idea how long she's been here?"

"Eight hours or so."

I nodded. "Makes sense. She'd probably have been spotted sooner if she'd been killed earlier. So after…what?… eight p.m.?"

Talbot didn't answer. Then we both caught sight of Darlene walking toward us with her forensics equipment.

Talbot smirked and spoke: "Here's the rest of your playgroup."

As Talbot began to walk away, he spoke again.

"Your turn to poke around."

CHAPTER 31

WHENEVER DARLENE CAME to a crime scene, I thought—for a split second, at least—that we'd contacted her while she was on her way to the airport. She was always pulling a large wheelie behind her. But, of course, Darlene's wheelie was packed with boxes of test tubes, piles of testing papers ready to be smeared with unpleasant body fluids, portable microscopes, scalpels, scissors, medical film. She was a takeaway laboratory.

Darlene headed straight for the victim. Justine met her, and I took out my iPhone and started walking in the direction of Wentworth Avenue.

I tapped the codes and double codes to get into Worldwide Private files. Then I tapped out "Elspeth Lombard" in Private Sydney's cryptic code. I pulled out my iPad to learn today's daily substitute code name for "Australia." Today the word for "Australia" was "Nova Scotia." *Got it.*

Elspeth Victoria O'Mara Lombard was the daughter of Norman O'Mara, a wealthy mining entrepreneur in Western Australia. The lady had married well also. Her husband was CFO of Buttress Finance Group—a big, global player. Made a name for himself on the Australian stock exchange in the early nineties, spent time in London at a British bank. Elspeth and Alex met in London, dated in London, and married in London.

Other background: the Lombards had two boys, nine and eleven, both at Cranbrook School. I lifted my eyes from the screen of the iPhone as I passed the end of the alley. Now I was on Wentworth Avenue. I saw a policewoman a couple of houses down. She was walking toward a squad car with two young boys. The Lombard kids, I figured. Jesus Christ! In one second, life can change from supergood to supersucky.

Leaning against a low wall, I gazed back at the screen and processed what I knew so far: *So, a second victim who's linked to the financial sector is found dead with fake banknotes stuffed inside her.* I wondered if Elspeth knew the first victim, Stacy Fleetwood…or, indeed, David Fleetwood? Must have. He was a senior cog at Citigroup. Thorogood would have all these answers in no time.

What other links could there be? I started to think laterally. Called Greta.

"Hey," I said gently.

"Is that Craig? Hi."

"Look, I'm calling about the latest—"

"Yep," she said. She was clearly trying to keep herself together.

"The dead woman is Elspeth Lombard." I heard a sudden intake of breath. Paused for a second. "You know her?"

There was a delay. "Um…not that well, Craig. But, yeah, I knew her."

"I'm trying to find links, Greta. Links with…"

"Okay…" Another sharp inhalation. "Er…let me…let me think. Alex, her husband…he knows David well—David Fleetwood."

"Through work?"

"Yeah, and socially. They're neighbors. They play tennis together. Stace…she played there too. Same club as us…down the road. And…er…the gym. Yeah, Elspeth goes to my gym… and Stacy's…"

"Okay."

"You think this is some sex thing, don't you?" she asked.

"No, Greta. I don't."

Silence.

"I'm sorry," she said after a moment. "I'm just…"

I kept quiet for a few beats. Then: "Can you think of anything? Anything unusual? Anything going on? I don't mean merely tossing the keys into the bowl."

"What *do* you mean, then?"

"Elspeth's husband is in finance. So is David Fleetwood. They work for different companies, but could the husbands maybe be working together on something?"

"Craig. I have no idea." She paused for several seconds. "All I know is that Stacy and Elspeth were just nice, normal women…right up until the times that somebody killed them."

"I understand, Greta. I understand," I said.

And then, in a manner totally uncharacteristic of her, she said, "Well, if you understand, then why the hell don't you and my husband try to *do* something about it?"

CHAPTER 32

FOUR KEY PLAYERS. The best and the brightest. Darlene and Mary and Johnny and Justine and four other folks sat waiting for a status meeting to begin in Private Sydney's conference room.

When I looked around the table, I saw four tired, drawn faces and eight, tired bloodshot eyes. These people were beyond the best and the brightest. They were passionate people. They cared deeply. They didn't need any hand-holding. I got to the point right away.

"All right. I know that everyone's up to his or her bum in work. And I know that we're not making a boatload of progress...yet. But, at the risk of sounding like some asshole of a coach, we're going to solve everything, every fucking thing. Take a deep breath. Let's get up to date, and then let's get to work."

A few nods, a few *okay*s.

"First, the Ho murder. Darlene has isolated DNA samples, but they don't tally with any records. Ho Meng is convinced the police can't help, and he's certain the Triads want him to coordinate a smuggling operation."

"There's this," Mary said. "Ho Meng is sure the Triads are out for revenge. He firmly believes that's why they've targeted him, killed his son. He believes they murdered his wife soon after the family arrived in Australia, a dozen years ago. I'm inclined to go with his hunch. Mainly because it's a lot better than just a hunch."

"So, Mary." I turned in my chair. "You have to dig further, and you've got to get him to dig further. That's key. Ho thinks he knows the gang, we have some DNA, but that's it. We need names, background. We need to know where the gang hangs out. For the moment, Ho refuses to work with the cops, and frankly I don't feel comfortable with that."

"Can we force him to cooperate with the police?" Johnny asked.

"No, we can't." I scanned the faces around the table. "But perhaps the police themselves could exert some… oh…influence on Ho."

Johnny smiled slightly. He caught on quickly.

"I don't mean to make this a family affair, but, Justine, could you talk to your brother-in-law about approaching Ho?"

"I already have."

I said "Good," as if I really meant that I was glad she'd spoken to Brett before I asked her to talk to Brett.

Then I turned to Darlene.

"Okay, Darlene. What's your latest?"

She looked down at a short stack of papers. Cleared her throat. "Dead woman: Elspeth Lombard, forty-one. Multiple stab wounds, fatal one to the heart. Tortured, face disfigured. She must have died pretty quick. I've found no prints, no alien DNA other than background stuff. The lab says there's no sign of sexual assault per se. But that's a stupid, academic argument. I would certainly call photocopied banknotes in a vagina an absolute form of rape and sexual assault."

Darlene paused, as if daring someone to disagree. No one did, and Darlene continued.

"There are, though, some long hairs that don't match Elspeth Lombard's hair. I found those on her dress. Doesn't necessarily mean much. She could've picked them up walking along the road, or at a restaurant, even a hair-cutting salon—any number of places. And, of course, as I just said, the banknotes—just like the last time—are photocopies."

"The victim's husband, Alex Lombard,

is CFO of Buttress Finance Group," I said. "So I'm wondering if there's a link with big-time corporate money." I looked at Justine and then Johnny.

"Obviously, our first touchstone has to be money, doesn't it?" Johnny replied. "Both husbands work in the financial sector. Banknotes placed ritualistically."

"But what about the elephant in the room? The fact that the money is fake?"

"In *both* murders," Mary added.

"But you don't have to be a genius to know it can't just be a coincidence the husbands are in finance *and* the two dead women were both abused the same way," Johnny insisted.

"Unless the killer is trying to trick us," Justine commented.

"Yeah, okay, anything is possible." I took a deep breath. "But money *is* the most obvious link we have at the moment, isn't it?"

"No," Justine said, and she said it quite emphatically.

"No?" We all looked to her.

"The most obvious link is . . . geography. The two women lived a street apart in Bellevue Hill. That's a stronger link than the financial one."

"You really think it has more to do with the fact that the victims lived in the same suburb?" I asked.

"Obviously I do," Justine said. "I just said that."

The room now held a bit of "who's the boss?" tension.

Justine continued. "If you're killing women in a neighborhood where most of the husbands are physicians or circus clowns, you're bound to conclude that their murders have to do with their husbands' being physicians or circus clowns. But I'm proposing that it's just the other way around. The killer is killing women who happen to live in an area where most of them are married to finance guys."

I had to admit: Justine's point was very well taken, and it would do me no good to debate it. The others on my team were waiting to hear what I thought. I spoke.

"So this asshole is actually killing women randomly, except it doesn't seem random. They live within a few streets of each other…Bellevue Hill is teeming with banker types, stockbrokers. It's that sort of area."

"I've experienced this sort of thing in L.A.," Justine interjected, and swept her eyes around the table. "The guy could be going for women with the same hair color—Stacy Fleetwood and Elspeth Lombard were both blond. He could be targeting women of a particular age. Fleetwood was thirty-nine, Lombard forty-one. It could be someone at their gym, the tennis club, the local coffee shop."

"Okay. So, basically, what you two are saying is that we're next to nowhere, because the financial link could well be absolutely wrong," Johnny shot back.

"Guess we are," I said.

"I think we've got to pursue the financial angle, but I think we should give equal weight to the neighborhood theory," Justine said. "And any other smart theory that we hatch."

She was doing her best to make me look good, and I did my best to try to appreciate it. I was actually about to congratulate her on her helpfulness when my cell phone rang. Only five or six people had the code to override my phone's Off position, and apparently—I looked at the caller ID name—Mark Talbot was now one of them. His boss must have shared my code with the asshole.

"Gisto, do a download on the e-mail marked 'Confidential Outdoors.' "

"Wait a minute," I said. Then I turned to Johnny and said, "Mind if I borrow this?"

He didn't have time to say yes or no. I flipped his MacBook Pro open, signed on to my backup e-mail, the one entitled "Case Confidential Only," and began scanning down to find "Confidential Outdoors."

"What's this about, Talbot?" I asked as I waited for the pages to load.

"The security company for your fancy building, Matrix? They've some images of the guys who killed Chang."

"But the killers snatched the hard drive from the guard booth," I said.

"Turns out they had another camera just outside the exit gate of the garage. Separate system. You got it there yet?"

I told him that it was still loading, and I took a few seconds to give his info to the others in the room.

Mary seemed to speak for everyone when she said, "That's the best news we've had all week."

The page was now loaded. I found the e-mail. I pressed Download. A snowy, scratchy video of a car stopped at the garage exit started to roll.

"Stop it at twelve seconds," Talbot said.

I pressed Pause at twelve seconds. The

camera held a nice medium close-up of the front windshield of the car. A driver's head. A passenger's head.

I played with a few video-control buttons to try to get better detail resolution. Finally I yelled into the phone.

"Talbot!" I screamed. "These faces are all blurred. They're blobs of gray. I can't even see a nose or an eye or a mouth. There's not a lab technique on earth that'll get us an ID on these guys."

I heard an unmistakable little chuckle from the other end. Then my cousin said, "It's the best we could do."

A moment later he hung up.

CHAPTER 33

THE PHONE IN Darlene's lab rang. She picked it up.

"This is Mickey Spencer. I'm stranded in your reception area, and it's lonely. The charming Cookie has disappeared. There are two VIP extensions here. Mr. Gisto's and yours. The choice was easy. Always call the girl."

"I'll be right out, Mr. Spencer," Darlene said, while she thought to herself: *Mickey Spencer. Mickey Spencer. I'm about to meet Mickey Spencer!* In that same split second she remembered her childhood bedroom and the two huge Mickey

Spencer posters that hung next to her bed. She remem—

"Please. It's Mickey."

Cookie the receptionist must have stepped out for a moment, Darlene thought. Receptionists always had to "step out for a moment." A cig, a pee, a secret cell-phone chat with their man. So Darlene left her lab and walked the few yards to the reception area. There was Mickey Spencer, seated behind Cookie's desk. Right behind him was the massive, unsmiling Hemi.

"G'day, young lady. I've actually come to see Johnny." Mickey stood as he spoke.

"Right. I'm Darlene...as you already know. Johnny—ah, Johnny isn't here."

Why is the leading forensics expert on the continent of Australia feeling like a teenager? That was what Darlene was thinking.

But Darlene said, "Uh, Cookie stepped out, I guess. If you'd like to wait—" she said, pointing to the chic reception-room furniture.

"Perhaps someplace a bit less public," Mickey said. Then he added, "Funny how Johnny *and* Cookie are both MIA at the same time."

"Isn't it? Come on to my place," she said.

"So what do you do here, then?" Mickey asked as they walked toward the lab.

"Forensics." Now Darlene was acting as if she were quite used to finding people like Queen Elizabeth, Pope Francis, and even Mickey Spencer in Private's reception area. Fact was, now that the initial surprise was wearing off, Darlene was becoming a good deal calmer.

"Wow! This lab is so bloody cool. I love *CSI*. Do you watch that show?"

"No," Darlene replied. "I see enough bits and pieces of dead people during the day."

Mickey stared at her, then shook his head slowly and grinned. "That is just the most insane job, Darlene!"

"Really? I think being a rock star is pretty insane."

"If you say so."

Hemi stood next to Mickey. He only moved when it was absolutely necessary, and his blank expression never changed.

"Can I get you something? Coffee?" Darlene offered.

"No, thanks," he said. "And I'm sure Hemi is well hydrated." Another pause, then: "Tell you what, though. Show me some interesting lab stuff. The creepier the better."

For a moment Darlene looked surprised. "Yeah, sure."

"You know, my parents always wanted me to become a doctor or a scientist, something like that," he went on. "I got stung by the rock 'n' roll bug, but I always regret not going to college or anything. I love science...Don't know much...but..." He laughed.

Darlene wasn't certain if he was telling the truth or if this was merely a part of his "I'm just an ordinary guy" act.

"You could still do it. You're not dead yet," Darlene said. She immediately wondered if she should have used a different expression.

"You reckon?" Mickey chuckled. "Yeah, I can see it now." He put a hand up, indicating newspaper headlines. " 'Former Rock Star Now Leading Forensics Expert.' "

"What are you working on this very moment?" Mickey asked.

"Amusing you," Darlene said. Then she quickly added, "Sorry, couldn't resist."

"You are a cheeky one," Mickey said, but he was obviously quite enchanted by her.

"Okay. Truly. Right now I'm investigating a kidnapping and murder."

"Wow!"

"We have some security-camera images of the suspects, but they're really not good. I can't make anything out."

"So what can you do about it?"

Darlene led him over to a flat-screen.

"I'm trying to enhance them with some new software I have."

Mickey gazed around the room. "Looks like pretty high-end stuff."

"Yeah, it is," she replied proudly. "State-of-the-art. But these stills are just too degraded."

"I can help," Mickey said.

Darlene lowered herself into a chair in front of the monitor. "You can?" She couldn't keep the skepticism from her voice.

"Well, not me personally. But I know a really great computer guy. A genius, in fact."

Before Darlene could reply, Mickey cut across her. "No, listen. The guy's amazing. This stuff "—he swept a hand around the room—"is cool, don't get me wrong, but in the recording studio, I use some really high-tech gear too. And my buddy—well, he works for me, actually—is *the biz*."

Darlene took a deep breath and put up her hands. "Well, great, Mickey. I'd appreciate any help I can get. What's your colleague's name?"

"Software Sam. I'll send him over."

"We pretty much know all the big-deal audio video people here, and in Tokyo and Beijing. I appreciate it. However—"

Her sentence was interrupted by a muffled sound coming from the doorway where Hemi was standing, filling the space.

"Park it elsewhere, Hemi," Mickey said. Hemi moved left, and Johnny entered. He shook Mickey's hand.

"Good to see you again, dude. So, what's new?" Johnny said.

"I was worried for a moment that Ricky Holt might have eliminated you," Mickey said.

Johnny laughed, a bit too heartily, Darlene thought. Then he said, "Let's head to a conference room."

"Right on," said Mickey. Then he turned to Darlene. "Thanks for the tour, and thanks for listening to me babble."

"That's okay. Maybe next time I can show you some of the machinery in action."

"That'd be grand," Mickey said. And Darlene was slightly surprised when Mickey gave her a peck on both cheeks.

She was even more surprised when he said, "I'll give Software Sam a call and have him come by to see you."

"That won't be nec—" Darlene began.

Mickey interrupted and spoke to Hemi. "Make a mental note for me to call Sam," Mickey said. Then, with a laugh, he added, "Hemi makes a 'mental note.' That's a good one."

CHAPTER 34

"I WANT TO help you, Johnny," Mickey said. They were in the small conference room near Johnny's cubicle.

"Well, the best way to do that is to try to remember every detail," Johnny said.

"Such as ...?"

"Well, if Ricky Holt wants you killed because you're apparently worth more dead than alive, that means either he's absurdly greedy or has money problems."

"Well, course it's 'cause he's greedy, Johnny. He's a businessman. Only thinks about dollars and cents."

"Yeah, but we also discovered that he'd filed for bankruptcy in the States. Did you know that?" Johnny asked.

"Sure thing, I knew it. Didn't really think it was important."

Johnny gazed into Mickey's pinkish eyes and counted silently to three before responding. "Not important—huh. Well, I hate to be condescending to a music legend, but around here, it's what we call 'motivation.'"

He glanced over at Hemi, who had sunk into a sofa at the back of the room, same fixed expression as always.

"God, this is all so fucked up!" Mickey exclaimed. Then he put his head down for a moment. "You got a drink, man?"

Johnny left the room for a few seconds and returned with a bottle of Johnnie Walker Double Black and a glass. He handed them to Mickey, who stared at the label.

"Is your last name Walker?" Mickey said with a laugh. Johnny just plowed ahead.

"Do you know anything about Ricky's finances, Mickey?" Johnny asked. "He must know all about yours. Does it only go one way?"

Mickey took a swig from the bottle, held it at arm's length. "Good shit."

"I'll take your word for it," Johnny said. "I don't drink."

"Lucky you."

Johnny raised an eyebrow.

"No, I mean it. Wish I didn't have to . . ."

"Holt's finances? Before my hair turns gray."

"I'm not an accountant, man. I don't know much about my own money, let alone my manager's money!"

Johnny rolled his eyes. Maybe he had overestimated this guy. Maybe Craig was right, and Mickey was drug addled. Or, put another way, maybe he was as dumb as his jokes.

"But he must have made a fortune," Johnny tried again. "How could he have ended up bankrupt?"

Mickey said nothing, just took another swig.

"Look, Mickey!" Johnny snapped. "How do you expect Private to help you if you don't tell us everything you know about the man?"

The rock star looked up and held Johnny's stare. "Yeah," he said finally. "You're right."

Of course I'm bloody right, Johnny thought, and he waited for the singer to sing.

"Ricky had a major problem. Blew fifteen mil…apparently."

"How?"

"Compulsive gambler, is what everyone said. But, look, dude, we all have our demons. I've never seen Ricky bet so much as a dime since I've known him. Got him drunk a few times, and he's told me straight that gambling is a mug's game. Says he put an end to the beast when he moved to Oz. Went into therapy, the lot. Gave up his old vices, doesn't even smoke weed now."

"And you believe him?"

Mickey considered the bottle again. "Well, I'll put it this way, Johnny." He lifted his eyes. "It's up to you to prove it if my manager has been lying, isn't it? And if he has been lying, then it will lead us to what we need—motivation!"

CHAPTER 35

THE GUY HAD described the car with absolute precision: 1971 Torino GT, bright lime, jacked-up rear wheels, avocado-and-red flame job along the sides. Mary recognized it immediately. How could she not? It was just like the anonymous writer had said in his e-mail to Craig.

He said he had important info about the murder of "that Chinese kid" but refused to come into Private's HQ. He'd meet someone from Private on the edge of Prince Alfred Park in Parramatta. Mary was that someone.

Mary crossed the hot gravel. She saw

the guy in the driver's seat—bleached blond mullet, navy-blue baseball cap, shades, cigarette, a mountain range of acne on his cheeks. His e-mail didn't mention that he'd have a very big Rottweiler in the back.

The guy leaned over, pushed open the door. The dog growled.

"Shut up, Thor!"

Mary kept her eyes fixed on the dog and slipped into the passenger seat.

"He's cool," the guy said. "Knows who's boss. Don't you, Thor?"

Mary moved to the edge of her seat.

"Buckle up. We ain't staying here," the man said, and fired up the engine.

"Five nights ago—Friday. I saw a kid that fit the description of that Ho boy in the paper. I found out you guys are investigating. Didn't wanna go to the pigs—hate 'em, but I felt I ought to say something. Other thing is, I hate the Chink gangs even more than I hate the pigs. It's the Chinks, right?"

Mary kept silent.

"I saw a car pull up about eleven at night. I was with a chick." He gave Mary a wolfish grin and turned back to the road as they took a corner, past some ravaged tenement blocks.

She gave him a hard look. "You saw this from your window?"

"Yeah, the Chinks were staying in an apartment a few floors beneath mine. I'm on the ninth."

"Can you describe the car?"

He looked affronted. "Course I can, I'm a bloody mechanic. 'Ninety-six Toyota Corolla. Piece o' shit. Blue. Faded rear bumper had an I LOVE MACCA'S sticker on it. They dragged the kid from the back. His hands were tied behind him. He was gagged, but he was squirming like a sonuvabitch. So they kicked him in the balls. I heard him grunt, poor little bastard."

"What did the two Chinese men look like?"

"That's the thing. I only caught a glimpse." He spun the wheel hard, left. "It was dark, right? The council hasn't fixed the streetlights. Besides, those Chink dudes all look the same, don't they? Usual shit—short, skinny, long black hair. One was wearing a leather jacket. I thought that was odd, as it was about a million fucking degrees outside, even that late."

Mary pursed her lips, looked away at the sidewalk flashing by.

"You got the number plate?"

"Oh, yeah. I left it for a bit, then I went downstairs."

"You did?"

"Told you. I hate 'em. That's why I'm here."

"Okay."

"The plate number was GHT…ah… two three R."

"Sure?"

"Absolutely."

"Well, thanks," Mary responded. "Anything else?"

"Yeah. I'm pretty sure they were in apartment sixteen, third floor."

"*Were?*"

"They left a couple of nights ago," he said quickly, then pulled the car to the curb, turned in the road, and headed back to the park.

"How do you know that?"

"Saw 'em, didn't I?" He glanced over at Mary. She caught a glimpse of the dog, dribble dangling from its chops. The guy accelerated down the street, screeched left, and the park lay directly ahead. "I checked with the block manager, Harry Griffin. I know 'im."

"You certain?"

"Of course I'm certain—Christ!"

"What's the full address?"

He paused for a beat, oddly reluctant. Then he pulled back into the lot. "Newbury House, Seventeen Canal Street. And that's all I got for you."

Then he said, "What have you got for me?"

Mary reached into her small leather duffel bag. She removed ten Australian hundred-dollar bills and handed them to him.

He took them, smiled, and made a loud kissing noise.

"Anything else for me?" he asked. He was looking directly at her breasts.

Mary said no way. She stepped out of the car, delighted that the 1971 Ford Torinos were built before the invention of driver-controlled door locks.

CHAPTER 36

AS SOON AS the scuzzball with the Rottweiler drove off, Mary called Darlene. She arranged to meet her an hour later at the address the guy had given her. Mary also told her to come prepared for forensic work. Then she called Parramatta Council. Within two minutes she had learned that Newbury House was serviced by a private cleaning company called R&M Cleaners.

Their address was barely a thousand yards from where she'd parked. Plus, the traffic was light.

The office was open, and as she approached the door to the left of an Aces Charcoal Chicken shop, a small group of Asian women in overalls were walking quickly down a flight of stairs. A van was parked at the curb. It had R&M CLEANERS written on the side.

Mary paused on the sidewalk to let the women pass. She glimpsed the plastic ID each of them wore attached to the straps of their overalls. That was all she needed. Twenty minutes and a trip to a passport photo booth in a stationery store later, she had a duplicate ID that she was sure would pass a cursory inspection. Then she drove on to Newbury House.

The block manager's office stank of cigarette smoke. The manager sat behind a small desk strewn with papers and an overflowing ashtray close to where he rested his left elbow. He was studying a racing paper.

"R and M Cleaners," Mary said confidently. He looked up from the paper, scrutinized her.

"Council sent me. Special clean for apartment sixteen," she said.

He looked puzzled for a moment. "You got ID?"

Mary pulled the fake from her pocket and held it out.

"Where're your overalls?"

She lifted her duffel bag and tapped it.

He shrugged and stood up, plucked the keys for the apartment from a rack on the wall behind him, and tossed them to her.

"When you're done, drop 'em in the box outside."

"Will do." Mary walked out, turned left, headed for the elevator. Emerging on the third floor, she saw Darlene waiting by the door to number sixteen, already prepped in plastic overalls and holding her metal box.

Mary unlocked the door to reveal a place where drunk college kids would feel right at home.

"Probably best if you leave me here for an hour, Mary," Darlene said as she began to pick through the trash lying everywhere.

"Leave you alone around here? You mad?"

"All right, but if you're going to nose around with me, at least put these on." She plucked a pair of latex gloves from her box of tricks.

Mary walked into the tiny kitchen as Darlene busied herself with a pile of rotting sandwiches and crumpled paper napkins on a cigarette-burned coffee table. The carpet stank of feet, cigarettes, and fried food.

The power had been cut off. In the kitchen the only light came from a tiny window over the sink. She heard a crunching sound underfoot and could make out dried noodles scattered on the cheap tiles.

She walked back into the living room. Darlene was bagging some cardboard cartons of the spoiled food and a few wooden chopsticks. She emptied two ashtrays into another bag and sealed it.

"Pretty bloody disgusting," Darlene remarked, looking up.

"Total arrogance," Mary replied. "Think they're above the law."

"In some places they are."

"Not in my city, they aren't!" Mary said, and turned to search the bedroom.

Moments later she called from the doorway. "Darlene?"

"Yep?" She stopped a foot inside the room. Mary had opened the curtains. The sheets were caked in dried blood. "I think we've found the operating room where they performed Ho Chang's eye surgery."

CHAPTER 37

THREE MONTHS AGO

JULIE O'CONNOR WAS a chubby little piece of flesh: flabby all around, with straggly blond hair, courtesy of a bottle of Nice 'N Easy lifted from the local 7-Eleven. People often say about fat people: "She'd be pretty if she'd only lose a little weight." Nobody ever said that about Julie O'Connor. The hair color didn't help. The thick layers of Maybelline didn't help. When she read in *People* magazine that David Bowie's ex-wife, Angie, had shaved off her eyebrows, Julie thought, *Hey, why not me?* So off came the eyebrows. And, of course, that sure as hell didn't help.

Bruce Frimmel was 265 pounds, with thick arms, vibrant red hair. He was one of those guys who had done everything to his face that a person could do: he had a Van Dyck beard, muttonchops, a haircut long on top and short on the sides, a diamond stud in his right ear.

A friend of theirs said that when big Bruce and fat Julie walked down the street together, they looked like the number ten.

The two of them lived in a public-housing apartment in Sandsville, in the western suburbs—no-man's land for any respectable Sydney-sider. It looked pretty much like a cheap motel room, just a little bigger. Two rooms: a cramped living area with badly stained green carpeting, a kitchen in the corner; next to that a bedroom, a bathroom with a cracked toilet, a cracked washbasin, and a shower curtain decorated with a pattern of hula dancers and palm trees. There were bars on the windows, bars on the front door.

Julie loved Bruce. Bruce tried to love Julie. But everything had gone wrong.

They wanted a kid, desperately. But nothing was working. So a month ago Julie had gone for an op, an op that, like their lives, also had gone spectacularly wrong. She developed an infection in her fallopian tubes, and within two weeks she was left infertile, completely barren. She would never have kids of her own.

Bruce had taken it badly. Very badly. Julie finally understood how badly one Tuesday at dusk, when she came home from the supermarket across town and heard Bruce slamming drawers in the bedroom.

"What ya doin', babe?" Julie asked. No answer.

When Julie walked into the bedroom, she saw Bruce bending over her own ratty, pink suitcase on the bed.

"Did we win the lottery?" She chuckled nervously.

Bruce ignored her.

"Tickets to Honolulu, Brucie?"

He turned, face hard.

She sank to the bed.

Bruce tossed some T-shirts into the
case. They landed on top of his footie
DVD and *Muscle Car* mag.

"I'm movin' out," he announced,
hands on hips.

"Moving out…Why?" Julie's face was
twisted. Then the inevitable question:
"Someone else?"

Bruce nodded. He made to sit next to
Julie on the bed but decided against it.

"Who? That slut from the video
store?"

Bruce looked down between his trainers at the dirty carpet. Shook his head.

"Who, then?" Julie's voice was far too
calm. Then she screamed…

"*WHO*…?"

Bruce was shocked for a second.

She pulled herself up from the bed,

rushed toward him. He was a lot taller than Julie and a lot heavier. Still, he stepped back. He thought she was about to cry. He'd never seen her cry, and an odd thrill rattled through him. A strange moment of pride that even he was a little ashamed of. But she didn't cry. Not at all.

"You piece of shit!" Julie howled, and she went straight for his throat, digging her nails into his skin.

Before the huge hulk of a man could pull her off, she'd drawn blood—deep nail drags across his neck. Then she smacked him across the face. It stung, but it also knocked him back into reality. He hit her, made contact with something hard, Julie's jaw. He almost lost his balance but caught himself, straightened, and really went for her.

She stumbled, landing hard on her back. Her head hit the rough carpet. Bruce dived on her, swung his fist round, and smacked her in the face.

"Useless bitch!" he screamed. "Can't even do the business…Well, thank

Christ! Who'd wanna have a kid with you?"

He smashed his fist into her face again, pulled himself up. Looked down at the red mess, blood streaming from Julie's nose, her lip split open.

Bruce turned back to the suitcase and clicked it closed, all the while listening to Julie's rasping breath and the blood gurgling in her throat.

CHAPTER 38

BRUCE SLAMMED THE front door. Julie lay semiconscious, blood drying on her face. Ten minutes later she was awake. She turned her head to the left, and blood streamed out of her mouth. She felt as if she were a drowning victim who had just been saved. The blood was the salt water she had swallowed. But where was the ocean? Where was the lifeguard? She spat more blood, and instead of thinking about Bruce and the horror of what had just happened, she began thinking of her father.

She saw him in her mind's eye. Her

dad, Jim—he would have sorted Bruce out. God, how she had loved her father. Loved him as much as she'd loathed her mother, Sheila.

Jim had been a cop in Sandsville. Julie's favorite memory of him was the day he took her to work and showed her around the Sandsville police station. She was ten and very proud of her dad.

They'd gone to the forensics labs in the basement. A man in a white lab coat had shown her the machines and racks of test tubes, told her about fingerprinting and a new thing—DNA profiling. She hadn't really understood any of it. But the next day she took a book from the local library. It was called *Forensic Investigation,* and she read every word of it.

Julie had begun to hate her mother when she realized Sheila didn't love her dad. There was another man. Julie wasn't sure if Jim knew about him, but she sometimes overheard her mother talking to her boyfriend on the phone when her dad was out. And once she followed her and saw her kissing a heavy-set man with a brown beard. How could any sane woman prefer this guy to her dad?

The day she'd seen her mother kissing her boyfriend had been the worst day in Julie's life. That is, the worst day until the really worst day arrived. The day two cops came to the door of their house and told them her dad had been killed on duty—knifed trying to stop a burglar. Jim was thirty-four.

CHAPTER 39

LATER THAT NIGHT Julie began to plot Bruce's murder. It all seemed perfectly logical, perfectly justified. She planned everything as meticulously as a not-very-meticulous person could. And as she took one step after the other on the road to killing her ex, she started to enjoy herself.

Bruce was a sex pig. She believed that all men thought with their dicks, but Bruce's sex drive was more powerful than most. So he was a bigger pig than most. She knew she could use that knowledge to her advantage.

Bruce always had a liking for what

he called "class ass." So Julie went to an Internet café and, using the "class ass" name Sabrina, sent an anonymous e-mail to Bruce's phone. She was sure he wouldn't reply right away. But she'd give him some time. "Sabrina" would send another e-mail.

At first, he was cautious, but after half a dozen messages all sent from different computers, Sabrina ensnared him, and the exchanges became more explicit. For Julie it became a delightful task. It was like writing pornography. She described what she would do *to* him and *with* him and *for* him. The asshole stepped into the trap. Soon he was begging to meet her in person.

She coaxed and teased like a pro, made it clear she liked rough sex in scary places. The more depraved, the more it turned her on. She had Bruce salivating.

She called in to him at an Internet café in Balmain and typed a message: "I want you ... TONIGHT!"

"When? Where?" he responded al-most immediately.

Finally "Sabrina" gave him the place and time: nine p.m., at a condemned house she had already staked out. The burned-out house was fenced off with wire mesh. On the perimeter of the property stood a large notice board that detailed the new development planned for the site. Another sign told the public to keep out.

The windows and the front door were boarded up, but Julie was prepared. She made short work of a couple of planks, securing one of the side windows in the living room. The room was strewn with crack pipes and newspapers and pigeon droppings.

She'd found everything she needed in the local hardware store, and she ar-ranged it neatly in the corner of the dilap-idated bathroom—a powerful battery-powered lamp, a hammer, and a new knife. She surveyed her purchases.

Then she waited patiently. Maybe it

was a half hour. Maybe he wasn't coming. Finally she heard someone approach the door. She recognized Bruce's sounds—the heavy tread of his feet, the irritating whistling—almost as if she were hearing him come into their own apartment.

"Sabrina?"

She didn't reply.

"Sabrina?" There was a nervous edge to his voice.

"In here," she called from the bathroom down the hall, and flicked on the battery-powered lamp. She stood behind the half-opened door.

Julie let him take two steps into the room, crept up behind him, swung her new hammer low, and said one word: "Bruce." He made a half turn, and Julie smashed him behind the right knee with the hammer.

He yelled and stumbled, grabbing the edge of the tub. She leapt on him and brought the hammer down hard on his head. Then once more. Her reward for her efforts was a bubbling little fountain of blood coming from the back of his skull.

The ridiculous haircut of his was soaked with his blood. He looked up into her eyes as she stood over him. He slowly twisted his head and saw her face.

"Julie!" he gasped. Then nothing.

She knelt beside him and pushed him over, onto his back. His eyes rolled back and up into his head. They were the eyes of a dead man.

Julie didn't know whether to laugh or cry. So she did both.

CHAPTER 40

GREASE-STAINED CARGO pants, heavy brown boots, a black, sleeveless top, and a bandanna. That was Mary Clarke's wardrobe choice when she walked into the Golden Wheel restaurant and bar in Campbelltown.

As soon as she took a seat at the bar, the place turned completely silent, the way it happened in saloons in old Western movies when the really bad "bad guy" or the really good "good guy" walked in.

"Coke, please. No ice if it's cold."

"Maybe you in wrong place, miss?" the Chinese bartender said.

Mary smiled sweetly. "Pretty sure I'm not. Now...Coke?"

The bartender walked to the fridge. Mary scanned the room. Thorogood had given her a pretty thick file to study on the new Triads. So she recognized a few of the Golden Wheel patrons from those classified photographs and documents.

Mary knew that there were two main gangs, one more important than the other. Latest intel was that they tolerated each other because each was run by two blood brothers who'd fought. At the moment the two brothers were on good terms. So apparently, the gangs were working together—for the moment.

The bartender broke Mary's concentration. "Six dollar."

She put the coins on the counter, lifted her glass, and continued surveying the room openly, even brazenly. One of the brothers was here, she noted. Lin Sung. A homely bastard with a pie-shaped face and eyebrows that should be trimmed with a lawn mower. She had studied his mug shot, sent over from Hong Kong that morning. When she'd seen his picture and Craig told her the guy was one of the two brothers leading the Sydney Triads, she'd joked that the poor bugger had obviously inherited the ugly genes. Then she saw the image of Sung's brother, Jing, and laughed out loud. Lesson learned: you don't have to be pretty in order to be a gang leader.

She felt a familiar ripple of power as she looked around. They knew she was either a cop or a PI. Most of the guys in this bar were not dumb. And even the dumb ones had street smarts. Whatever their intellectual makeup, they all had the brains not to lay a finger on her, at least not yet.

A man got up from a table. It was Lin Sung. He was all smiles, wire thin, snappily dressed, if you happen to go for shiny fabrics and narrow ties, circa 1979.

"Do I know you?" he asked. "What're you drinking?" He flicked a glance at the bartender.

"I don't think you do, Mr. Lin," she said. "And I'm enjoying this Coke— don't need another, thanks all the same."

Lin gave a slight bow. He looked at her again. "Well, then," he said, "is there anything else I can do for you?"

"Nope!" Mary said, smacking her lips. "Just here to have a Coke. Seemed like a nice place...from the outside."

Lin straightened up. The fake smile disappeared. He turned and walked away.

She drained her glass of the much-discussed Coke, pulled herself off the stool, and walked toward the restroom. What she actually wanted was a look at the outside rear of the building.

She opened the bathroom door and locked it behind her. Filthy basin, filthy toilet. But there was a square frosted-glass window covered in wire mesh. The window was stuck closed. She moved her right palm over the frame, found just the right spot, hit it with the flat of her hand. The entire window frame collapsed. She climbed through.

Outside, behind the bar, it was pretty much as anticipated—stinking, over-loaded bins, empty steel beer barrels, a fish skeleton ground into the dirt.

There was a small, falling-apart shack across the alley. She crossed the alley and tried the door handle. Locked. A solid kick knocked it in; the bolt snapped, clunked to the ground.

A storeroom piled high with large car-tons. Chinese writing on the sides. She hoisted one down, plucked a Swiss army knife from her pants, and slit it open. It was filled with bags. She moved one aside, sliced along the seam. Rice spilled out, over her heavy boots. She closed the knife and slipped it back into her pocket.

A sound from outside. She ducked down beside a tower of boxes.

Two men, both Chinese, walked through the space that had been created by the missing door. As soon as the men stepped closer in, Mary charged at the doorway. Once outside, she heard some-

one shouting in Chinese. One of them
tackled Mary. She fell. She used her elbow with all her strength and hit the
temple of the guy who had tackled her.
The aim of her elbow was absolutely
perfect. The guy was out—guaranteed
concussion. Also guaranteed was that
the other guy would be right on top of
her.

She was correct about that. And the
guy on top of her was Lin Sung, his
hideous face broad with a big grin and a
mouth full of bad teeth.

Suddenly she felt a sting of pain in
her left hand. Mary ignored the pain and
managed to use that same hand to re-
trieve the knife from her pocket.

She had her knife out in three seconds. As she snapped it open, she managed to slash Lin Sung in his left forearm. It was barely worse than a scratch,
but it grabbed his focus for a moment.
That gave Mary the few seconds she
needed to shove him off, turn him over.
Now she was kneeling on his shoulders,

pinning him against the gravel. His smile had not disappeared. He looked at the point of Mary's blade. It was a fraction of an inch from his Adam's apple.

Lin lifted his head a little.

"Who killed the Ho boy?" Mary asked quietly.

"Who is Ho?"

Mary touched his skin with her knife.

"The Ho boy?" Lin said, slowly and sarcastically. "I'm afraid that name is new to me."

"I would slash your face open," Mary hissed, "but it would do your looks a favor."

Lin chuckled. "What do you people say? Sticks and stones—"

"Who killed the kid?" And now she cut his throat just enough to let some blood drip out. It was as gentle as a nick from shaving.

Beads of sweat broke out on his forehead.

"You'll kill me before I speak," Lin Sung said.

Mary stared him out for ten, twenty seconds, becoming more and more aware of her own pain, her hand throbbing. She flicked a glance downward and saw a puddle of blood. She stood up. She looked down at Lin Sung and sneered at him.

The moment she turned to walk away, Lin Sung, still on the ground, began to laugh.

CHAPTER 41

YASMIN TRENT'S THICK, black hair was held in place on top of her head by two ruby-studded antique combs. Casual chic—that was her intention, and it worked. Other jewelry? A small, gold Rolex on her left wrist, a bunch of skinny Elsa Peretti silver bangles on her right. Roberto Cavalli jeans. Her beige shirt was made from a fabric that was just one small step away from being see-through.

She walked briskly from the SupaMart through the parking lot. Her small tote held a container of no-fat cottage cheese, a spit-roasted chicken, a small box of green tea, and a bottle of Tanqueray gin. She had no reason to notice Julie O'Connor, standing a few yards from the rear door of Yasmin's Toyota Land Cruiser.

Julie was in a remarkably good mood. In a local stationery shop an hour earlier, she'd overheard women talking about the "serial killer." It was exhilarating, but, of course, it made her realize that people were on the alert now. So she had to take extra care in her next murder.

Yasmin Trent touched the remote. The car beeped and flashed. Yasmin pulled herself into the driver's seat, placed her tote on the passenger seat, and shut the door. Just at that moment, Julie jerked herself into the back.

Yasmin started to turn. Then she screamed.

"Don't move, bitch!" Julie hissed, and Yasmin felt something hard and cold at the nape of her neck. She screamed again.

"Once more and I'll slice your cute little head right off."

Yasmin shut up.

She could see the figure in the back reflected in the mirror. It was a woman in a hoodie: bleached, crispy blond hair protruding from under the fabric, no eyebrows.

"Drive," Julie said quietly.

Yasmin froze.

"Okay," Julie said a little louder. "I get that you're terrified. But you *will* turn the key. You *will* pull out of the parking lot. And you *will* drive along the road, or I'll not only slice you up, I'll come back for your kids."

Julie saw Yasmin turn the key, and the engine fired up. The car pulled out of the parking space. Julie kept the eight-inch blade she was holding tight up against Yasmin's slender, tanned nape.

"Good," Julie hissed. "You're being a very good little girl."

CHAPTER 42

YASMIN DROVE. SHE was amazed at how driving was such a reflexive endeavor. In her sweating, shaking, stomach-turning fear, tears continuously filled her eyes. But she noticed that her driving was impeccable. She stopped at the stop signs. She slowed down at the yellow lights.

A few minutes into the drive, she found enough saliva in her mouth to actually speak: "What is this all about?"

There was still a noticeable tremor in Yasmin's voice. That pleased Julie, and by not replying she only heightened Yasmin's anxiety.

They were driving down the main thoroughfare from the eastern suburbs toward the business district. Following Julie's instructions, Yasmin pulled the Land Cruiser off at the next junction, through a toll booth, and onto the freeway, heading west.

"Please—what is this about?" Yasmin asked.

"It's about you and me."

"What's that mean?"

Julie chuckled. "Power, Yasmin. Everything in life always ends up being about power."

"I don't..."

"You don't understand? Maybe you're as stupid as you look."

Yasmin had no response. She just kept driving.

"So, the Rolex, Yasmin?" Julie said slowly.

Yasmin touched her wrist involuntarily.

"How many blow jobs did that take, eh?"

Julie could see Yasmin's horrified face

in the mirror. "Everything costs something, Yasmin. Come on. Everything costs—"

"Is that it? You want my watch? Here, have it ..." She reached for the clasp.

"No, you stupid bitch! I don't want your shitty watch."

The car swerved. So much for reflexive driving.

"Careful, honey," Julie mocked.

The car swerved again. This time it was deliberate, and Julie knew it. Julie pushed the tip of the blade a fraction of a millimeter into Yasmin's neck, making her squeal.

"Stop doing that shit ... NOW!"

The car swayed once more. Horns blared. The Land Cruiser crossed lanes. Yasmin cut in front of another car. More horns, screeching tires.

Julie pulled her mouth up close to the woman's ear. "Here's some info: If you don't stop this shit, I will take your twins from preschool. I will take them somewhere very private ..."

Yasmin slowed the car immediately. It felt to Julie that Yasmin was about to pull over onto the shoulder of the road.

"Don't…"

Julie watched Yasmin's face in the rearview mirror, white as dead flesh. She was staring fixedly at the road ahead.

Then Yasmin spoke very softly.

"You're the killer!" Yasmin said, shocked at her own discovery.

Julie felt a stirring of pride in the pit of her stomach. "Just drive," she said. "Not much farther now. It'll all be over very soon."

CHAPTER 43

A VARIETY OF disgusting objects lay across the steel-top counter—a pile of filthy and bloodstained sheets, takeout cartons that reeked of rotted food, cigarette stubs, crumpled Kleenex, used condoms, empty booze bottles, plastic pieces of what was once a TV remote— all collected from the deserted Triad apartment in Parramatta.

Mary pulled up a chair as Darlene turned away from the monitor to face her.

"I could tell by your tone on the phone that you're disappointed," Mary said.

Darlene shrugged. "Look, perhaps we were hoping for too much."

Then she noticed Mary's bandaged hand. "What happened there?"

Mary glanced down. "Oh, a stupid accident. Don't ever use the pointed end of a knife to remove an avocado pit."

"I'll keep that piece of advice on file," Darlene said. Then she handed Mary a pair of rubber gloves. Mary stood up next to Darlene.

"The blood on the sheets matches Ho Chang's, of course. His prints are all over the bedding, on the food cartons, chairs in the kitchen."

"What about other fingerprints? The guys who abducted him?"

Darlene paced back to her workstation, Mary in tow. She tapped at a keyboard. The image on the monitor changed to show several sets of prints.

"I've found four distinct sets in the apartment, excluding Ho's. I've also separated out three samples of DNA."

"That's great . . . yeah?"

"Not really. One set of prints and one DNA sample belong to a plumber who'd worked in the apartment a few weeks back. He had a record—petty theft in 1990; meant he was in the database. Another set of prints belongs to the manager's wife, Betty Griffin."

"Could she or her husband be involved?"

"She died last month. Ovarian cancer."

Mary snorted. "And the other two?"

"According to my analyzer, the DNA comes from two different Asian males."

"And?"

"That's it—no matches in any databases. Same for the prints."

"So we've narrowed it down to—what?" Mary declared. "About a billion men?"

"Actually, I think it's closer to two billion."

CHAPTER 44

"TO OUR WONDERFUL Greta!" Brett Thorogood said as he lifted his crystal flute of Veuve Clicquot. He clinked it with Greta's and those of their two closest friends, Deborah and Barry. The doorbell rang.

"I'll get it," Greta's son, Serge, called as he ran from the playroom, his younger sister, Nikki, close behind.

"That should be Christine, the babysitter," Greta said.

"You're so lucky, having a regular sitter," Deborah said.

"Christine's great. She works in the bakery at the local SupaMart. The kids love her. I think Brett likes her a little too much too."

Brett laughed and said, "Wait till Barry sees her. He'll love her too. She looks like the sweetest cupcake in the bakery."

"You sound just like an idiot school-boy," Greta said.

Footsteps echoed from the hallway. Serge came in with a woman who was clearly not a "cupcake" or, frankly, any kind of fantasy confection. Greta stared at the chubby woman with spiked blond hair and mottled skin. She was clearly confused.

The woman stepped forward and held out her hand. "Hi, I'm Julie—Julie O'Connor."

Greta noted the SupaMart uniform and badge.

"Christine went home sick from work. She called you, right?"

"No, she didn't," Greta said.

"Oh. Well, I've got a lot of experience. I sometimes babysit with Christine. We're old friends—"

"It's not that," Greta said, as pleasantly as possible. "It's just, I don't know you…"

Julie let out a gentle sigh. "Okay…I understand." She turned to leave.

Brett stood up, touched Greta's arm, and whispered in her ear. "We're stuck, darling. It's your party—we have to go now, but we can't leave the kids on their own."

"Just a sec, Julie," Greta said. "Just let me try Christine." She plucked her cell from the table and hit the speed-dial number. It rang five times, then went to voice mail. Irritated, she snapped the phone shut.

"We'll be good," Serge said.

"It's not that, sweetie."

"Too late to get anyone else," Brett muttered.

"Okay! Okay!" Greta put her hands up resignedly. "Julie . . .?"

Julie's face was expressionless.

"I apologize. Would you…?"

Julie smiled sweetly. Then she spoke.

"I understand your concern. But, listen. Don't worry."

"I'll try not to," Greta said. "Our cell numbers are on the pad near the kitchen phone, and the children should be in bed by nine and—"

"Calm down, Mrs. Thorogood. Have a good time," Julie said.

Brett smiled and said, "We will. Sorry for the confusion."

CHAPTER 45

SHE CLOSED THE door. Nikki and Serge were standing together, eyeing their babysitter.

"Where'd your eyebrows go?" Nikki asked.

Julie tilted her head to one side, touched her face, and feigned surprise. "Aaaggghhh! They've vanished!"

Nikki didn't even smile.

Then Julie said, "Truth is...I woke up this morning, and they were gone!"

Nikki was highly skeptical. "No, you didn't!"

"Honest, I did! I think my cat ate 'em

while I was sleeping." Julie clapped her hands together. "So, what's your favorite game, kids?"

The siblings argued about whether they should play Pocket God on the iPad, Xbox, or the interactive game that came with the latest Harry Potter DVD. In the end, the wizard won out.

Julie indulged them for an hour. Then it was bedtime and Julie's turn for some fun.

She wanted to see the inside of a house like this, see how the bitches around here lived. Earlier that day, she'd told Christine, the Thorogoods' regular sitter, that Greta had dropped into the store and asked her to pass the message on that she had to cancel tonight. Then she'd slipped Christine's cell from her pink work smock.

The plan had worked, and now that she was here, she was stunned. She'd never been in a house like this and couldn't get her head around the fact that only four people lived in it. It did nothing for her state of mind.

She walked through the main living room. She picked up crystal vases and marble ashtrays, elaborately framed photographs and silver bowls. She settled the things back down carefully. The stairs beckoned. She spun round and walked up the wide metal-and-glass steps to the first floor. The kids were asleep.

She took the second flight of stairs to the top floor, a single, expansive area devoted to the parents. A vast bedroom decorated in navy blue and white, a wall of windows looking out onto the ocean. A marble bathroom bigger than most people's living rooms. *Holy shit.* There was a shower that could fit about twenty people in it. *Two* tubs, *two* basins, and a whole separate room for the toilet.

The wife's closet had endless shelves, along with a sliding ladder attachment; that way the bitch could get to the Jimmy Choos on the very top shelves.

Julie slipped between the rows of clothes. She ran her hand along the parade of dresses and skirts and suddenly realized that they were hanging and separated according to color.

Beyond the walk-in was a small room. Well, what the hell was this? Greta's personal dressing room. A counter, a chair, necklaces hanging from stands, makeup set out in precise rows. It looked like the cosmetics counters in the fancy boutiques on Castlereagh Street North.

Julie squatted down and chose a lipstick—Guerlain Rouge d'Enfer—applied it carefully, and studied the result in the mirror over the counter. She gave herself an approving little nod, found the Dior mascara, and stroked it on. Then she walked back to the clothes.

She took her time, sifting through the garments carefully. She read the labels—Lanvin, Chanel—chose a bright-red dress, slipped into it, and managed to zip it up halfway. It was ridiculously tight on her, but she didn't care.

She picked a pair of leopard-print Blahnik shoes, crammed her toes into them, and strode into the bedroom, where a mirror occupied half a wall.

"I was born for this," Julie said to herself, and did an ungainly twirl, almost falling off the shoes. That was when she saw Nikki Thorogood staring at her. Nikki's mouth was wide open.

Julie reacted with incredible speed, whirled on the girl and grabbed her before she could take a single step back. She brought a rough hand to the girl's mouth and pulled her backward against her body. Nikki's petrified squeals were muffled by Julie's fat fingers.

"Shut up," Julie hissed in Nikki's ear. She twisted the little girl to face the mirror. She slipped the stiletto off her right foot and lifted the bladelike heel to the kid's throat. "Tell anyone about this, Nikki...and I will come for you in the night and I will kill you very, very slowly. Do you understand me?"

The kid was too terrified to move.

Julie tightened her grip. "DO YOU UNDERSTAND ME, NIKKI?"

The girl nodded, and slowly Julie loosened her grip.

CHAPTER 46

THE PLACE WAS a great choice. Icebergs is, I gotta say, the best restaurant in Bondi Beach. Plus, it sits on the side of a cliff, has an amazing, panoramic view back across a spectacular stretch of ocean. The Thorogoods had booked the entire restaurant for Greta's birthday party. One hundred people.

Justine was a knockout in a tight-fitting red cocktail dress. She'd put her hair up, Audrey Hepburn–style circa *Breakfast at Tiffany's*.

I felt her tug on my sleeve and lead me toward a balcony across the room.

I plucked two champagne flutes from a surf dude who had put on a waiter's uniform for the night.

"Greta seems to be having a fun time," I said. She was dancing with Brett, and they both looked happy. Greta was laughing at something her husband was saying close to her ear.

"It's good to see. Hasn't been having a good time recently, since the murder."

We stared out at the darkness, broken up by the lights of Bondi. In the sky was a sliver of moon, something my late wife used to call a "toenail moon."

"Ice," Justine said suddenly.

"What?"

"We seem to have a thing for ice, you and me. You took me to the ice bar, and now I get you here at Icebergs." She produced a gorgeous smile.

"Where next?" I said. "Would you like dinner somewhere with 'ice' in the name?"

"Ah," Justine said, "I'm flattered, but…"

"Sorry."

"No, don't be. But..."

"You're taken...Jack's a very lucky man."

There was an awkward silence. Then she said, "So, Craig... How did you end up here, doing what we do? This crazy job?"

"Nice diversion, Justine!" I laughed. "Long story."

"Got time." She took a sip. I turned, leaned on the balcony. "I was born in England. Mum died when I was twelve. I never knew my father. I was sent to Australia to live with my uncle and his family."

"And why PI work?"

"Ah—well, that was thanks to the love of a good woman."

She raised an eyebrow.

"I studied law, but when I actually got round to practicing it, I found it bone-dry. Let me rephrase that: it sucked. Then I met my future wife, and I fell in love. Becky was the one who pushed me into trying something I really wanted to do. So I set up my own little company: Solutions, Inc."

It was clear from Justine's nodding that she knew something of what had happened to my family.

My cell trilled. I recognized the Private Sydney number. "Darlene—you're working late."

"Sorry to call you at the party. Just had a call from police HQ. I thought you'd want to know. There's been another murder."

CHAPTER 47

"REPETITION IS SO bloody boring," Darlene said.

We were standing over the dead woman. Same story—face disfigured with multiple cigarette burns, skirt up and over her legs, a roll of fifty-dollar bills in place—only this time the victim was an almost-black-haired brunette.

"This bastard's getting me down," Darlene said as she knelt beside the body.

The victim's blood had pooled on the concrete beneath her, drenching her clothes red. I'd already learned that her name was Yasmin Trent: forty-one, mother of three young boys, lived on Gervaine Road in Bellevue Hill, about five hundred yards from Stacy Fleetwood and Elspeth Lombard. Yasmin had been a homemaker, and her husband, Simon Trent, was a dentist. So that pretty much wiped out the finance-motive theory. Most dentists made a nice living, but they didn't have the dough that the finance guys had.

There were two Police Forensics officers working on Yasmin Trent's body. They'd mellowed toward us recently. I guessed they realized we weren't going away. Plus, they'd benefited from Private's lab resources.

As I looked around myself, I caught a glimpse of Mark at the wheel of his car, the door open. He was talking to a sergeant.

I left Darlene and nosed around. It was a patch of waste ground behind a gas station in Sandsville, in the western suburbs. The late-evening traffic was light on the freeway beyond the forecourt. The place was scrappy and grimy. A rusting car stood to one side. A few weeds poked through the concrete nearby. A dead palm stood close to the rear wall of the gas-station building.

The MO had altered here slightly. It was another new, disconcerting aspect to this case. Killers rarely change their MO, even subtly. Like the other victims, this dead woman was from the eastern suburbs. She'd probably never even been to Sandsville before today, maybe just seen it on TV when channel 9 news carried an item about a knifing or a house blaze to the west. It was only, what? Thirty miles from here to Bellevue Hill? But the two places might as well have been in different solar systems.

So the question was: what was Yasmin Trent doing here? Was she killed here or in Bellevue Hill? This was different. On the other hand, some of the MO was the same—facial disfigurement, multiple stab wounds to her back, vaginal

ATM-in-reverse. We were looking for one sick motherfucker.

I felt a tap on the shoulder, turned to see Darlene. She had her box of forensics equipment in her left hand.

"That was pretty quick."

"I've learned what to look for. That part of it's predictable. The hard work comes later. But you know what, Craig? There's something not quite right about this."

"You mean the body being here in Sandsville?"

"No, it's that, but it's more than that. I can't put my finger on it. But I will. I've got very sensitive fingers."

CHAPTER 48

HO DAI WAS thinking about sleep. He was almost dreaming about sleep. He'd just gotten back to his apartment after leaving his father's house, walked into his tiny kitchen, gotten a glass of water, and heard a sound. He drank the water.

The noise came again. Footsteps? Voices? Whatever the noise was, it was also soft. But it was real. He saw two shadows pass by a glass wall close to the front door. Then Ho Dai watched as the handle turned and was released.

He padded across the floor and into the bedroom, reached the built-in wardrobe, pulled himself inside, and eased the door shut. It was dark, but he knew where he kept the gun his father insisted he have. He felt the grip just as the intruders made it through the front door and into the tiny hall.

Dai pulled the weapon down from a shelf and pointed it directly ahead. He heard someone enter the room.

"Mr. Ho," a voice said, "we know you're in here."

"I have a gun!" Dai shouted. "Open the door and I'll shoot."

A bullet thudded through the door and smacked into the wall a foot to Dai's left. He felt his bowels loosen, just managed to control himself. A second bullet sent shards of wood flying in the dark and crunched into the wall at the back of the wardrobe. It was so close that splinters flew into Dai's arm, making him cry out.

"Open the door a crack and drop the gun outside, or we'll shoot again," said the same male voice.

Dai stood rigid, trying to think, trying to rationalize.

"I'm counting to three. One..."

Dai was wreathed in sweat and breathing hard. He couldn't win. Whatever he did, he was dead meat.

"Two..."

He could barely make himself move. Had to force his arm forward. The door opened an inch, two inches. He tossed the gun onto the carpet and slammed the doors outward, propelling himself into the bedroom. He tripped, crashed to the floor, and felt the cold barrel of a gun on the back of his neck.

CHAPTER 49

ANTHONY HILARY WAS feeling horny as hell. Everything had been arranged with Karen. He would surf at six a.m. with his buddies, Chad and Frankie, and then he would meet her at the empty old house he'd found the day before. When he'd first suggested it, Karen was reluctant, but he'd eventually persuaded her.

"I can promise you the most comfortable and cleanest sleeping bag in Sydney," he'd told her with a grin.

"Oh! I'm touched!" she'd responded. "I must remember to mention that to my parents when they ask me why I'm leaving for school an hour early." But then she had shaken her head and smiled. "Okay, Ant. Seven a.m."

The surf was excellent this morning, but Anthony's mind wasn't on it. He wiped out way more than usual. Even Frankie and Chad noticed.

"Dude, what's with you? You totally wasted that wave."

"Yeah, sorry, man," Ant responded. "Look, I'm gonna bail."

"What?"

"Can't focus. I'll put the board in your car, all right, Frankie?"

His friend waved and slipped back into the surf.

Half an hour later, Anthony was standing outside the house on Ernest Street, in Bondi, watching the shifting morning light on the roofs across the road. He didn't normally do this sort of thing. He and Karen were good kids from the same coed school. He loved her, and he believed she loved him.

They were seventeen. Yeah, he rational-
ized, some kids their age were parents
already, but he and Karen could never be
alone together, were watched over 24-7.
It pissed him off to no end.

Karen was fifteen minutes late, and
Ant was growing increasingly frustrated.
When she arrived, he just managed to
stay cool.

"Okay, lover boy," she said, sidling up
to him and reaching on tiptoes to kiss
him full on the mouth. He looked down
at her tanned face and dark curls. He
was hard almost instantly.

"Come on," he said, and took her
hand.

The front door was broken, and, although it looked like it was closed se-
curely, it actually hung halfway off its
hinges. Ant escorted her down a narrow
passage to the second room on the left.
She could hear music drifting along the
hall and glanced at her boyfriend as she
recognized the tune, Angus and Julia
Stone's "Big Jet Plane."

Karen stood at the entrance to the room, holding Ant's hand. He knew he had her. She was entranced. He had cleaned it up, swept the floor, made a bed of sleeping bags. The curtains were drawn; two dozen candles glowed. An iPod played softly through a portable speaker system. The song ended and was followed by Karen's favorite, Alicia Keys's "No One."

"Oh, Ant. This is the best." She turned and kissed him again, sliding her tongue between his teeth and producing a low moan in the back of her throat. Ant felt he would burst there and then. He swept her up, lowered her gently onto the soft layers of the sleeping bags.

It all went a little faster than Karen would have liked. But it was still good for her. Ant was pretty sure of that. When it was over, they lay together, looking joyfully and contentedly up at the shabby, pitted ceiling, as if they had just finished making love in the Sistine Chapel.

"Back in a sec," Karen said softly. She pecked Anthony on the cheek and pulled herself up. "Bathroom."

"Hey, take this." Ant reached into his bag for a large bottle of water. "No main supply."

Karen looked pained and then crouched down to kiss Anthony again. "That's very thoughtful," she said.

He watched the girl's naked form in the candlelight and threw his head back onto the makeshift pillow. He really thought that this was the high point of his life. That things could never be better than this.

Then he heard Karen scream.

CHAPTER 50

INSPECTOR MARK TALBOT felt like shit, and days like today, the ones that started out really bad, were almost impossible to bear.

He'd woken up at six a.m. with a sore head from a big night out with two of his tequila-loving buddies and had dragged himself into the station by seven thirty. Forty minutes later the call had come in—another grisly find. This was getting to be fucking ridiculous. Back in the car.

The traffic was terrible all the way to Bondi, and around eight o'clock it turned stormy—black clouds rolling in

over the ocean. He switched on the ra-
dio, pushed the button for classic-rock
FM, and felt better as Steely Dan's
"Reelin' In the Years" filled the car.

"All right, what's the story?" Talbot
said as he got out of his car and followed
a sergeant into the empty house, the rain
crashing down around them.

"Best see for yourself, sir."

Talbot dashed into the hall, his jacket
soaked. The forensics people were everywhere. Huge spotlights blazed, pow-
ered by a portable generator. None of it
did his head much good. At the end of
a corridor, there was a bathroom, with
two officers in plastic suits crouching
down. The tub, toilet, floor, and white
walls were splashed with dried blood.
A lab guy was photographing the scene.
Talbot saw a line of dry red-black dots
leading from the room, out toward the
kitchen and the rear of the property.

The stench hit him as he entered the
yard. The smell of death. He knew it
well.

The blood trail stopped on the west side of the back garden. There was a large stain on the patio, close to the fence. His team had already lifted the paving stones and dug away some soil. Talbot, hand over his mouth, could see part of a corpse—a woman, faceup in the dirt.

He waved over one of his sergeants who had been standing on the other side of the shallow grave. "The basics," the inspector said, his voice phlegmy.

"Young guy called us about seven thirty. By the time we got here, the place was deserted."

"What was he doing here?"

"Looks like the kid was a vagrant— evidence someone had slept in the front room last night."

"Obviously wasn't here long. Probably nothing to do with the crime. Needs checking out, though."

The body was no more than a couple of feet beneath the surface. Three men in blue forensics suits lifted the dead woman out of the opening and laid her on plastic sheeting.

Talbot and his sergeant took two paces toward the body.

One of the forensics officers leaned in and brushed away some soil.

Most of the woman's clothes were decomposing. Her flesh barely clung to her bones.

"Dead for weeks," the forensics guy muttered, his voice muffled by his mask.

"Clear the soil from her pubic region," Talbot said.

The officer moved the brush down the dead body, swept away the sand and grains of soil. Some skin and flesh came away with it. A roll of fifty-dollar bills had been wedged into her vagina.

"You want me to call Private?" the sergeant asked.

"No, I don't think so," Talbot said. "Not this time."

CHAPTER 51

COOKIE POKED HER head around the door and into Private's lab. She was afraid of disturbing Darlene. So Cookie quietly said "Knock, knock" in that cute voice that some young women can use without sounding like idiots. Cookie always got away with it.

"What's up?" Darlene asked. She saw a tall, skinny guy with hair like a giant bird standing just outside the room, trying to peek inside.

"Er...this is—what did you say your name was again?" Cookie asked, turning and deliberately obstructing the doorway.

"I-I-I'm S-S-Sam," the man said.

Darlene looked at him blankly for a second, and then the name registered. "Software Sam? Mickey's friend?"

"The very s-s-same."

Cookie glanced at Darlene, then at the tall guy, and stepped aside.

"Mickey reckons you're a whiz with computers," Darlene said, leading Sam into the room. Cookie disappeared with an equally cute "Bye-bye."

He was gazing around, taking it all in approvingly. "Yeah...I-I-I am. So, w-w-what's your problem?"

"Look, I don't mean to be rude, but this equipment—well, it's all pretty new. Most of it's one-off stuff, custom made. I wouldn't expect you to be able to help with it."

"I could g-g-give it a go."

Darlene studied him. *You look ridiculous,* she thought. *But, then, so did Einstein!*

"Okay. I'm having a problem, primarily with my image-enhancing software."

She led him across the room.

"I'm working on some blurred images from a security camera." She pointed to a large Mac screen, sat, and tapped at the keyboard. Sam stood beside her chair,

A pair of indistinct faces came up.

"Th-th-they're the o-o-originals, r-r-right?"

Darlene looked up. "No, Sam, but they're the best I can get."

He whistled. "What s-s-software package you using?"

"It's a custom-made one from a friend of mine in L.A. He calls it FOCUS."

"Yeah, well, it's c-c-crap, isn't it?"

Darlene produced a pained laugh.

"C-c-can you open up the p-p-program for me?"

Darlene shrugged. "Okay." She brought up the appropriate screen. Then she offered her chair to Sam.

The screen filled with symbols and lines of computer code.

"I'll c-c-clone this first," Sam said. "As a b-b-backup." He tapped at the keyboard with extraordinary speed. Darlene watched as the algorithms and rows of letters and numbers shifted. Sam paused for a second, peered at the screen. Then his staccato key stabbing started up again.

Two minutes of concentrated effort, and the visitor pushed back Darlene's chair. "Th-th-that sh-sh-should do it," he declared.

"What did you do?"

"B-b-boosted the r-r-response parameters, r-r-realigned the enhancement s-s-software to concentrate on th-th-the contrast and the w-w-warmth c-c-components."

Darlene returned to her chair and clicked the mouse a couple of times to bring back the main screen. She opened the FOCUS software package, clicked on the image from the security camera, and pressed Import. A new screen opened, showing a crisp, sharp image of two Asian men, the picture so clear you could almost count their eyelashes.

"That's incredible!"

"I-i-it is pretty c-c-cool, i-i-isn't it?"

Darlene stood up. "I'm so sorry I ever doubted you."

"No prob." Software Sam looked a little embarrassed. "Oh! Almost forgot. M-M-Mickey gave me these." He held out a bunch of invitations. "H-h-half a dozen p-p-passes to his b-b-birthday bash, tomorrow night at the V-V-Venue."

"Fantastic!" she said.

"I'll be there too," Sam said.

"Fantastic!" Darlene repeated. And she meant it.

CHAPTER 52

I WAS STARING at the monitor on Dar-lene's desk.

"That's just amazing!" I exclaimed as I studied the piercingly clear image of the two men who'd killed Ho Chang.

"I'd like to take credit for getting it up," Darlene said, "but it was Mickey Spencer's buddy, some guy they call Software Sam."

"Yeah, Cookie told me he'd been here — kind of a weirdo."

"A weirdo genius. Actually, he's kind of a cutie," she said. Then she quickly added, "In a weirdo-genius sort of way.

You know, when the class nerd takes off his glasses, and suddenly you think, *He's not so bad.*"

"I know the situation. As the class nerd…but not with the same results you describe," I said.

"Oh, yeah. I'm sure," Darlene said. Then she got back to business: "So, what do we do now? We going to share this with the cops?"

I contemplated the image. "Oh, I don't think so…Not yet, anyway."

Darlene gave me a quizzical look— quizzical but satisfied.

"If we do that," I went on, "someone will blab, and these bastards"—I waved a hand at the monitor—"will disappear. No, this is ours, Darlene. At least for the moment. You been able to do anything with it?"

"I've tried. Spent all afternoon attempting to match up facial characteristics with databases all over the world. Not getting very far. Same old problem. The Triads bribe the authorities in Hong

Kong. So nothing's on record. If there's nothing on the two men, then the CIA, MI Six, the Australian intelligence agency can't get a handle on them. These guys have no DNA records, no fingerprints or photo presence at all. As far as the investigative agencies are concerned, they don't exist."

CHAPTER 53

I WAS IN the New South Wales police pathology lab with Darlene. She was looking over the rotting remains of the dead woman discovered in the old house in Bondi. And I was looking at Darlene. No matter how long I was in this business, I could never get used to the human forensics.

As I watched her work methodically, I felt myself growing more and more pissed off. Private hadn't learned about the corpse for at least five hours after it was found. Even then, it was only because Darlene heard about it thirdhand from a friendly cop at police HQ.

The victim was Jennifer Granger, thirty-eight, from Newmore Avenue, a street that happened to be perpendicular to Wentworth Avenue in Bellevue Hill, where Elspeth Lombard had been found. It was within spitting distance of the other victims' homes. Although, of course, in Bellevue Hill, nobody spat on the street.

"I spoke to one of the sergeants at the station in the CBD," I said. "Jennifer Granger was reported missing three weeks ago, on December fifteenth." Darlene didn't look up, but she couldn't have missed the anger in my voice. "Who reported it?"

"Her husband. She was supposed to be on a girls' weekend in Melbourne but didn't show. Her girlfriends didn't tell her husband, a gynecologist called Dr. Cameron Granger, until Sunday morning."

Darlene lifted her head at that bit of info.

"Two of them knew Jennifer was having an affair. So they figured she had used the weekend as a cover, without telling them. The same two women tried to text her. When they got no reply, they phoned her cell. No response. Straight to voice mail. We've followed up on the calls; their story holds up."

"Probably dead at least twenty-four hours by then," Darlene said.

I stared at the mess of rancid flesh that smelled of newly applied chemicals. I tried and failed to imagine her as a beautiful, wealthy woman engaged in an affair.

"What's the husband been doing all this time?" Darlene asked.

"The sergeant at the station told me that Dr. Granger called them at least twice a day," I said. "Went to the station half a dozen times, offered a reward of ten grand for any info. That was all in the first week after she vanished. One of Jennifer Granger's friends finally enlightened her husband about the affair. But he still kept up the pressure on the cops. In fact, he doubled the reward."

Darlene raised an eyebrow. "Here's an informed guess: this poor woman is the first victim."

"Is that based on anything empirical? Apart from the fact that she died three weeks ago?"

"No, just a hunch. The murder is a bit different from the others... done with less confidence."

I tilted my head and frowned.

"The murderer got the woman to come to him... in a derelict house, away from Bellevue Hill. Now, though, he literally goes straight to the victims' doorstep."

"What about Yasmin Trent?"

"I'm convinced she was snatched. Probably close to where she lived. The cops found her car fifty yards from her body."

It was my turn to look surprised. "I didn't know that."

"They checked the odometer. The last journey in the car was thirty-one miles. Precisely the distance from Bellevue Hill to where Yasmin Trent's corpse was discovered in Sandsville. I reckon our killer is beginning to feel the heat in Bellevue Hill and mixing it up to keep us off his scent."

I was about to reply when the door opened and Mark Talbot walked in.

"Just passing." My cousin smirked.

"We need to talk," I said.

CHAPTER 54

"WHY THE HELL didn't you tell us about this woman?" I said.

Mark and I had walked to a deserted storage area at the back of the building.

"One of my officers caught a pick-pocket in Darling Harbour this morning, Craig," Talbot said. "Should I have told you about that?"

He moved toward me, stepping right into my personal space. But he apparently had a whole routine starting.

"Oh, and that pesky graffiti artist who keeps spray painting a wall just off George Street in the CBD? Got him too. Sorry...forgot to mention—"

"Maybe you think you're being clever, Mark," I said calmly. "But we have a deal with the police, don't we?"

"*You* have a 'deal' with the deputy commissioner."

"And you have to abide by it."

Talbot came even closer. He was about my height. We were eye-to-eye.

"This morning I used my professional discretion."

"No, you didn't. You did this deliberately, to screw me over. And you just showed up here to continue the fun."

He shrugged. "Well, yeah, maybe I did."

"Thanks to you, we lost five hours of precious investigation time."

He laughed. I could feel his breath. "Just listen to you . . . you fucking smart-ass. 'Precious investigation time'! Who the hell do you think you are? You're a PI. You can fool the deputy commissioner, but not me."

"I'm very disappointed."

"You what?"

"I'm disappointed."

He leaned in, his eyes narrow. "Disappointed! You cocksucker! Who do you think you're talking to?"

I went to gently push him back. And that was when he took a swing at me.

I blocked his fist, and he stumbled back a step, went for me again, his right arm swinging round.

I had learned something about fighting since I was eight years old. Better yet, Mark wasn't in the best of shape. I dodged his fist so easily, it was embarrassing—which enraged him more. His left fist came up, slower, but at an oblique angle. It grazed my shoulder. I grabbed his wrist and bent his hand back.

"Don't, Mark!" I said in his ear.

His breath was on me again, his mouth close to my left cheek. I bent his hand a little more and sensed him shift position, his right knee moving up toward my groin. I turned my body away, and his knee hit me in the hip. It hurt.

Still gripping him with my left hand, I swung round, sending a right hook to his face.

He fell back and landed heavily on the floor, blood streaming from a cut just below his left eye. He began to get up.

"Stop!" I hollered, but he wouldn't listen.

"Asshole! You always have been…!" He growled, got to his feet with surprising speed, and rushed me. I whirled round, elbow out, and he ran straight into it, nose first. I heard the cartilage crunch. He spun, hit the floor again, lay still for a few moments. In the police business we call that "victim supine."

I heard him groan, crouched beside him, keeping my guard up. He glared at me, blood streaming from his nostrils. His left eye was already puffed up.

I offered him a hand, but he spat at it. His saliva landed on the floor between us.

"Suit yourself," I said. Then walked away.

CHAPTER 55

WHEN I RETURNED to the morgue, I tried to look cool, calm, and composed. I was none of these.

"You all right?" Darlene asked.

"Yeah, fine."

"You look like you've been in a fight."

"You're perceptive. You should go into detective work."

She smiled and then pointed to Jennifer Granger's corpse.

"It's very similar to all the others," Darlene said. "Face burned and cut, stabbed in the back repeatedly. The same money dump—*fake money* dump. No sign of sexual assault. No DNA."

"You're leading up to the word 'but.' Aren't you?"

"But . . . ," she said. Then paused.

"You've found something, haven't you?"

This time she grinned. "You should go into detective work. I've found a partial print on one of the photocopies."

"Oh."

"Which convinces me even more that Jennifer Granger was the first victim. The killer was less practiced. He made a mistake."

CHAPTER 56

I KNEW THAT the dark-silver fabric suit Dr. Cameron Granger was wearing was expensive. I knew because I'd seen it up close in the Armani store the week before. I knew because I also found the price tag hidden in the suit and examined it. Then I stepped quickly away, as if the mannequin wearing the suit were about to mug me.

Let me put it this way: Cameron Granger was *annoyingly* handsome. The square jaw, the big head of blond hair, the blue eyes, the broad shoulders. He had the whole annoying deal going for him. Plus, I knew that he had a big house in the eastern suburbs, a tennis court in the backyard, and, I guessed, a million-dollar yacht moored somewhere where other people were paid to take care of it.

I had to quickly remind myself that Dr. Granger had just learned that his wife was cheating on him . . . and dead.

He indicated a plush, gray suede sofa, sat on one end, with me at the other. He looked suitably morose.

"I've been to the morgue. Been briefed. Given my report to the cops."

"If you don't mind my saying so, you seem very contained, Doctor."

"What can you do? I've had some time to absorb it all. After Jennifer failed to show up with her friends, I assumed she'd either run off with her lover or she was dead."

I appraised the man again. Was he using fake courage to overcome his grief?

"You had no idea your wife was involved with another man?"

"Oh, right…What more traditional motive for murder is there than being cuckolded?"

I held his eyes, and he looked away.

"Again, I apologize, but it strikes me as a little odd," I said. "Why would a wife risk losing such a lavish lifestyle by messing around?"

Granger surprised me by simply shrugging. "You tell me, Mr. Gisto. Maybe she thought she'd never be caught."

"When did you see your wife last?"

"I went through this with the police." He sighed. "I kissed her good-bye in the hallway of our home. Waved as she got into her car. She was leaving for the airport—or so I thought—to see her girlfriends in Melbourne."

"Then, later, you got a call from one of them."

"Yes, Maureen Miner, over thirty-six hours later, actually. She'd tried and failed to reach Jen by phone…Got worried…Stupid bitch."

"You sound pretty angry. Wasn't this Maureen Miner doing you a favor?"

"Oh, yeah! The sisterhood, keeping my wife's infidelity a secret. Great favor. Remind me to send her a thank-you note."

"Right," I said evenly, thinking about all the times men had closed ranks and kept their buddies' secrets to themselves. "Well, you obviously would like the killer brought to justice—you've doubled the reward."

"I doubled it once again earlier this afternoon."

"Is there anything you can think of that might help us…and the police?"

"Look, Mr. Gisto, I've told the police everything I know. I saw Jennifer leave the house. I assumed she was doing what she said she was going to do and met her mates in Melbourne. I didn't hear a thing until Maureen called. That was three weeks ago. Maybe you should speak to the guy Jennifer was fucking."

"As a matter of fact, Dr. Granger,

we've found the guy. My colleague is with him right now."

The doctor didn't even ask the name of his wife's lover. All he said was:

"Really? Do me a favor. Give him my very best wishes."

CHAPTER 57

THIS GUY WAS Jennifer Granger's lover? This was Nick Grant?

You've got to be kidding.

This was the guy who stole Jennifer Granger from her husband?

Justine studied the man sitting in front of her and wondered how any woman could find him attractive. He was about fifty years old, tall, almost sickly thin. He wore a brown vest and cargo shorts. On his left arm was a full-sleeve tattoo of grapevines interspersed with hearts and stars.

Nick Grant had agreed to meet on

"neutral ground"—a pub on Napoleon Street in Bondi.

"Look," he said, fixing Justine with a confident gaze. "Me and Jen—it was a casual thing, right? She was getting quotes for an extension on her house in Bellevue Hill. Took a shine to me right off." He gulped his beer, gave Justine a faintly flirtatious smile. Then his expression turned serious. "I was sorry to hear what happened..."

"She was with you the weekend she was murdered?"

"No! That's just it. I hadn't seen her for weeks. As I said, it was casual. I think we only did it four times, five max. She'd arrange everything and call me up with half an hour's notice. Tell me to put on something clean...that she was in room one thirty-one at the Four Seasons, or room four twenty-eight at the Hyatt, wearing nothing but high heels." He grinned stupidly. "Well, what do you do?"

"And the weekend of December fourteenth? You were in Sydney?"

"No." Nick Grant shrugged. "I weren't."

"So where were you?"

"In Melbourne."

"Melbourne?"

"Yeah—you look surprised."

"No, no, go on."

"Rugby piss-up. Me and the lads. We went to see the Waratahs at AAMI Park. Fantastic game—and afterward! Sunday…whoa…a complete blur. Took Monday off. Went back to work Tuesday. We're on a big job in Mona Vale." He nodded toward the Northern Beaches.

"So when did you hear that Jennifer Granger had gone missing?"

"One of her friends called me out the blue. I didn't know what the woman was talking about at first. She was another stuck-up bitch—sorry. I mean she was—oh, fuck! You know what I mean!"

Justine stared at the man.

"This woman," he went on, "Maureen? She said Jennifer hadn't shown up for a girls' weekend. 'Why you telling me?' I said. Apparently, Jen had mentioned my name and the company I work for, and this Maureen tracked me down. Cheeky bitch. I got a bit pissed off with her. Told her she'd better not tell anyone where to find me, especially Jen's bloody husband."

"And nothing else happened?"

"No. Not another word till this morning."

"So when was the last time you saw Jennifer Granger?"

Nick Grant took another gulp of beer and pondered the tabletop.

"Well, let me think … Must have been two weeks before the Melbourne weekend. Yeah—early December, at the Sheraton."

Justine shivered. She couldn't help but recall the gruesome and sad, shallow grave where Jennifer Granger was found.

She suddenly found herself saying aloud what she was only thinking.

"What a terrible mess some people create for themselves," she said.

Nick Grant didn't seem taken aback.

"I'll agree with you there, ma'am. I'll agree with you there."

CHAPTER 58

THE PARTIAL PRINT from Jennifer Granger's body appeared two feet wide on the flat-screen. Darlene studied the lines, what analysts call whorls and loops. She remembered a stat from college—there's a one-in-64-billion chance that any two people share fingerprints.

The partial on the screen looked completely unremarkable. It was, perhaps, two-thirds of a full print, limited in value but better than nothing.

Darlene double-clicked the mouse and highlighted the image. Then she moved the picture to an icon on the screen. The file disappeared, and a box came up with the words GLOBAL DATA-BASE ANALYSIS IN PROGRESS. Beneath this, a line, a tiny red dot to the left, and the words ESTIMATED TIME REMAIN-ING. 42 MINUTES—the time it would take for the Private computer system to compare the partial print with prints in every database throughout the world, some two billion records.

Darlene pushed her chair back, ran her fingers through her hair. She was extremely frustrated. Here she had some of the best forensics equipment in the world, yet she'd spent three days drawing a blank on four connected murders. At the back of her mind, something was nagging at her. It'd been needling her for at least twenty-four hours, but she couldn't pinpoint it.

She got up and walked across the lab to a bench. She'd filed every piece of data she had on the four murders. Most of the info was on the computer, and there were a few coded written reports kept in a filing cabinet. Here on the bench stood ninety-six test tubes in twenty-four racks. Each one was meticulously labeled. Each one contained something from the murder scenes.

She scanned the racks. There were slivers of cloth, particles of soil, fragments of body tissue, blood-soaked fabrics, hairs. Hairs! She moved the racks forward, one after the other, taking care to keep everything in the correct order. Then she saw what she was looking for—a test tube containing a single, whitish-blond human hair.

Darlene then walked to a powerful drive that stored all the crime-scene photos. She tapped the mouse and brought up the photo collections from the past three days. Clicked a folder entitled "Yasmin Trent." Scrolling down, she stopped over image number 233. A smile spread across Darlene's face.

CHAPTER 59

DARLENE DROVE STRAIGHT from her lab to the house in Bondi where Jennifer Granger had been found. She knew the Police Forensics team would still be there, and she wanted one more search around the place herself.

Darlene recognized the cop who met her at the front door. He gave her a warm smile. "Darlene," he said. "Back again?"

"Can't keep me away from a good murder scene, Sergeant Heller," she said, reading his ID badge. He was young and good-looking. She'd spotted him at the earlier murder sites, and she was certain that he'd noticed her too.

"It's Brian," he said, and he followed her through the hall. They stopped at the door to the bathroom, the bathroom that was obscenely splattered with blood.

"The murder was committed here," the sergeant said.

"You don't say!" She laughed. "So, I heard you got tipped off by a vagrant who slept in the front room last night."

"That's what we thought at first. A young guy called us early this morning. We followed up, traced the cell call. Turns out he's a seventeen-year-old high school guy. He and his girlfriend snuck in here for an early-morning quickie. They're both respectable kids from good families. But they picked the wrong spot. Their parents were real pleased to hear about it all."

Darlene gave a small laugh, and then she and Sergeant Heller walked outside, into the intense afternoon sun. Darlene saw four men in boilersuits digging up the lawn and the overgrown flower beds at the rear of the house. Two others were sifting through the soil, searching for further clues.

Then Darlene heard a cry from one of the diggers. She ran across the yard.

Two of the men were bending over an opening in the ground. Darlene skirted the edge and crouched down. Decayed human bones. Patches of white caught the light of the sun—a forearm, protruding from the dirt.

The forensics guys ran over, saw the bones, and squatted down beside Darlene.

"Keep digging, real gently," one of them said to the men with the shovels. They had to clear the soil near the arm with different spades, spades no bigger than a soup spoon.

The grave was shallow, not even three feet deep, and soon the outline of a large man could be seen. A few patches of gray-brown flesh remained on his dead bones. Strands of red hair clung to his skull.

CHAPTER 60

IT WAS PAST six p.m. and Johnny was leaving the office when the phone rang. A young female voice told him she was calling from Bonza Records and inviting him to a "VIP concert" starring Mickey Spencer, starting at eight thirty that night.

He just had time to get home, get changed, and get a cab to the concert site—a rather macabre place called the Old Quarantine Station, near Manly.

The cab pulled into the drop-off lot. Johnny paid the driver and walked toward the noise. He knew this place from when he was a kid. For more than a century—since it was built, in the

1820s—it was the place where visitors to Australia were quarantined before being allowed into the country. Thousands had suffered terribly in this place. Decades ago it'd been turned into a theme park, "the most haunted place in Australia."

Close to the old shower block and the mortuary, the original boiler house had been converted into a swish restaurant and conference center. Johnny emerged onto a cobbled courtyard lit up by massive lights on rigs. Directly ahead stood a stage strewn with musical equipment, men in black jeans and T-shirts testing mics. There were perhaps a hundred people milling around in front of the stage. Most were wearing suits, drinking champagne, chatting animatedly.

Johnny walked over to a waiter carrying a tray, took a glass of orange juice. A leggy blonde approached with a clipboard. Johnny gave her his name.

"Ah, yeah!" she said. "I was the one who called you earlier. I'm Melanie." She extended a hand.

"So, what's this all about?"

"Promo for the suits. Even stars as big as Mickey need to lay on a show for the execs and the sales guys."

Johnny nodded. "Strange choice of venue."

"Oh, we like to be a bit different!"

There was a sudden hush as the strains of a famous classical piece Johnny couldn't put a name to flowed from the speakers on either side of the stage. A man wearing a cream linen suit and a Mickey Spencer T-shirt walked out stage right, radio mic in hand. It was Ricky Holt.

"Ladies and gentlemen…welcome." His voice was deeper and softer than Johnny had imagined. He smiled at the crowd, pointed at someone at the front, laughed good-naturedly. "Thanks for coming along. It's a sort of celebration of Mickey's birthday tomorrow. Now, Mickey's well and truly wired, and he is RARING TO GO! So, please, give it up for my boy—Mickey Spencer."

The lights died, the entire stage turned black. A drum rhythm started, and a bass guitar came in. Then the lights burst on, thousands of watts of color. And there was Mickey Spencer, dressed entirely in white, crouched, microphone in hand. He screamed and the music came crashing in.

The crowd, lubed on expensive champagne and free cocaine, went wild. The song rocketed along, growing more and more powerful as it went.

Johnny had seen videos of Mickey Spencer, of course. His latest song already had a million hits on YouTube, but seeing him live and only fifty yards away was something else. He looked around and saw Mel nodding appreciatively. Then he turned back to the stage. Johnny was hardly able to believe that the demure, shy character he'd met at Private could transform himself into this massive personality, this rock god, parading in front of them.

CHAPTER 61

I WALKED INTO Darlene's lab and caught her doing crazy little dance steps. She looked as excited as a little kid waiting for Santa Claus.

So of course I couldn't resist saying, "What's happened? The newest edition of *Forensics Now Journal* arrive early?"

She gave me a smile and tilted her head to one side. "Just got back from the house in Bondi. And guess what? There's a second body in the garden."

"Really? Are you shitting me?"

"Nope. It's a man. From the level of decomposition, I'd say he's been dead two, maybe three months. Severe facial disfigurement, multiple stab wounds. Sound familiar, Mr. G.?"

Since being a detective meant always playing devil's advocate, I said: "But it's a totally different MO—a male victim. It doesn't make sense."

"I've taken samples. Police Forensics is all over it. There must be some link. Has to be the same killer."

Then Darlene said: "There's some other news."

"Lay it on me."

"I think I *have something* on this killer."

Darlene walked to her desk and picked up a file.

"Something was always bugging me about these crimes."

"Yeah, you said something in Sandsville—Yasmin Trent's murder."

"It came to me a couple of hours ago." She pulled a test tube from the pocket of her lab coat and held it out.

I took it and lifted it to the light.

282 • James Patterson

"A strand of hair?"

"Specifically, bleached-blond hair. Found on Elspeth Lombard's blouse."

"She was blond. Not one of hers?"

"I don't think so. I've just had it under the scope. A particular bleach was used. Every brand is slightly different. This is a cheapie—slightly higher peroxide level than the more upmarket dyes. Doesn't sound like the sort of stuff a woman like Elspeth would use. Also, see how a good third of the hair is dark? The woman this hair came from doesn't keep up with her color. She let it grow out. Again, doesn't fit Elspeth's profile."

"I don't see what—"

"Okay . . . The thing bothering me was that when I first arrived at the scene of Yasmin Trent's murder, I ran off a couple hundred shots on my camera and must have subconsciously noted a strand of blond hair lying across the dead woman's arm. I was distracted by something and had to talk to one of the cops for a minute. By the time I got back, the

Police Forensics guys were packing up, and I set to work."

"You'd forgotten about the hair?"

"I don't think I really registered it consciously."

"But the camera did."

Darlene pulled a photo from the folder. It showed a magnified white-blond hair lying on a piece of dark fabric.

"And Yasmin was a brunette," I said.

"She was." Darlene took back the photo. "I called Forensics straightaway. I got one of the good guys there, Gabe Ruggie. He's okay, seems to like me. He checked their files. Sure enough, they have a blond hair from the Yasmin Trent murder scene."

"Holy shit!" I said.

"Yep. They profiled a DNA sample from the hair. Couldn't match it with any database. They sent the profile over."

She pulled a piece of paper from the folder and held it out. It was a chart showing the analysis of the sample.

"And this," she said proudly, "is the profile I have from the hair on Elspeth Lombard's body." She handed me a second sheet. The two charts were identical.

"Hair from the same person."

"Absolutely no doubt—and the DNA does not match either blond victim. Oh, and one more thing."

"Okay. Lay it on me."

"Absolutely no Y chromosome in the profile."

"You mean…"

"You got it, Sherlock. The DNA is definitely from a female. Our killer is a woman."

CHAPTER 62

JOHNNY TOLD ME what had happened to Mary at the Triad place. So when I ran into her on my way to my office, I just nodded to her, and she followed me inside. She had to know—Mary was as bright as investigators come—that I wasn't happy.

Just to make sure, I started by saying, "I'm really pissed with you. What the hell were you thinking?"

She sat down and tried hard to keep her bandaged hand just out of sight.

"Information gathering, Craig. I went into places worse than that friggin' Triad dump all the time in the force."

"You could have gotten yourself killed. You might as well have put up a billboard on George Street: 'Triads—we're after you!'"

"They knew already. Word travels fast in this city. Besides, that's exactly the desired effect. I wanted to give them the shits!"

I let out a deep sigh. "Okay." I put my palms down flat on the table. "It's done. So how's the hand?"

"Just a scratch."

"Yeah, right! A sixteen-stitch scratch!"

"God!" Mary exclaimed. "Can't a girl keep anything secret around here?"

The phone rang.

"Mr. Gisto?"

It was Ho Meng, and he sounded a little out of breath.

"I need you to come here to my home immediately. There has been . . . a development."

CHAPTER 63

JOHNNY WAS WALKING toward the exit gate at the Old Quarantine Station where the cabs were lining up when he heard someone call his name.

He turned just as a huge, black Mercedes SUV pulled up. Mickey Spencer had his head out the window, a big grin on his face.

"Jump in, mate."

Johnny peered inside. Even more impressive than the car was the girl in the backseat next to Mickey. She had legs as long as the Mercedes and her exquisite face wore a perfect model's pout. The ubiquitous Hemi was in the front passenger seat, next to the driver.

"I'm good, Mickey."

"Dude! You're coming to the after-party, right?"

"Party?"

"My place. Come on, hop in." He spread his arms. "Plenty of room."

"Okay."

The car pulled away as the door closed, and Johnny landed on a seat facing Mickey and the girl. There was an ice bucket in the middle of the floor, two uncorked bottles of champagne inside. On the fold-down tray was a mirror with half a dozen lines of coke. Johnny noticed white powder on Mickey's upper lip.

"Johnny, meet Katia, my girlfriend. Katia, this is Johnny, a good friend of mine."

The girl looked at him seriously, didn't move a muscle. She had jet-black hair cut in a severe bob with a high, straight fringe; huge, dark eyes;

and of course those comes-with-the-total-package cheekbones. She was dressed entirely in black except for what looked like a miniature sword about an inch long on a pink ribbon at her incredibly pale throat.

"I know you don't drink, Johnny, but do you…?" He nodded toward the cocaine.

"No, thanks, Mickey."

"How dull," Katia said, but she did smile after she said it. She had one of those nonspecific "Continental" accents.

"Each to his own," Mickey said. "Katia is a brilliant guitarist, Johnny. She's Russian and was in a band in Moscow. They were called Khuy."

"Which translates as 'penis,'" the girl said. Again the smile.

"Isn't that fuckin' great, man? I fell in love with her when I learned that. Six months ago…longest relationship I've ever had!" He turned to the girl. "And I love her."

Katia leaned in to kiss Mickey. They

stayed glued together for a few minutes while Johnny looked out the window at the buildings flashing past.

Finally Mickey pulled away, wiped his mouth, and refilled his and Katia's glasses.

"So, man, you like the show?"

"I was knocked out," Johnny replied earnestly.

"Excellent. Excellent." Mickey downed the champagne in one gulp. "Well, I think you'll enjoy the party even more." And he gave one of his huge smiles.

CHAPTER 64

MICKEY'S SYDNEY CRIB was a penthouse in Woolloomooloo. Gray, minimalist, clean lines, massive windows looking out toward the harbor, all at a $10-mil price tag.

By the time the SUV got there, the place was packed. About two hundred people in various states of weird wardrobe—women and men in midthigh shorts with neon-bright T-shirts, assorted piercings and tats, fur boas and garden-party hats. Insanely loud music (apparently John Mayer and Justin Timberlake were back).

Mickey and Katia vanished, and Johnny wandered around the huge room. This was his night to consume a few pints of orange juice.

He was still somewhat amazed that Mickey Spencer was actually his buddy, his client, his source for coke, should he ever want it.

Johnny knew he was a poor kid from a middle-class suburb. But Johnny was determined—through school, through his job at Private—to leave the crummy background behind him. And meeting Mickey—well, that had been a totally unexpected piece of great luck.

He felt a tap on the shoulder and turned to see Katia.

"Can I speak with you?" she said.

"Sure."

She led the way across the main room. During that little walk Johnny caught a couple of famous faces from TV and YouTube, as well as a lot of not-so-famous but very beautiful young women. Johnny also noticed Ricky Holt talking to Mickey on the far side of the room. Katia motioned toward one of the balconies just as Johnny saw Holt hand Mickey a small package wrapped in brown paper.

"I'm sorry I was so rude earlier," she said. "You know, when I was on 'automatic bitch' and said 'How dull.' "

Johnny shrugged. All he could really think of was how lovely and refined her voice was.

"I didn't mind…Plus, I *am* pretty dull."

Katia smiled at his self-effacing charm.

"I didn't realize you were the guy from Private. Mickey's been singing your praises." Johnny's face could not conceal his pleasure at that comment.

"I'm very concerned for him," Katia went on.

"Because of this Club Twenty-Seven business?"

"Of course."

"He's convinced that Ricky Holt—" Johnny said.

"I'm very aware of that," Katia said. "But...Oh, I just don't know...Let me be straight about this. I'm worried that Mickey's losing it."

"Drugs?"

"Everything, Johnny. Everything. It's almost as though he has some weird death wish."

"So you think Ricky Holt has nothing to do with it?"

"You're the private investigator."

He fell silent, looked back to the room, filled with people. There was a sudden commotion. A woman ran over from a doorway in the far wall. She was shouting something, but Johnny couldn't hear anything over the thumping music.

Katia's instincts kicked in immediately. She barged her way through the packed room. Johnny followed her. Some drinks went flying.

The music stopped abruptly, and a hundred threads of conversation stopped with it.

CHAPTER 65

THE SILENCE IN the room was being replaced by the low, nervous rumble of voices as people pointed and whispered. As soon as they arrived in the room, Johnny followed Katia through a door into a cavernous marble bathroom. Three men stood around Mickey, who was lying faceup on the floor. A fourth man was kneeling beside him. He opened a metal briefcase.

"Fuck—*yob*—*govno!*" Katia screamed, madly mixing her languages. She also knelt on the floor.

Mickey was semiconscious. He was

drenched in sweat. Foam surrounded his lips. His arms and legs were twitching.

Katia spoke to the man who was opening the metal case: "Dr. James, you've got to…"

The doctor ignored her.

Katia went to grab Mickey.

"Please!" the doctor snapped.

Dr. James pulled a syringe from the case and squeezed the plunger a fraction of a millimeter. A tiny squirt of liquid now dribbled from the tip of the syringe. Then the doctor leaned forward.

With one shockingly violent movement, the doctor plunged the syringe into the middle of Mickey's chest, right through to his heart.

Mickey jolted upright. Then, as the doctor withdrew the needle, the rock star slumped back, his eyes snapping open. He rolled to one side and vomited.

It was then that Johnny noticed the package he'd seen Holt hand to Mickey minutes earlier. It was open on the floor, a used syringe and an empty vial lying on a piece of brown paper.

CHAPTER 66

HO BLENDED IN almost perfectly with the cream-colored sofa he was sitting on. He was dressed in a beige polo shirt and wore silky cream-colored chinos that could have been made from the same material that covered the sofa. He stood up, shook my hand, but it was apparent that he was upset.

"What's happened?" I asked.

"Dai has disappeared. I called his cell and home numbers half a dozen times. I went to his apartment and rang the bell. No response. I let myself in. There were signs of a struggle. A gun had been fired into the wardrobe."

"Any blood?"

Meng shook his head. He looked down and stared at the plush, white living-room carpet.

"And you haven't contacted...?"

Ho looked up. "No, Mr. Gisto, I haven't called the police."

I sighed. "There's more, isn't there?"

"A ransom note. Same as before. Either I do as they say, or my son dies."

The unforgiving light from the ceiling halogens made the man appear almost transparent.

"Is there an ultimatum?"

"Midnight tonight. I say yes, or Dai..."

I nodded.

"And there was this."

He leaned over to an ornately carved side table and picked up a small cardboard shoe box. He removed the lid and handed the box to me. I looked inside and— *Holy shit!* I saw an ear nestled on a bed of bloodied cotton.

"We can't fuck around, Ho. Forget about us trying to catch the two goons who kidnapped and killed Chang. We have to get the police involved and go much higher up in the gang hierarchy."

Ho closed his eyes for a second.

"This has gone too far for Private to deal with alone," I said. "And, actually, by not going to the police, you're in danger of breaking the law yourself."

He sniffed at that, but a moment later he nodded and said, "I know."

CHAPTER 67

TENSE. ANGRY. CONFRONTATIONAL.

If you could ignore those three poisonous moods darkening the atmosphere in the briefing room at Sydney Police HQ, we were just a bunch of guys hanging out together.

Ho Meng and I were sitting around a steel table with the very pissed-off Brett Thorogood, the very pissed-off Mark Talbot, and a senior detective in charge of the police assault force, Evan Freitas.

Deputy Commissioner Thorogood was commanding proceedings from the head of the table. It was no-nonsense all the way.

"Mr. Ho," Thorogood said, looking directly at the man. "You know these gang people better than any of us. Do you have any idea of the identity of the men?"

Ho sat completely still. He now possessed some sort of Zen-like calm that I thought covered a seething anger and horrible pain.

"As you are aware, the lead operatives in Sydney are the Lin brothers, Sung and Jing," Ho said stiffly. "They are Four Twenty-Sixes."

"Which means?"

"The Triads have clear distinctions between ranks and positions in the gang. They are each given numbers based upon the I Ching numerological system. The leader of the Triads is Four Eighty-Nine. His name would be 'the Mountain' or 'the Dragon.' I believe the gang in Sydney is a fragment of the Noonan, perhaps the most powerful of the Triads. The Dragon, the Four Eighty-Nine, is a man named Fong Sum. I met him once in Hong Kong. He's there now."

"So he'd be like a don in the Mafia?" Talbot asked.

Ho nodded slowly. "There are many differences, but, very broadly speaking, yes, he would. He controls a global network. The Sydney gangs are just a small part of it."

"And the Lin brothers—how many people work for them?" Freitas asked.

"That I do not know for certain."

"Ballpark?"

"I would estimate perhaps forty to fifty foot soldiers in the city," he responded.

"Foot soldiers are the rank and file, right?" Thorogood asked.

Ho nodded again. "They are known as Forty-Nines. I would suggest the men who abducted Chang and Dai would have been their best Forty-Nines, men who are working their way up in the pecking order. This would have been a big job for them."

"As this whole heroin project is for the Lin brothers too," I remarked.

"Indeed," said Ho.

"Okay," the deputy commissioner said. "So, do we have a consensus as to what to do next?"

Ho apparently felt he was a participant in answering Thorogood's question. He spoke: "I have come to the conclusion that the only chance we have of saving my son is to convince the gang that I will do what they want."

"And that could provide us with a platform for a sting operation," I added.

Mark looked at me with suspicion. "Us?"

"We're here to do anything we can to help," I said directly to Thorogood. "But we can't do it unless we're armed—like the rest of you. My assistant Mary Clarke and I are licensed to carry firearms."

"I appreciate your contribution," the deputy commissioner responded. Then, to make sure his opinion registered, he looked directly at Mark Talbot and said, "I think we can work together on this."

CHAPTER 68

WITHIN A FEW hours, the cops were at Ho's house, setting up tracking equipment for landline and cell calls. There was quite a crew from our side too. Talbot, Freitas, and Thorogood were there to babysit, and I had Mary and Darlene with me this time.

At eleven p.m., Ho Meng made first contact from his landline.

He tried to keep the call going, but the foot soldier at the other end wasn't dumb. The call ended before the police expert could locate the caller within less than a square mile. Ho gave the anonymous Triad member a mobile number. The guy clicked off before saying when he would respond. We just had to wait.

"We brought along some technology that might help," I said. Mark gave me a contemptuous look (surprise!), but Freitas and Thorogood were full of hope and full of enthusiasm.

Darlene walked across the room with two small boxes. She stacked them on a low table and opened the lid of the top box. Then she plucked out a cell phone and removed the back cover. "Put your SIM in here," she said to Ho Meng. "When they call you, we can get a better trace on them than with the conventional gear." And she flicked a glance at the police operator, with his suitcase-sized tracking unit, which he'd set up on the sofa, close to the home phone.

Darlene picked up the second box and pried open the lid.

We could all see inside. A white pad

with a black dot the size of an aspirin on top. "A microtransmitter," she said.

"We can place this anywhere on Mr. Ho's body, and it'll pick up conversations and relay them to a receiver. You'll be close by in a van, right?" Darlene asked the cops.

"I'll be with the assault unit," Freitas replied. "Inspector Talbot will be in the van."

I glanced at him. Mark looked away.

"Okay." Ho nodded. "So, what happens now?"

Thorogood looked up. "We're ready when they are. Just need them to give a call."

CHAPTER 69

JULIE HAD FALLEN asleep in front of *Australian Idol* and, as usual, was dreaming about her father. In her dream, all the bad things had never happened. Dad was still alive. Julie had finished school, gone to college, become a police forensics officer.

"Geeeeee-na Esposito!" the announcer shrieked, and as the *Australian Idol* winner was announced, the audience began to cheer, and Julie woke to the strains of Sheila Watson singing "I Dreamed a Dream."

Gina Esposito wept for joy as the horror of Julie's own life came rushing toward her. She closed her eyes again, and there was her mother, screaming at her. When Sheila felt that Julie had misbehaved—that was when the torture began. She had kept her locked in her bedroom for days, a plastic bowl for a toilet. Sheila gave her only beets to eat.

Later, the torment increased. Sheila would tie Julie to a chair in the kitchen, gag her, and burn her arms with cigarettes.

On her eleventh birthday, her first since her father's death, she received nothing. Then, just before bedtime, Sheila tied her to the chair again and told her if she made a sound, she would put her feet in the fire in the living-room fireplace. Then she pulled out one of Julie's incisors with a pair of pliers.

This treatment continued for four years. She could never say a word for fear of torture. She hid the scars and the marks, made excuses for every lost tooth, every bruise. Then, one day—she knew it would happen one day—Julie snapped.

On the evening of her fifteenth birthday, Julie knew she would be in for her traditional "gift." As Sheila busied herself, getting ready to go out, Julie slipped a medium-size kitchen knife into the back pocket of her jeans.

Her mother appeared in the doorway to the kitchen. She was wearing a clownish amount of cheap makeup. There were two lengths of cord in her left hand.

"In the chair."

When she didn't move, her mother began to smile. Took a step toward her.

Julie pulled the knife from her pocket and swung it around. She stopped just before her mother's face.

The woman screeched, the smile vanishing instantly.

"You...! In the chair!" Julie hissed. And when her mother didn't react, she moved the knife a tiny bit closer.

She tied Sheila with the cords, gagged

her with a tea towel, then brought the knife to the center of her forehead.

Sheila was shaking, her eyes filled with terror and hatred.

Julie moved the knife another fractional distance closer, scoring her mother's flesh. The woman screamed under the cloth, but it came out as nothing more than a muffled hum. Julie heard a rush of liquid and saw her mother's urine flow over the front of the chair and onto the floor.

"You never made me do that, even once, you useless bitch!" the girl announced proudly. She pulled the knife away and pocketed it again, turned, and walked out.

CHAPTER 70

LIN SUNG CALLED at midnight.

Ho managed the call deftly, and I could understand how he'd been such a successful cop in Hong Kong and snowballed his skill into a lot of money with his businesses in Australia. He was quick and smart and calm.

Darlene's app was able to pinpoint the caller in under ten seconds. It was impressive but actually not much help. Lin was calling from a pay phone outside Luna Park, in North Sydney.

"We would like to meet you," Lin said, his voice coming into our individual earbuds. Of course, his words also went straight to a digital recorder.

"You will have my son?"

"Not this first time."

"Then there will be no meeting."

Silence from the other end. I held my breath.

"You are hardly in a position to negotiate, Mr. Ho."

Ho paused for a moment. "I entirely disagree."

Lin gave a small laugh. "Don't begin a game of bluff with me, Mr. Ho."

"I'm not bluffing." Ho's voice was stony.

Another, longer pause.

"Very well. We'll bring the boy. But we will only consider an exchange if all our conditions are met. Do you understand?"

Ho said nothing.

"I'll assume that is a yes, Mr. Ho. And if you invite a third party to our meeting, your son will be killed before your eyes."

When Ho still did not speak, Lin said. "Blackball Reserve, forty-five minutes."

Then Lin hung up.

CHAPTER 71

WE WERE ON the freeway, ten minutes short of Blackball Reserve near Manly, when the agreed-upon rendezvous was changed. I was in my car, Mary in the police surveillance vehicle with Mark, and, next to him, a plainclothes officer, driving.

Two hundred yards ahead of them was Ho. He looked like a lonely old man, driving alone in his Bentley.

My cell clicked on to speaker. It was Mary.

"New destination," she said. "A green-and-chrome warehouse near the airport."

We all turned off at the next junction and headed south. I couldn't see the Bentley, but I had to assume that the cop directly ahead of me could. My car was fitted with a police tracker set to a broad range of frequencies. I could hear their comments, and I knew that Central Control had quickly redirected the chopper assault team to the new location.

We reached the place in thirty minutes, pulling up a hundred yards away from the warehouse. I parked behind the surveillance vehicle and ran over to join them as Ho's car vanished into the shadows. Mary opened the sliding door, and I climbed in. Mark and an operative were at the controls. We could hear every sound Ho made through the tiny transmitter.

"Assault Officer One," the operative in the van said. "This is Control, come in." AO1, I knew, was Evan Freitas.

"Control. We're in position. AOs Four, Five, Six, and Seven are in a small room across from the main warehouse building. I'm with AOs Two and Three, on the opposite side. I have visual contact with Mr. Ho's vehicle."

A screen on the wall of the control room of the van lit up with a night-vision video feed from AO1's helmet. It showed a fuzzy image of Ho's Bentley entering the derelict warehouse, lights ablaze. It stopped, Ho dimmed the lights, and the image improved dramatically.

Now a black Mercedes with tinted windows, registration LS1, entered through the north end of the dilapidated building. It crunched over the pitted floor, strewn with pieces of metal and crushed concrete. It stopped about fifty yards short of the Bentley.

Ho stepped out of his car. He took a couple of steps toward the Mercedes. The car's engine was still running; both rear doors were opened. Two men slipped out. They were slender, black-haired figures. The slightly taller one of the pair was Lin

Sung. He was dressed in his usual vintage, narrow-lapelled shiny black jacket and skinny tie. His brother, Jing, was wearing a blue tracksuit and white trainers. They walked slowly toward Ho as the driver clambered from the front of the Mercedes to stand by the hood.

"It's a pleasure," Lin Sung began, and put out a hand, which Ho ignored.

"Where is my son?"

Lin Sung chuckled and flicked a glance at his brother. "There is great value in patience, my friend."

"I'm not your friend." Ho looked from one brother to the other. "I'm here to make a deal with you, as we agreed."

"Yes, and—"

"I want my son released. Then I will cooperate."

Sung sighed, cackled.

"You find it funny?" Ho asked coldly.

"You don't?" the younger brother butted in. His voice was oddly effeminate, completely at odds with his macho stance.

"Ho's hanging tough," I whispered to Mary, who was standing beside me in the police van.

"Hope he doesn't overdo it."

I turned back to the screen and saw Lin Sung take a step closer to Ho. "We have the boy," he said slowly. "But we need assurances. Surely you understand that? If we return him to you, what is to say you will cooperate?"

"You have my word."

It was the younger brother Lin Jing's turn to produce a half-assed laugh. "Ah! Your word!" he said, nodding. In an instant his mirth had vanished, and he pulled a gun, a Chinese Type 64, from his waistband. Lin Sung saw it and glared at him, but he didn't flinch.

Ho looked from one man to the other.

"In case you couldn't tell, this isn't going well," Mary hissed in my ear.

Freitas's voice came through. "Hold positions. No one move till I say."

Sung deliberately moved closer to his brother and slightly in front of him.

"We are all reasonable men," Sung said, and he tilted his head slightly as he stared at Ho Meng. "I understand you want your boy back, but you have to put yourself into our position, Mr. Ho." Then he turned and snapped his fingers at the man standing by the hood of the Mercedes. He walked to the back door and opened it.

"You may see your son."

The driver leaned in and helped Ho Dai climb out. The young man's hands were tied behind his back, and he was shaking, weeping, petrified. He had a large, bloody wound where his left ear had once been. He saw his father and began to speak. "Say nothing!" Lin Jing barked, then whirled around to Ho again, his gun raised.

"There. Your baby's safe. Now we talk."

"What is it you want from me?"

"At last—" the younger gangster exclaimed, but his brother cut in over him.

"Your business provides a perfect cover for one of our...trade plans."

"Drugs...You want me to get heroin in," Ho said.

Sung smiled, nodded.

"And in return?" Ho flicked a look at his son, who was still standing by the car, the driver gripping his right arm.

"When you have proven your worth, he will be released."

Ho gave Sung a venomous look. "No deal," he said, and started to turn.

"You motherfucker!" the younger brother screamed. He was clearly about to squeeze the trigger of his Type 64.

"Go!" yelled Freitas.

CHAPTER 72

SENSORY OVERLOAD PLUS. Shouts from the assault team, yells and thuds from the warehouse floor. On the screen, a smudge of movement through the night-vision lens.

Ho fell to the floor. I couldn't tell if he'd been shot or had hit the ground to avoid a bullet. Sung reached toward his brother. Ho rolled to one side as the younger brother fired a second bullet. Sung was yanking Lin Jing's arm down when the assault team, in full body armor, burst into the warehouse from two different directions. They screamed as they ran. Their Enfield SA-80s were leveled and ready.

The younger Lin reacted instinctively. Pumped up, he jumped for cover, into a pile of metal drums. Then he fired at the approaching cops. Before he could reach the barrels, he was ripped open by at least three different shots. He crumpled in a heap.

Sung whirled round, ran, and reached the Mercedes. Dai and the driver were crouching behind the car. The driver had pulled a gun. The kid looked like a puppet—big cartoon eyes, limp limbs. Sung reached cover and pulled out a semiautomatic, Bulgarian-made Arcus 94.

Lin grabbed Dai, and we all heard the gangster yell something that I couldn't understand.

"Hold your fire," Freitas's voice boomed through the speakers.

On the screen, I could see the fragmented image of Lin Sung rising slowly from a crouching position. He had the semiautomatic at Dai's temple. The driver shuffled out of the scene. He slipped behind a huge pile of rusting plant machinery. Then Ho Meng stood up slowly, apparently unharmed. He started to walk toward his son.

"Let my boy go!" he yelled.

Lin Sung ignored him and took a step forward. Then he opened the driver's door with one hand and shoved Dai inside the Mercedes. Lin Sung moved in beside him. They disappeared from view behind the tinted windows.

Ho reached the car, but it made no difference. The car roared away. The cops had their machine guns raised, jumping aside as Lin accelerated toward them. The car skidded on the uneven floor, drifted for a second, tires screaming. Lin got it under control and apparently pushed his foot to the floor.

I didn't wait another second. I slid open the door of the surveillance vehicle and ran across the gravel to my Mas. Then I hit the remote.

CHAPTER 73

I SPUN THE car backward on the gravel, turned into a pitted lane beside the warehouse, and shot away.

I couldn't see the Mercedes, but I knew I was racing in the right direction. It was the only route to the perimeter fence. Careering round a bend topping sixty, I hit a yard-wide hole in the tarmac, bounced out, the suspension stretched to the breaking point. I almost lost my grip on the road as the rear end came out. I just pulled it back.

The entrance to the freeway was about fifty yards ahead. I caught a glimpse of

Lin's car as it shot through the gates and accelerated up the on-ramp. I dodged a stupid fucking pothole, swung left, then a hard right. I opened up the engine and tore onto the M5, headed west.

The Mercedes was quick, but the Maserati was quicker, and, driven by someone in my crazed state of mind, it was *fantastically* fast. Lin may not have been aware of my advantage, but he learned of it immediately when I began to gain on his car, halving the distance between us in less than thirty seconds. The M5 freeway was almost deserted, and I had the Mercedes in my sights, only twenty-five yards ahead. The speedometer read 120.

Lin took the next exit. His car went screaming onto Princes Highway, toward Rockdale. It was a smart move: a slower road, a greater chance of urban traffic, plenty of turnoffs. It leveled the playing field—a little.

At 1:15 a.m., the street was pretty much empty of traffic. Lin pulled the

Mercedes onto a side street, took it wide, and just missed an oncoming car. I screeched after him, barely missing the same car.

It was a narrow suburban street— rows of modest houses, parked cars to the left. Lin jumped the lights. I slowed and checked, followed him over the junction. He took a right, a left. More residential roads, a church, a McDonald's. I caught a sign for a sports field and saw a line of trees.

Lin waited until the last possible second, roared into a narrow lane just before the park. I braked and flew around the corner.

The Mercedes had disappeared from view. Then I realized I'd shot straight past it. Lin had taken a hard right off the road and pulled up onto a rutted track at the edge of the field.

I reversed and caught some movement in my rearview mirror. Lin was out of his car, gun in hand. He rushed around to the passenger door. He

326 • James Patterson

yanked it open, dragging Dai to the ground.

I stopped, slipped out, kept low. The Merc was ten yards away. Lin was pulling Dai up, the barrel of his gun at the kid's temple.

I was in the shadows, but Lin certainly knew precisely where I was crouched. He could have taken a pop at me, but then he risked losing Dai. "Stop!" he shouted into the night. "Or I'll kill him."

I pulled back and crept behind a line of bushes. He couldn't be sure where I was now. I moved fast. Lin and the boy dropped out of view for a few seconds. Then I found an opening in the bushes and saw that they hadn't moved.

I picked up a stone, tossed it to my left. Lin whirled around. He had his free arm around Dai's throat.

"Stop the stupid games," Lin said, an edge to his voice now. I was getting to him.

I moved to his very far right and could

see the back of his neck, wet with perspiration. I leveled my gun to his head.

"Let him go."

Lin spun around.

"Let. Him. Go."

"No!"

Some instinct told me I'd pushed him
too far. I fired, and his gun went off simultaneously. Lin flew backward. The
hood of his car broke his fall; a cloud of
red exploded from his head. Dai jolted,
screamed, and collapsed to the ground.

I rushed over, expecting the worst.
Blood was running down Dai's cheek,
dripping from his jaw. He must have
thought, like me, that Lin had been
about to shoot. Dai had moved just in
time. Lin's bullet had just grazed the
boy's temple.

I pulled Dai to his feet. He was shaking fiercely. I untied the cord around his
wrists. He started to cry, tears streaming
down his cheeks. He put a hand to his
face and came up with bloodied fingers.
I thought I might start crying myself.

We could hear sirens. "It's okay," I said, and I knew I was saying it as much for myself as I was for Dai. "Just a scratch. You're going to be fine, Dai. It's all over, buddy."

CHAPTER 74

I SPENT ABOUT three hours at police HQ. I didn't get the "full hostile" interrogation treatment, but no one was especially easy on me. As expected, Mark Talbot could barely contain his pleasure at witnessing the intense, relentless questioning.

Finally everyone agreed: I'd handled the situation with Lin as wisely as possible.

Brett Thorogood summed it up: "An impossible situation with a reasonable outcome. In fact, I'd say the good guys won." Mark looked disappointed.

From HQ I drove directly to Private. I

was exhausted, and I was filthy. I showered and shaved in the workout locker room and stepped back into the grungy clothes that had traveled with me to hell and back. Then I called all parties into the conference room. Frankly, they didn't look much fresher or better rested than I did.

I started the meeting by holding up the front page of the *Sydney Morning Herald:* SYDNEY SLASHER CLAIMS NEWEST VICTIM.

"We're getting nowhere fast with this one, gang," I said.

Then I looked at Darlene. "Anything, Darlene, anything? Please tell me you've got something."

"Only what I said yesterday afternoon. I'm sure the killer is a woman."

No one said anything. They'd all been updated on Darlene's DNA findings.

"Not conclusive, though," Mary said. "We know the victims were all acquainted. The blond hairs could have come from a mutual friend."

Darlene looked at the table, nodded.

"But what if they *were* the killer's? Let's run with that for a sec," I said.

"There's no match in the database," Johnny commented.

"Means nothing. Maybe the murderer had never committed a crime until..."

"All right," Justine said suddenly. "What if she happens to be a 're-spectable' bleached-blond friend of the dead women and part of the same social circle? Maybe the motive was some relationship mess or simple jealousy."

"The wife of a banker or a corporate suit gone crazy?" Darlene looked interested. "Maybe it *is* a sex thing. An eastern-suburbs mum taking revenge on women her husband's slept with?"

I raised my hands. "Hang on, let's calm down!"

"Actually, I don't believe that," Darlene backtracked.

"Why?"

"For a start, the hair was not recently bleached. There was significant regrowth. That in itself suggests the woman doesn't pamper herself. How many wealthy women walk around with weeks of roots growing out?"

"Search me!" Johnny said.

"And my sister insists Elspeth and Stacy weren't messing around," Justine said.

"Besides," I added, "the banknotes don't fit the theory, do they? The very fact that the notes are fake suggests the killer isn't a rich woman living in the same area as the victims . . . unless that's a trick."

"Oh, for God's sake!" Mary exclaimed. "We're going round in bloody circles!"

"No, no . . . rewind," I said. I was suddenly excited. I stood up and started pacing close to my chair. "Let's say it's *not* a trick and the killer *is* poor—a woman from outside the area. She can't afford real fifty-dollar bills. She photocopies them. Yes!"

I surveyed the room, the faces of the team. For a moment they all looked confused.

Then I remembered something. "Darlene, you told me the other day that the fakes are high-quality photocopies. What if our killer photocopies the notes at a shop instead of at home? And what if...what if the murderer, this woman whose bleached hair strands have been discovered...what if she doesn't live in the eastern suburbs but works there?"

"I'll get on it—visit all the copy shops in the area," Johnny said, as excited as me. "I think you're on to something, Craig."

Then Mary spoke: "Hey, Johnny, before you start making the rounds, let me give you some advice."

"What's that?" Johnny asked.

"It'll be a lot easier to drive if you remove your lips from the boss's ass."

For the first time in a long time, everyone in the room laughed.

CHAPTER 75

JOHNNY MAY HAVE been an ass-kisser, but he was a hardworking ass-kisser. He was back in my office with a report by midafternoon.

"There are five copy shops within a three-kilometer radius of Bellevue Hill. First three shops I visited drew a complete blank. Guys there had no idea what I was talking about when I asked them if any odd-looking, memorable, cheesy blond women had been in. Made me feel bloody stupid, actually!"

"What about the other place?"

"Fourth shop was on New South

Head Road, about a mile from Bellevue Hill. The manager was a nice guy. Said he'd seen one particular woman come in a few times during the past three weeks. She didn't look 'odd,' exactly, just miserable, run-down. But get this. He described her. Above-average height, well built, bleached blond."

I rubbed my hand over my chin and stared at Johnny silently. "And the fifth shop?"

"Jackpot! A very sweet girl running the place."

"Yeah, yeah…"

"She'd seen the same woman at least twice during the past month."

"Don't fuck with me, Johnny. You've got more. I can tell by your tone."

"The last shop keeps surveillance-camera records for a month at a time. I gave the girl a hundred bucks, and she ran off a copy of the disk for me."

He handed me the disk as if he were presenting an award. I slipped it into my PC.

It was a crappy-quality recording but good enough. It showed a woman coming into the copy shop, moving from the counter to a self-service machine. She placed something indistinct on the machine's tray and watched as half a dozen copies emerged. She then paid for them and left.

"Quite a powerful-looking woman," I said.

"And piss ugly," Johnny said.

I exhaled.

"Sorry!"

"Can't see much of her. But she definitely has bleached her hair."

"First thing I noticed," Johnny replied.

"Take it through to Darlene. See if she can do anything with her magical imaging equipment."

CHAPTER 76

DARLENE WATCHED THE short clip taken at the copy shop. Johnny was leaning on the back of her chair, peering at the screen over her shoulder.

"It's pretty bad quality," she mumbled.

Johnny said nothing.

"But, thanks to my new buddy, Software Sam, I might get something out of this. It works just as well for video as it does for still images."

She ran her hands over the control panel of the image enhancer. Then she turned back to the computer keyboard and moved her fingers over the keys.

The screen went blank for a second. Then the film spooled back to the start. Darlene tapped another couple of keys. The clip was now 500 percent clearer.

The woman lumbered into the shop. She was wearing a shapeless navy-blue sweat top, handbag on her right shoulder. The camera caught her straight on. She had a wide face, a flat nose, small eyes. Her shoulder-length hair looked greasy. It was blond. Not dyed well— a bottle from a pharmacy, badly applied a long while ago. She wasn't wearing makeup. Then Johnny said something that Darlene had already noticed: "She doesn't have any bloody eyebrows."

"Not the prettiest specimen," Johnny said, a little more diplomatically this time. "What would you say? Five seven, five eight? Hundred and seventy pounds?"

"Five nine, one seventy-five."

"I bow to your superior skills," Johnny retorted.

"I used to work in a carnival," Darlene said.

Johnny laughed at the joke. Then they continued to watch the video of the woman walking towad the photocopier.

"Can you close in on her just a little bit more?" Johnny asked.

Darlene played her fingers over the keyboard, slowed the film, zoomed in and adjusted the enhancer to sharpen the picture. She was pushing the software to its limits.

Tugging the mouse gently, she moved the center of the image to see what the woman was placing on the copier. They both noticed she was wearing latex gloves. She plucked a sheet of paper from her bag. It was impossible to see what was on it.

Darlene slowed the film, then let it creep forward a few frames. The first insert began to emerge. She shifted perspective, closing in on the paper spewing from the copier. It appeared slowly. She moved in closer still. She toggled the controls on the enhancer, prayed the software would hold up.

And there, in the plastic collection tray of the photocopy machine, lay a sheet of paper containing the image of four fifty-dollar bills.

"Excellent job," Johnny said.

"We still don't know who she is," Darlene commented. "I'll get this over to the police. They may know something about Blondie."

CHAPTER 77

IT WAS A good thing that the door to my office was wide open when Darlene came rushing in. If it had been closed, she'd have crashed right through it like Bugs Bunny.

She was waving a paper file.

"Better sit down for this one, Craig," she yelled at my empty desk.

"I am, Darlene," I called from the sofa.

"Just off the phone. Sergeant Tindle called. They've ID'd the remains of the man at the Bondi house."

She sat at the other end of the sofa from me, opened the file. "Name's Bruce Frimmel."

She handed me a photograph of the man from police records.

"He'd served time. Assault charge five years ago. His DNA was on file. He vanished two months ago."

"And you reckoned the guy in the garden had been dead for two to three months."

"Police Forensics also identified two distinct sets of blood splatter in the bathroom at the house. One is Frimmel's, the other is Granger's blood. They were both killed in the same room."

"Interesting."

"Just wait. It gets much more interesting. Bruce's girlfriend, Lucy"—Darlene glanced at the file again—"Lucy Inglewood—was questioned when Frimmel vanished. She told the police he had crossed a few people. There was a biker gang in Blacktown he'd upset, and a few months earlier, he'd broken up acrimoniously with his last girlfriend, who he'd lived with for a few years."

"The police looked into these, I take it?"

"They interviewed everyone who'd known Bruce Frimmel. Sergeant Tindle worked with Inspector Talbot on it. They talked to twenty-odd of Frimmel's associates and those close to him, including his ex, Julie O'Connor. The sergeant called me, Craig, because he had just seen the security-camera stills of the woman in the copy shop I sent over this morning."

She plucked two sheets of photographic paper from the file and handed one to me. "This," she said, "is the best image from the security camera."

I stared at the photo of the woman walking toward the copier.

"And this is the woman Sergeant Tindle interviewed two months ago? This is Julie O'Connor?"

I glanced at the second photo, then held the two images side by side. "We have our killer," I said.

"And you know the best bit? According to police records, as of two months ago, she was working at SupaMart in Bellevue Hill."

CHAPTER 78

"GOOD NEWS," MARY said as we pulled up outside the SupaMart in Bellevue Hill.

"Yeah? What?"

"Iceberg lettuce is on sale for a dollar nineteen a head."

I wasn't in the mood to laugh.

I checked my watch. It was just past noon as Mary moved from the car to the sidewalk. She pulled on her shades and waited a moment for me to step out of the car and lock it. I led the way to the store, keeping the keys in my hand.

The manager's office was at the back.

A girl standing on some steps was stacking spaghetti-sauce jars on shelves and pointed the way.

"Take a seat, take a seat," the manager, Jimmy Kocot, said.

Obviously bored, I concluded. *Slow day in Bellevue Hill.*

"We're looking for Julie O'Connor. Understand she works here."

"Julie? Yeah, she does. Should be here now, but she isn't."

"What do you mean?"

"Didn't turn up for her shift this morning." He frowned. "What's this all about? You cops?"

"No," Mary said. "We're from an investigative agency. We've had a call from one of Julie's relatives," she lied. "Miss O'Connor's aunt way over in Perth has died. The family want to reach Julie."

"So maybe the old girl's come into some money," he said.

"Maybe."

"Well, of course...I understand... Mustn't assume anything."

"No," I responded. "Would you have Julie's address? Maybe a phone number?"

Kocot looked doubtful for a few moments. "That might not be possible. There's a certain confidentiality—"

"Sure," Mary said in her sweetest voice. "It's just, the family is *desperate* to get in touch with Julie. She apparently left her relatives in Perth under a cloud, years back."

"I didn't know that," Kocot responded. "Might explain a thing or two."

"What makes you say that?" I asked.

"Well, I like Julie, but she's never been the most...communicative of my staff. Never made friends with the others. She's a bloody good worker, though— that's why I kept her on." He paused.

"Okay—I can't give you her phone number, 'cause she doesn't even have a phone. But the address..." He turned toward a mini filing cabinet on top of his desk. Flicked through the cards. "Yeah, here it is: Six Neptune Court, Impala Road, Sandsville. Let me know what the outcome is, will you? It'd sure help me to know if Julie will ever be coming back."

CHAPTER 79

JULIE WAS SITTING on her beer-stained, grease-stained, bloodstained sofa, watching TV with the sound off. It was the only way to watch, because the voices of everyone on *Australia's Next Top Model* could make you crazy with anger.

Beside her lay her scrapbook and a notebook. She picked up the notebook first. She kept this in her overalls pocket at work. In many ways, she had the perfect job for her purposes. Working at the checkout of SupaMart in Bellevue Hill each day, she was on the lookout for potential victims. Each day held a constant parade of spoiled wives of successful eastern-suburbs bankers, brokers, and doctors. These women floated into SupaMart, Gucci-clad and dripping Tiffany, to buy zero-fat milk and goat cheese with their private-school-uniformed brats. To them, Julie was either invisible or an object of contempt. Julie loathed them.

But Julie also had access to their personal details. She had their credit-card data, she caught their names when they bumped into their snooty friends and had a little "chat" at the checkout. She noted everything she heard. The same women came in several times a week. A month of listening and note taking, and she knew a great deal about Samantha, Sarah, Donna, and dozens of others including Yasmin Trent, Stacy Fleetwood, Elspeth Lombard, and, of course, Jennifer Granger, the wife of the bastard who'd started it all.

She returned the little notebook to the

top pocket of the lumberjack shirt she was wearing, and she picked up her scrapbook. She'd devoted a double page to each of the murders, numbered them.

1. JENNIFER GRANGER
2. STACY FLEETWOOD
3. ELSPETH LOMBARD
4. YASMIN TRENT

Beneath these names: descriptions of each murder, recounted in her scratchy misspellings. Interspersed with the words, Julie had pasted in pictures of adorable, chubby babies, pictures she had cut out of magazines.

In the middle of the scrapbook, she sorted and listed the info she had scattered in her notebook. Here she neatly arranged rows of credit-card numbers, addresses, friends' names, husbands' jobs, husbands' offices, kids' schools. All of it had been routinely transferred from the notebook.

Now she turned through the pages of the scrapbook slowly, as if it were a religious manuscript. She studied all the information she'd transferred over the months.

"Tabatha," Julie said aloud. "Married to Simon, a 'very handsome' broker at Stanton Winslow. Address: Eight Frink Parade. Four kids—shit! Busy girl!"

Turning the page ... "Mary—ah, nice Catholic girl, Mary. Irish ancestry, no less. Works for a local charity—Homes for Rejected Pets. How lovely! Two kids, Fran and Marcus. Husband, a spinal surgeon at Royal North Shore Hospital ... Tempting, very tempting."

She flicked to the last page. A newspaper article about the murder of Jennifer Granger. She had read it hundreds of times. She had memorized it.

Then Julie's eyes drifted downward as she slowly read all the factual material she had collected on this woman.

"Well, hey ... will you just look at this," she said in a whisper. "Just look at this. I'd almost forgotten this one ... Oh, this one would be perfect!"

She leaned forward, the scrapbook on her lap, turned back to her pages listing the murdered women, flicked to a fresh page, and wrote: "NUMBER 5." Then a name.

CHAPTER 80

"SOMETIMES I CAN'T believe how easy all this has been," Julie announced to the empty living room.

Then she looked up at the silent television screen and at once felt a shiver pass through her body. She immediately ramped up the sound.

"...the body has been identified as that of Bruce Frimmel," the newsreader said. The TV camera held the image of the dead man—an old police mug shot, Julie imagined.

The newsreader continued: "Frimmel is thought to have disappeared in

November. A police spokesman said he was most likely killed around that time."

Julie jumped up at the sound of tires screeching outside. She ran to the small window of her first-floor apartment and saw a white car pull up near her entrance in the small scruffy courtyard.

She snatched up the scrapbook, ran into the bedroom, and tossed it on the bed. Then she scrambled to the bottom of her wardrobe, where she had her backpack already prepared with the things she knew she might need when it was time to leave fast.

Back in the living room, she moved to the filthy kitchen area. She picked up a book of matches, darted back to the bedroom, struck a match, and held the flame over the end of the scrapbook.

At first the paper resisted. It felt to Julie that seconds were passing as minutes. She had to quell her rising panic. The match expired without the flame catching. She'd just singed the edge of the flimsy cardboard cover.

She struck another match and steadied her fingers by gripping her wrist with the other hand, blowing gently on the flame.

It caught. Julie couldn't wait a second longer. She dashed out of the room and into the tiny hallway, leaving the door ajar. She could hear footsteps on the stairs, a woman's voice. She puffed her way up to the second level, then around the bend and onto the next flight. Now she was on the top floor. Only the flat roof was above her.

She leaned over the railings and saw two people, a man and a woman, approach the door to her apartment. The pair moved along the wall and then disappeared. Julie turned, rushed as best she could up to the final flight of stairs, and pushed the exit door onto the roof.

The roof was supremely quiet—just the hum of traffic from the main road, the occasional squawk of a lorikeet. Julie crouched beside a utility pipe running along the edge of the roof, felt around for the brick she knew was there, found it, shifted it, plucked up the key.

She ran back to the shed, slotted the key into the lock, pulled open the door. The inside of the shed was so dark that it seemed almost black, but her eyes adjusted quickly. She scanned the shelves of jam jars filled with nails and screws, tins of paint, rolls of wire, bits of plastic tubing, and a bench scattered with tools. On the floor stood a five-gallon plastic drum with FLAMMABLE written in large, black letters around the middle. She locked the door from the inside and crouched down, keeping her breathing shallow and listening for approaching feet.

CHAPTER 81

MARY AND I had the brains to drive to Sandsville in her smartly sensible Toyota instead of my stupidly extravagant Maserati. In case you don't know what most of Australia knows—Sandsville is, hands down, the most dangerous section of Sydney. It's all gangs, all the time.

The seedy apartment blocks of Neptune Court looked as though they could collapse at any moment. There were three buildings clustered around a scrap of land, the grass worn to nothing. The music of the place was people fighting, babies screaming, occasional gunshots, rap tracks blasting.

Of the three buildings, we knew we were headed to the one that housed apartments one through twenty.

The door into Julie O'Connor's block was closed, but the steel-reinforced glass had been smashed in. I climbed through the hole, Mary half a second behind me.

Number six was on the first floor, but I smelled the smoke before I was half-way up the flight of stairs. Mary moved ahead, leaned on the wall next to the door, then swung inside. I was right behind her. She turned into the living space, swept the room, then moved to the only other part of the apartment, a tiny bedroom.

Yellow flames swirled up from black-ened sheets. The fire was small—a pile of papers, but smoke had filled the room. We grabbed a pillow each and smacked at the fire. Then I found a quilt on the floor, threw it over the small blaze.

"This was just started," Mary said.

"Must've missed her by seconds. You keep looking around. I'll be back in a minute."

I ran onto the landing. No one. I noticed half the doors were boarded up. There were two more floors above this one. I ran up the first flight, saw no one, reached the top floor. There was a ROOF EXIT sticker on a door. The door had been pushed outward.

I eased out onto the roof. It was deserted. I spotted a workman's shed in one corner, paced over to it slowly, carefully. I tried the handle. It was locked.

I did a three-sixty, saw the black metal railings of a ladder descending from one corner. Walking across the roof, I peered over the edge. The ladder dropped three floors to the ground. No one. Nada. Nothing.

CHAPTER 82

WHEN I RETURNED from the roof, I found Mary sitting at a kitchen table that had been shoved up against the wall between the stove and an ancient fridge. She'd pulled on latex gloves and was leafing through a clutch of charred papers.

She fished out a second pair of gloves from her cargo pants. Tossed them over. The place still smelled of fire and ash.

"Tried to open a window," Mary said. "They're sealed tight."

"So we can just die of smoke inhalation? I don't think so," I said, and I picked up a kitchen chair and threw it against the small kitchen window. A shattering of glass and a bit of fresh air.

"Checked every cupboard. No one in any hidey holes. Looks like Julie O'Connor left in a hurry. No sign of a handbag or purse. Found this on the floor over there," Mary added, picking up a five-dollar bill.

"Probably dropped it after snatching notes from this," I replied, indicating a jar lying on its side on top of the fridge. A few coins had been left inside.

"So, what do you think she was burning on that bed?"

"It's hard to tell." Mary nodded at the crispy papers. "Some of it's just burned to nothing."

She pointed to a pile of black, fire-ravaged paper. Then she carefully sifted through a few pages of what looked like some sort of photo album or scrapbook.

"Don't want to damage it more; Darlene'll kill me," she said. She opened the pages meticulously to the only section of the book left untouched by the fire. It was about the halfway point in the book.

I walked around and looked over her shoulder. There was a scorch mark running across the paper. "Personal info, descriptions," I said. "That set of numbers halfway down the left side." I hovered a finger over the damaged papers. "List of credit-card numbers—always sixteen digits, batches of four."

Mary nodded, turned a page carefully. "She's headed each double page with a woman's name."

I felt a tingle pass up my spine as I saw the name at the top of the first pair of pages. "Elspeth Lombard."

"All right," I said, "let's bag this stuff. Get it to Darlene." I glanced around to search for a plastic container. That was when we both heard, from the hallway, the sound of smashing glass.

Then an enormous explosion.

CHAPTER 83

"IT'S A GODDAMN Molotov cocktail!" I yelled.

Mary was up in a flash, her chair flying across the kitchen floor as she ran for the bedroom. I scanned the room desperately and spotted a plastic trash bag resting against a garbage can. I picked up the bag. Moving as fast as I could, I poured the contents of the bag onto the floor, then tucked the bag inside my jacket.

Before I'd finished, Mary was back in the living room, clutching a pair of blankets. The flames from the hall had spread, fiery tendrils reaching toward the ceiling. A ratty sofa close to the hall end had caught fire; the cheap foam melted quickly and added to the choking stench.

Mary ran to the sink, pulled on both taps, twisting them to the max.

"Got to wet the blankets!" she hollered, and then threw the blankets under the running water. Then she threw one of the sodden blankets to me. Following Mary's lead, I ducked my head under the stream of tap water. Then I wrapped the wet blanket around my shoulders, across my front, letting the bottom edge flap against my shins.

"Go!" I bellowed, and without wasting another second, I ran straight for the flames and the hallway.

The fire had engulfed about half the living room. I could sense Mary a foot behind me as we stumbled into the hallway.

The heat from the fire hit me like flames from hell. I knew I had to keep running. The floor was scorching the soles of my shoes.

Gripping the blanket, I reached for the latch and twisted. It was locked.

Panic rose up in my chest. It was getting harder and harder to breathe. I turned to Mary. I'd never seen her scared before. Then we both reached the same decision at the same moment and charged forward, slamming into the door together.

I heard the wood splinter and managed to stagger back. My chest was screaming at me. It was like walking barefoot on hot coals, but I knew that if I didn't keep going, we would both die.

We ran for the door again. A pain shot across my shoulders and up my neck. The door gave, but only opened a fraction. We charged a third time. The door fell outward, and I collided with Mary as we crashed onto the concrete landing.

We pulled ourselves up, but I tripped on the blanket, falling heavily against a door the other side of the landing. I did my best to ignore the pain. I threw the blanket aside and felt something hit me hard across the face and chest. I looked up and saw Mary leaning over me, beating out a line of fire across the front of my shirt.

CHAPTER 84

MARY AND I staggered out onto the dirty ground between the buildings. I was leaning forward, hands on my knees, gasping for air. A couple of teenagers ran to us. Mary was coughing from deep in her gut. Then she turned away and vomited.

Now the first teenage guy approached, and I suddenly felt a terrible pain in my jaw. I stumbled back and caught a glimpse of the other kid as he jumped on Mary's back.

Before I could take in that the bastard had hit me, he swung his fist again. I dodged it, lashed out, and caught him on the side of his face.

I heard a crash from behind and saw the kitchen window of Julie O'Connor's apartment shatter outward, a great sheet of flame spewing out. The teenagers were distracted, and Mary had apparently recovered. She whirled round, a string of vomit running down her vest top. She threw the teenager from her back. The little bastard crashed to the ground face-first. I landed a second punch to the side of the other kid's face. As this kid fell, Mary connected her right boot with his balls. He doubled up, moaning.

I noticed that the bandage around Mary's hand was bloodied. "You okay?" I gasped almost inaudibly.

"I've had better days, Craig. You?"

I started coughing and couldn't stop for at least ten seconds. Fire alarms in the apartments began to wail.

Mary had taken her cell from her pocket. She nodded toward the car as she called 000. We ran toward our vehicle, leaving the two thugs groaning in the dirt.

I watched a group of people running out of the building.

Mary was giving instructions into the receiver as we crossed the patch of ground. I was limping like an injured footballer leaving the field, and I was totally in awe of how incredibly fit and powerful Mary was.

It was only as we got ten yards from the Toyota that we realized all four tires had been slashed.

CHAPTER 85

JULIE O'CONNOR CAUGHT the train from Sandsville and headed for downtown. She had close to two hundred bucks, a fortune to her. She also had a stolen basic green American Express credit card. But the most important thing she had was a plan.

She was still feeling the buzz, the thrill she had experienced—making the gas bomb from the materials in the shed, tossing it into the apartment, descending from the roof of her block using the metal ladder, and slipping away in the commotion.

Ten minutes into the train journey, she left her seat, walked calmly past a disgusting young family—cute wife, cute husband, cute twin girls—in the next aisle and into a corridor. She pulled open the door to the wash-room.

Yanking her backpack from her right shoulder, she let it drop to the floor. Leaning down, steadying herself as the train swayed, she found the plastic bag she'd put in the backpack earlier. She placed it on the small sink.

Pulling out a dark wig, a fake mus-tache, and a baseball cap, she arranged them on the side. Tugging on the wig, she tucked a few loose strands of bleached-blond hair beneath the edge, found the tube of glue she'd pur-chased, ran a line of it along the back of the mustache, and put the mustache in place. Then she tugged on the cap. Looking at her reflection in the mirror, she had to smile. Now, that wasn't hard at all.

She was wearing jeans, boots, and her lumberjack shirt. She felt around the inside of her handbag and pulled out the roll of banknotes—real ones: three fifties, plus a twenty, a ten, and a few coins. There was a second roll wrapped in an elastic band—ten fifties. This second batch, however, was photocopied money. She planned to use that later.

She then removed the AmEx card, a packet of mints, and her favorite baby picture, one she had salvaged from her scrapbook. It showed a real cute kid, about nine months old—a cuddly baby girl wearing diapers and pink shorts. She had an adorable fat tummy and was crawling toward the camera, a big smile on her face.

Julie stuffed all these items into the pockets of her jeans, opened the window of the washroom, and tossed out her handbag. Then she checked herself in the mirror again and brushed a stray bit of wig under the cap. Taking a deep breath, she leaned into the mirror, real close, her face filling her view. She bared her teeth. "You can do this, Julie O'Connor," she hissed. "You can do this... *baby!*"

CHAPTER 86

"JUST GOT THIS in from one of our cars out in the western suburbs, sir," said Sergeant Raj Petigara. He then handed a sheet of paper to Inspector Mark Talbot. "Thought you might find it interesting."

Talbot scanned the report and grinned, touched the Steri-Strip across his nose. Craig Gisto had almost been barbecued, then beaten up by a couple of teenagers in Sandsville.

"He's not, by any chance, in the Serious Burns Unit of the Royal North Shore Hospital?"

"Not this time, sir," the sergeant replied.

"Shame," Talbot remarked under his breath. He glanced at his watch. "Hell. I'm late." Then he turned and walked quickly down the hallway.

He crept into the conference room just as Brett Thorogood was about to start talking. Talbot found a seat and manned out Thorogood's glare.

There was a buzz of excitement in the room. Even Talbot could sense it. He could feel his own adrenaline pumping. This was why he'd joined the force: a manhunt—well, a womanhunt in this case.

"This is the suspect," Thorogood announced, pointing to a large photo of Julie on the whiteboard. "A snap taken when she started work at SupaMart."

Hideous bitch, Talbot thought.

"Don't have much on her," the deputy commissioner went on. "Name: Julie Ann O'Connor. Age: twenty-seven. Current address: Six Neptune Court, Impala

Road, Sandsville. No record. So far, so ordinary. Her father was a cop, Jim O'Connor—killed in the line of duty in 1996. She disappeared in 2000, off the radar until 2004. Cropped up in state records as a cleaner for a small engineering firm, Maxim Products, in Campbelltown. Our friends at Private have come up with some useful stuff."

Talbot felt a knot in his gut. He hated even hearing the word "Private."

The DC clicked a remote, and footage from the copy shop came on screen.

"Appears the woman is underprivileged, lives in a slum, works in one of Sydney's most affluent areas. Could be some sort of motive for her killings. She copies the banknotes using a couple of different copy shops near Bellevue Hill."

" 'Underprivileged'?" Talbot said, and looked around at the five other officers in the room.

"Yes," Thorogood replied. "Your point?"

He had none, just hated the term.

This bitch was dumb as dirt. Didn't she know you could photocopy at home on a printer? But she probably didn't have a computer—a big satellite-TV dish, for sure, but no computer. She'd probably never touched one, had no clue. Douche bag.

Thorogood was talking again. "Private has confirmed a positive DNA match to place this woman at two of the murder scenes for sure. She was also in a long-term relationship with the male victim, Bruce Frimmel, whose DNA was found at the same murder scene as Jennifer Granger's. Both victims' bodies were found in the yard of the house on Ernest Street in Bondi."

Talbot wanted to retch. That word again. "*Private.*" Thank God it had been one of his boys who'd put the pieces of the puzzle together, matching the photos of the O'Connor bitch.

"Any idea where this Julie O'Connor is now, sir?" It was Chief Inspector Mulligan, Talbot's immediate superior. He was leaning back in his chair, arms folded across his chest.

"Good question. We don't know. But we will. I've just heard that about an hour ago, Craig Gisto and Mary Clarke from Private almost got O'Connor at her apartment in Sandsville. Ended up, though, that they were lucky to escape with their lives. The bloody woman fire-bombed the place with them inside."

Mark exhaled loudly. Thorogood gave him an odd look.

"I'm pulling out all the stops," the deputy commissioner went on. "Closing the airports, putting up roadblocks ringing the city, every spare man out on the streets. We'll get her, and when we do, she'll live out her days in a ten-foot-square cell."

Not if I get her first! Talbot thought.

CHAPTER 87

JULIE GLANCED AT her plastic Little Kitty digital watch: 5:03 p.m. It sure as hell had been a busy day.

She had left the apartment hours earlier, thrown in the Molotov, slashed the car tires. Now she was walking around town dressed as a man, feeling increasingly confident. No one seemed to notice. She just blended right in... blended right into the pool of humanity around her. She, of course, knew she was not *like them,* not *like them* at all. She was a different breed from the people she brushed shoulders with, different from all these assholes she stared at, all these people who merely glanced at her and moved on with their normal, happy lives.

They all had homes to go to. In those homes were people who loved them, people they loved. They had lives, careers. Julie had nothing, and it made her feel…it made her feel…For the first time in her life, she actually felt liberated. Free. Totally free. She was her own powerhouse. Why, look at that, there was some dumpy, middle-aged rich broad who smiled at her, a flirtatious sort of child. Yes, she was free. She could do anything. *She* could even be a *he*.

CHAPTER 88

"OKAY. I KNOW I'm interrupting."

Johnny looked up from his desk and saw Katia, Mickey Spencer's girlfriend, standing at the entrance to his cubicle. She looked even more stunning than she had the night before. She was dressed entirely in white—a long, flowing skirt that reached almost to her ankles, a snug-fitting white T-shirt, and the miniature sword on the pink silk ribbon, still around her neck.

"A visit from you could never be called 'interrupting,'" he said. *Mr. Suave is in the building.*

She gave him a faint smile.

"How's Mickey?"

"Oh, he's absolutely fine."

"He's fine?"

"Up by two this afternoon, and off to a rehearsal at three."

"But…?"

Katia gave him a broader smile. "You're pretty naive, aren't you, Johnny Ishmah? That's so sweet."

God, he hoped he wasn't blushing.

"I don't know much about the rock world, but I'm not totally naive."

"Mickey has an incredible constitution, *and* he keeps Dr. James close by. Ricky insists upon it. Mickey can't stand the doctor. Thinks he's grossly overpaid."

Johnny perched himself on the edge of the desk and decided to go for what he wanted.

"Last night, you were starting to tell me what you thought about this Club Twenty-Seven thing."

She paused, bit her lip. Then she spoke.

"You know, Johnny, I really don't know what to think anymore. I've been with Mickey for six months. He was a user when I met him. He drinks heavily. But...you know...he's a rock star... That's what rock stars do, isn't it? The difference now is—well, he's become a lot worse in the last two months."

"And you think that's because he's approaching his twenty-seventh birthday? Or do you think Holt is pushing him into killing himself?"

Katia folded her arms and looked as though she were about to burst into tears.

"Look, Katia," he said, "last night I saw something."

She fixed her huge, dark eyes on him.

"The smack. I saw Holt give it to Mickey just before he went into the bathroom."

Katia exhaled through her nose. "Of course he did," she said, her expression cynical. "Mickey was a prize racehorse. He used to be Ricky's most valuable asset. Now, though, even with his career on the slide, Ricky's still Mickey's supplier for everything…especially first-class drugs."

CHAPTER 89

JOHNNY HAD BEEN tailing Ricky Holt for over two hours. After all that time, he ended up at Kings Cross—the cheesy stretch of strip clubs, ugly discos, and dumb-luck gambling haunts called the Strip.

Holt had ducked into a joint called the Roxy. Johnny stopped outside, nodded to the bouncer at the door, and pulled out his wallet as he walked in. He approached a woman in a significantly short black dress. The dress also had a significantly plunging neckline. She was sitting at the business end of the bar. Her legs were crossed, and a large sign above her nearby cash register stated simply: ENTRANCE FEE, $50. Johnny paid up.

Some nameless dance track, all bass drum and bubbling synthesizer, thumped away. Spotlights on a small, circular stage moved in a random pattern and sent splashes of color across a couple of girls wearing G-strings and nothing else. Several punters stood near the edge of the stage, looking up at the girls and the glare. A smaller bar near the stage was surrounded by UV lights.

Johnny scanned the room, but he had trouble making anything out. He moved slowly around the edge of the space, trying not to make himself obvious to the half dozen men sitting at tables. He did not see Holt. Then he caught some clear movement in the corner of his vision, a man slipping under an arch. A notice to the side said PRIVATE ROOMS.

Johnny made his way over, slowed as he reached the arch, and took a couple of paces into a narrow corridor lined with closed doors. At the end of that corridor stood an emergency exit. The door was open and led into an alley. The music was quieter here, just the thud of the bass drum. Johnny paced along the corridor and heard voices coming from beyond the exit. He recognized Holt's voice. Then he pulled in close to the wall and held his breath, straining to hear what was being said.

Then came a thumping sound, a groan—and suddenly Ricky Holt was flying toward the emergency exit. Ricky grabbed on to the door frame, trying to break his fall. Johnny couldn't help himself. Stupidly, reflexively, he jumped aside and into full view of the men in the alley.

CHAPTER 90

JOHNNY RAN THROUGH the doorway and cut right. He was suddenly in the alley, and he hoped he'd gain a few seconds' lead before the men realized what was happening.

The dark passageway was cratered with potholes and strewn with garbage. Johnny tripped on—what the hell was that? A battered television set. He almost went down but steadied himself and kept going. He glanced over his shoulder and saw two thugs running toward him through the shadows. Beyond them, the rear lights of a car.

Another dark alley was on his left. He turned and ran through the black back street. He could hear that the two men had reached the same left turn and were coming after him, gaining on him.

His eyes burned with perspiration. *Am I too young to have a fucking heart attack?* he thought. Somehow he found some backup energy. Where the hell had that come from? He ripped along the narrow lane and came out onto a brightly lit road. People were walking— couples, children, friends out with other friends. He saw restaurants and bars and shops. He could merge, maybe. He could disappear into all these faces. But these guys would get him somehow. It was their turf.

Directly across the street, a dark entrance to another alley. He darted across the road. Horns blasted. Johnny swerved, gained speed, and flew into the passageway. But the men pursuing him were just as fast. They crashed into the lane a few yards behind him. Johnny put on a final burst of speed, reached the end, a T-junction. He swung left and tripped. He hit the ground with a skin-scraping thud, and he heard a very un-pleasant snapping sound. His back. He knew something had snapped in his back.

CHAPTER 91

THEY YANKED HIM up and rammed him against a wall. One of them pushed his hand tightly at Johnny's throat.

"A little nosy, aren't we, kid?"

Johnny stared into the man's face. His head was shaved, and his eyes were like two dirty pools of water. His eyebrows could have benefited from a trim with a power lawn mower.

"I was just leaving the place."

"Uh-huh."

"Look, I'm not interested in what you were saying."

The other guy laughed and took a step

forward. The first man loosened his grip on Johnny's neck, grabbed his left arm, pulled it up hard behind his back. Johnny cried out in pain.

"A little word with the boss is in order, I think," the guy with the big eyebrows said. He pushed Johnny forward, and in a few moments, they were back on the busy main thoroughfare.

Two minutes later the two men had marched Johnny to the rear entrance to the Roxy. A big, black BMW stood in the lane.

Johnny struggled to get away, even though he knew he had little chance of that. The two men each had an arm on him. They came around to the side of the vehicle, and the one on Johnny's right opened the door with his spare hand, pushed down on his head, and shoved him into the car. Then he slipped in beside him. The other guy ran around and jumped into the driver's seat.

"You a little out of breath?" the boss

asked, turning to the henchman in the back. "Gave them a run for their money, eh, Johnny? *Johnny Ishmah?*"

Johnny stared at the boss. He had a flabby face, small, black eyes, and a big, nauseating grin.

"Jerry Loretto!" Johnny said, amazed. "It's been a long time . . ."

CHAPTER 92

"ALL RIGHT, YOU two—piss off," Jerry Loretto said, and so they quietly stepped, pissed off, into the alley and slammed the doors shut.

"Well, well!" the youthful-looking boss exclaimed. "Never thought I'd see you again, Johnny. What the hell you doin' here?"

Johnny had regained some composure, took a deep breath. "Could ask the same of you, Jerry. You watching one of your dad's places?"

Jerry snorted. "My own, you cheeky little bastard. I'm a big boy now!"

Johnny knew Jerry Loretto was only twenty-four, although his expensive outfit—striped suit and navy-blue polo shirt buttoned to the neck—looked like it would be perfect for a fifty-year-old undertaker. Yes, that was what Johnny thought, but he kept his observation to himself.

Johnny had known Jerry in high school. Not that Loretto had shown up at school very often. Even then he'd been a petty criminal—newsstand holdups and purse snatchings.

Johnny had studiously avoided the guy. Jerry Loretto was one of the school thugs, a thoroughly nasty piece of work, even at the age of eleven. But one day Loretto crossed the path of another tough kid from a neighboring school who had intruded into Jerry's "patch," selling weed and Ecstasy. Loretto had been jumped, knifed, and dumped by the roadside. Johnny had found him, and Jerry had begged him not to call an ambulance because he didn't want anyone to know what he'd been up to.

Johnny had helped Jerry get home, and after that, Loretto was his guardian angel.

"I'm here on an investigation," Johnny said, a little embarrassed. "I'm a PI."

"Holy shit," Loretto said. Then he burst out laughing. "Well, I guess that figures. Johnny, you always were a little pussy." Then Jerry added, "You're not investigating *me*, are you?" Johnny thought the question was half-joke, half-serious.

It was Johnny's turn to laugh, a nervous edge to it. "Nah, I'm keeping an eye on your buddy, Ricky Holt."

"That shyster?"

"He manages one of our clients."

"Mickey friggin' Spencer?"

Johnny nodded.

"So, what do you want to know about Holt, then, Johnny boy?"

"Well, he's obviously up to his neck in it."

"Up to his eyeballs, more like—up to here." And then Jerry indicated a level six inches above his head.

"Gambling?"

Loretto nodded. "Stupid bastard must be the worst punter in history, but he don't give up."

"How much does he owe you?"

Jerry frowned, then tapped his nose. "Client confidentiality," he said, and laughed loudly. "Let's just say *a lot*."

"And you've given him an ultimatum?"

"One he probably can't meet."

Johnny nodded. "He really is up to here..." He imitated Loretto's earlier gesture.

"Oh, yes, Johnny boy. I sure as shit wouldn't want to be Ricky Holt in three days' time."

CHAPTER 93

AT TEN P.M., Darlene was alone in Private HQ. She kept unsociable hours, always had. And she never saw anything odd about rising at noon and staying awake until seven a.m.

Darlene approached the large metal table that dominated the center of the lab. The powerful light above the table bleached the gray steel bench to a near-white color. Under that light, on the steel counter, lay remnants of Julie O'Connor's papers, the papers that had been salvaged from the apartment in Sandsville.

Darlene had already spent several hours sorting and sifting through the material. She'd compiled three separate groupings. Group one: useless ashes. Group two: almost-useless scraps of paper worth examining. And, finally, group three: a small heap of material that might prove helpful.

This last pile included about a dozen pages of a scrapbook. Wearing latex gloves, Darlene carefully turned these pages. It was a peculiar mess. Many of the surviving pages contained pictures of Julie holding babies. Then there were pictures of babies cut from magazines, ads for baby strollers, baby clothes, disposable diapers.

A few pages in, she saw a crude drawing of a nursery. Then, on the following pages, names. A long list, two columns to a page. At the top of the left column, Julie had written "GIRLS." Topping the right column was the word "BOYS." Under these headings were dozens of names, alphabetized, some crossed out and written over, many misspelled.

A set of double pages from the scrapbook had separated from the spine. She saw familiar names. One said, "WHORE NUMBER 3: ELSPETH LOMBARD"; the other, "WHORE NUMBER 4: YASMIN TRENT." Beneath this were details of the murders from Julie O'Connor's perspective. Crude in their descriptions, shocking in their candor.

Darlene carefully continued to leaf through the notebook. Then she suddenly stopped.

On the brightly lit counter lay another double page that had slipped away from the others. She could see three words: "WHORE NUMBER 5." Next to that heading was a deep brown scorch mark.

CHAPTER 94

A SHORT, STOCKY man with a mustache walked from the train station, south along Seymour Avenue, then right onto Sebastian Road. He'd walked this route many mornings before. It was 10:15 p.m. as he entered the parking lot of SupaMart. No one had seen anything odd or distinctive about the guy. It was, of course, Julie. And the only one who knew that... was Julie.

Julie strode straight past the entrance, the rectangle of glass fronting the store, then down a broad alleyway toward the other parking lot, the rear parking lot.

Hanging a left, she found the darkened doorway at the back of SupaMart. The door was bolted and padlocked.

Julie slotted a key into the padlock, turned it, found a second key for the lower Yale, twisted that, pushed—and the door swung inward.

She now stood in a corridor, flipped a light switch. A fluorescent strip buzzed and sputtered into life. Concrete floor, concrete walls, concrete ceiling. She pulled open another door, took three paces along the passage, stopped at yet another door. It was marked STOREROOM 1.

It was unlocked, the light on. It was filled with boxes of canned goods—soups and gravies and vegetables for the shelves in the store. At six a.m. tomorrow, a three-person team would arrive to take the goods out onto the shop floor. Later tomorrow, a threesome from a truck would arrive to unload their cargo of more boxes to be put into this storeroom, to be put on the store shelves. It was a long and boring cycle.

Julie knew that there was a concealed cupboard at the back of the third shelf up from the floor. She had spotted it weeks ago when she was sent to the storeroom to bring out a carton of Heinz beans. She yanked on the handle. Inside were a few items she'd put there two days ago—clothes, a sleeping bag, a thermos, and some basic toiletries.

Julie gathered the things up, unfurled the sleeping bag on the floor, and lay on it. She was used to sleeping rough. After walking out on her evil mother, she'd lived on the streets for four years. She'd been raped twice, had her skull fractured as she slept in a park, and almost died on the operating table. No, unlike the stupid, soft bitches she delighted in killing, she knew Julia Ann O'Connor was as tough as they came.

She leaned back against the wall and pulled out her crisp, new, virginal notebook. At the top of a double page, close to the back, a name. Beneath this an address, followed by a list of people—the woman's family and friends. Then a collection of phone numbers. Last, some notes, a set of things she thought might one day be useful information about the woman she'd targeted: "Favorite restaurants," "Gym address," "Phone number," "Habits."

Under "Habits," she'd written: "This whore likes to run. She runs and she runs…silly bitch. She runs around Parsley Bay, a couple of miles from her house. Always same time—early riser, this babe…six a.m. Easy!"

CHAPTER 95

THE YOUNG MAN knocked on the lab door.

"Come on in, Johnny," Darlene said.

"How'd you know it was me?" Johnny asked.

She looked up from her microscope.

"I just knew. I just know everybody's knock, everybody's footstep," she said. "I guess I was born to have this kind of job."

"Seems kinda creepy," Johnny said with a smile.

"Whatever," Darlene said without a smile.

She pulled back from the scope and said, "Take a look."

He peered into the eyepiece. "Means nothing to me."

"And not much more to me," Darlene said. "It's part of Julie O'Connor's scrapbook, but it's so badly charred, I can't make out the words. I'm getting really pissed with it, to be honest."

"Not surprised." Johnny paced over to Darlene's desk. He saw the small pile of invites Software Sam had left yesterday.

"I heard about these," he said, picking up the tickets. "Mickey Spencer's birthday party...right? Craig mentioned them."

Darlene nodded. "Yeah, that guy...Friend of Mickey's dropped them in. With all the stuff going on here, I'd forgotten."

Johnny stared down at the invitations. "It's tonight." He stared into Darlene's eyes.

"It is?" she said.

Then Johnny was surprised to hear

himself say: "Darlene? What is wrong with you? How could you forget? We should be there to watch out for him. What if Ricky Holt—"

He stopped shouting. He was panic struck.

"Shit," Darlene said.

"Come on. Let's get moving," Johnny said.

"Johnny Ishmah," Darlene said, "you're not asking me out on a date, are you?"

He flushed red. Maybe it was embarrassment. Maybe it was fright. Probably it was both.

"Oh my God! You're blushing!" Darlene said, hand to mouth. "How..."

"Don't say 'cute.' "

"All right... How *not* cute!"

CHAPTER 96

NO DOUBT ABOUT it. Darlene was a bundle of contradictions, and all those contradictions intrigued Johnny. To put it mildly. Darlene's job was messing with blood and body parts, but underneath her lab coat, she always wore Prada or Chanel. A trust fund from her rich dad bought her a three-bedroom apartment with a harbor view, but she drove a seventies VW Beetle that she'd lovingly restored herself. Johnny was definitely intrigued, but he also knew that Darlene was way out of his league. This "date" for the Spencer party was a totally unexpected bonus for the lad.

Darlene, with Johnny in the passenger seat, chugged the old VW through the exit gate of the garage. The security guard smiled and gave her a shy wave.

"Sweet bloke," she said to Johnny. "After his concussion he came to work in just a few days. He insisted."

It was 11:32 p.m., and the sidewalks of the CBD were crowded and busy. They passed a club on George Street called Ivy, a line out the door stretching two blocks.

Johnny leaned in toward the radio— an original sixties collectible. He pointed to the machine. She nodded, and Johnny nudged down the On switch. Classical chamber music came out from the speaker.

"You ever been to anything like this before?" he asked, picking up the invitations.

Darlene shrugged. "Once or twice, but it was a while ago."

Craig had told Johnny that Darlene had been a model for almost a year after graduating from college. Darlene had never brought up the subject. So Johnny assumed it hadn't been a good experience.

"How do you change channels on this thing?"

"You don't like Brahms? The dial."

Johnny slowly turned the knob. He passed through a jazz station, the ABC late program. Then some pop music came on. He went past it, backtracked, tuned it.

"Unreal!" He turned to Darlene.

"What?'

"Only Mickey Spencer's new single. Heard it on Spotify this morning."

"The song is happening all over," she said as she turned off George Street. The two of them fell silent for a few moments. They just listened to Mickey's new song.

She hung a right onto Castlereagh Street and then looked at Johnny. "Pretty catchy tune. Why so quiet? What's up?"

He was pale, staring at the radio. Held up a hand. "Sssh! Listen!"

The music swelled; Mickey repeated the chorus: *"I just wanna die at midnight in your arms. Like Jimi and Janis and Kurt... Club Twenty-Seven charms."*

"They'll be playing it at his funeral service if we don't hurry up," Darlene said.

CHAPTER 97

THERE WAS A long line to enter Mickey's celebration, but Darlene and Johnny simply showed their invites to the bouncer at the head of the line. He glanced at the papers, stared briefly at Johnny and significantly longer at Darlene. Then the big guy nodded to the double doors.

"How'd you get us in so fast?" Johnny asked.

She handed him the invitations.

"Look closely," she said.

Sure enough, he saw a tiny engraving at the top right: "Very VIP."

The club was huge. Music throbbed from powerful speakers. Lights swept and flashed. One vast wall was covered with an early Pink Floyd–esque display of psychedelic colors.

"We've got to find Mickey fast," Darlene said.

"Good luck with that."

It was packed. Darlene and Johnny pushed their way across the main floor.

They reached the bar. Johnny leaned in and tried to get the attention of the bartender.

Darlene wasn't paying much attention; she was busy looking for Mickey. She very quickly caught the eye of the bartender. She was good at doing that.

"Hello, darlin'," he oozed. "What can I interest you in?"

She switched on the charm. "Listen, I need to find Mickey. I'm a friend."

"Of course you are, sweetheart!"

She flashed her invite, and the guy changed his tone and his tune.

"Okay. Cool. So, how can I help?" the bartender asked.

"Where's Mickey right now?"

The man shrugged. "How should I know?"

"I *really* would like to know," she said. "And I *really* think Mickey would like me to know too. Okay?"

He straightened. "Upstairs in his suite. Two twelve. Second floor, far end...I'd use the stairs; the heads are taking the elevator straight to the washrooms on the second floor."

He smiled and tapped his nose.

CHAPTER 98

MICKEY'S PARTY WAS part heaven, part hell. Part porn flick, part Disney cartoon. Part fun house, part madhouse. You get it. The party was precisely what you'd expect of an international rock star on his twenty-seventh birthday.

Barely clad, kohl-eyed women tottered around holding flutes of champagne. A dealer sat at a corner table, dispensing his drugs as if they were Big Macs. A voluptuous female dwarf wearing nothing but a yellow bikini bottom carried a tray on her head. The tray was heaped with cocaine.

Meanwhile, Mickey and Katia held court in the bedroom. He strummed an acoustic guitar and sang one of his lesser-known songs. A spliff dangled constantly from the corner of his mouth.

Hemi overflowed an armchair located close to where the bedroom spilled out into a vast lounge. He'd positioned himself there deliberately so he could follow the action. He was drinking his usual sparkling mineral water. Hemi spoke to nobody, and nobody spoke to Hemi.

Mickey was on the last repeat chorus of his song when he saw Hemi roll forward and collapse onto the carpet in a wobbly, groaning heap. He stopped strumming immediately and turned blankly to Katia. She hadn't seen the big man crumple, but she, like everyone else, had heard the sound of him reaching the floor. She was the first up and across the room. Mickey came around the end of the bed, still holding his guitar.

Katia crouched beside Hemi, and together Mickey and Katia managed to roll the big guy over. He was out cold and began to snore. She raised her head to Mickey, and she burst out laughing. The rock star looked concerned for a moment, but then he started laughing too.

"Too much sparkling water, Hemi," he mumbled.

Katia stood up and came around to hug Mickey. "Let's get outta here."

He looked down at her, eyes swimming. "But it's my party."

"I want to take you somewhere quiet and lick you all over."

Mickey giggled stupidly. "Well, that's an offer I ain't gonna refuse...am I?"

"Ricky's room is empty. He's banned everyone—"

"Ricky's?" Mickey said, and he suddenly looked frightened.

"Don't worry—he's downstairs schmoozing. I got the key earlier from Reception. I wanted us to see in your birthday together...just you and me. I want to protect you. No one can touch you till after midnight."

CHAPTER 99

THEY STUMBLED DOWN the corridor. They fell. They giggled. They stood up again and stumbled some more. Once they reached the door, Katia slipped the key card into the lock slot, opened the door slowly, and then pulled Mickey inside. Ricky Holt was getting up from the end of the bed, a bottle of Wild Turkey 15 in his hand. He had a split lip and a line of Steri-Strips across his cheek.

"Ah!" Katia said.

"Don't mind me," Holt mumbled.

Mickey began to jabber. His words made no sense. He pointed at Holt.

"What's he saying?"

Katia shushed Mickey and guided him to the bed.

Holt looked at the bottle and frowned, turned it upside down. "Damn!"

"I'll get you something." Katia left Mickey sitting up on the bed, his head back on a mountain of pillows. He was gazing at Holt warily. A few moments later, Katia was walking back from the drinks cabinet with a bottle—this time Jack Daniel's—in one hand and a tumbler almost full to the brim with amber liquid. She handed the tumbler to Holt. He made a grab for the bottle. "No, no," she tutted, and crossed over to Mickey.

Holt pulled himself into an armchair and took a gulp of Jack.

"I've got a story," Katia announced. Holt looked at her blearily.

"Oh, I like stories," Mickey said, swigging from the bottle.

"There was a pope. I can't remember which one. It was a long time ago, maybe in the tenth century, sometime

420 • James Patterson

like that. Anyway, he wasn't a very popular pope, and so he decided to go on a tour of all the papal dominions, to try to buy the favor of his flock with indulgences. He reached Verona on Midsummer Day and was dispensing his promises and his money to the people of the city, when a woman who was known to be a witch stood up and yelled to the crowd that the pope would die on October second of that year. That would be just over three months later."

Mickey was looking at her, rapt as a child listening to Goodnight Moon. Holt had his eyes closed. His chin was on his chest.

"They arrested the woman, of course, burned her at the stake in front of the pope. But even though the witch was dead, the pope was terrified by her curse. He returned to Rome immediately and tried to put the memory of what had happened in Verona out of his mind. But it was no good. As October second approached, the pope became more and more agitated. On October first, he gave strict instructions to his staff and to the cardinals and locked himself in his private chambers. He would see no one, and he would not eat or drink anything until just after midnight on the morning of the third.

"The pope's servants followed his every wish, and as the clock struck midnight and October second passed into October third, the room was unlocked. The elated pope sprang from his bed, walked toward the servant, tripped, smashed his head against the leg of a table, and died instantly."

Mickey looked horrified and was just about to say something when Holt tumbled to the floor.

"Shit!" the rock star exclaimed. "We've lost another one!" He turned to see that Katia had taken off her pink silk-ribbon necklace and had the sharp tip of a very sharp, very small sword at his jugular.

CHAPTER 100

THE STAIRS STOOD at the far side of the dance floor, and that dance floor was packed with heaving, sweating bodies. Music seemed to make the walls and ceiling actually shake.

"Must be a back way," Johnny yelled into Darlene's ear.

She glanced at her watch. It was 11:55 p.m. "No time to look for it." She made for the edge of the crowd, forcing her way between the revelers and the wall of mahogany paneling. It was almost impossible to move.

Johnny took out his Private ID and

squeezed past her. Under the pulsating light show, he looked like a plain-clothes cop holding up his badge. The sea of humanity more or less parted before him.

He reached the stairs, and Darlene almost fell over him. "Neat trick," she said.

The first floor was dimly lit, the noise from below still incredibly loud. A red carpet led them down a corridor of guest rooms. They ran for the second flight of stairs.

It was quieter—no one around. Then they heard a sound—laughter, a girl squealing. Darlene glanced at her watch: 11:58.

On the far wall, a sign: SUITES 208–215, an arrow pointing left. Darlene turned on her heel, headed off, Johnny close behind.

The door to 212A stood ajar. They slowed, turned in, and almost fell over a couple of girls rolling around on the floor and kissing passionately. Three people were passed out on the bed. No Johnny. No Katia.

Darlene ran through the bedroom, into the lounge, and then on into the second bedroom. The bed there was a tangle of limbs, groans and moans audible above the music coming from a beat box in the corner.

After checking that Mickey wasn't one of the bodies on the bed, she did a one-eighty and charged back into the main bedroom.

"Anything?" she asked Johnny.

They tore along the plush carpet, careered around a corner, pulling up just short of an elderly Asian maid pushing a cart filled with toiletries. She was wearing earplugs.

"Whoa!" she shrieked. The woman pulled out the earplugs, grimacing at the noise.

"Sorry," Darlene said. "Can you help? We're looking for Mickey Spencer."

"Who?"

"The pop star?"

"Never 'eard of him," the maid said irritably. "Just can't stand this awful noise." She paused. "Oh, I know who you mean! It's his party . . . right?"

Darlene nodded.

"He's got suite two twelve B. Terrible mess he always makes."

"He's not there."

"I saw him a few minutes ago. He went off with a girl."

Johnny stepped forward. "Tall, skinny, black hair?"

The maid nodded. "They went that way. The older man—his boss."

"Mickey's manager?"

"Whatever you call him . . . His room is down there . . . two fifteen. That's where they were headed."

CHAPTER 101

DARLENE AND JOHNNY ran down the corridor. The door to 215 was locked. Johnny whipped a penknife from his pocket. Darlene stood aside as he flipped a blade out and slid it into the lock. He twisted it right, left, back, then left again. They heard a click, and the door opened.

Katia was on the bed, crouching over Mickey. She appeared to be almost supernaturally huge. Mickey was frozen in fright, rigid. He was eyeing the tiny but deadly sword at his throat.

It took Darlene and Johnny a second to

absorb it all. They saw Mickey's manager, Ricky Holt, unconscious on the floor.

"Okay, everyone, whatever's happening needs to stop happening," Johnny said.

"Yeah...what *is* happening?" Mickey said drunkenly. His brain obviously knew that it should sober up, but his voice hadn't gotten the message yet.

Katia blew the singer a kiss. "Dear Mickey," she said softly. "You see...just like all pop stars, you could never keep your cock in your pants, could you?"

"What?"

"You probably don't even remember her, do you?"

"Remember who, Katia?"

"In 2010 Fun Park played a gig in Moscow. You must remember that!"

"Yeah."

"After the show you met a young girl."

"I've met a lot of—"

"Don't!" Katia screamed, and she moved the blade in her hand forward a fraction of an inch.

Mickey's fists clenched. "Agh!"

"Don't be a fucking baby all the time." She was angry. She kept shouting.

Mickey took a couple of deep breaths. He was wet with sweat.

"You bitch!"

She blew him another kiss and smiled sweetly. "I mention the girl because—"

"What fucking girl?" Mickey turned his eyes to Johnny and Darlene and gave them an imploring look.

"That girl was my sister, Anais. She got pregnant. You made her pregnant."

"What! I didn't—"

"You didn't what, Mickey? Didn't screw her? I know you did."

"I had no idea—"

"She e-mailed you. She tried to get to you and 'your people.' Never a single reply. You discarded her, simply brushed her off."

Katia was staring down at the singer, her face contorted, eyes ablaze. Johnny and Darlene knew that the best they could do was to do nothing.

"Katia," Mickey pleaded. "Please...I didn't know. No one told me. Maybe I can help now...."

"She's dead, Mickey. Died having a backstreet abortion."

There was a stillness in the room. No one spoke.

"I'm so sorry—" Mickey began.

"Sorry?"

"I didn't know..."

"Anais suffered so much." As tears filled her eyes, she relaxed her grip.

Mickey moved in the bed and placed his hands gently on Katia's cheeks, wiped away the tears. He looked into her eyes.

"I really am...," he said.

For a moment, Katia began to respond, closed her eyes, went to kiss the pop star. But then her eyes snapped open. She seemed to explode. She shoved him back and brought the tiny dagger to his throat again.

"I could have killed you anytime, Mickey, but I wanted you to *suffer*. I filled your tiny brain with the idea of you joining Club Twenty-Seven a few months ago. You've been too drugged up to remember that. And I could also blame it all on Holt. I made you think that too."

Johnny and Darlene were scared. Katia had suddenly became frighteningly calm. A soft smile turned her mouth upward. Her teary eyes were suddenly gentle. She leaned back slightly. The calm was more frightening to Johnny and Darlene than the shouting and cursing. They could see her tighten her grip on the miniature sword.

There was a movement from behind the woman. Darlene and Johnny managed to stay still, to show no reaction.

Katia shot them a glance. "I have to kill him, you see," she said, now way too calm. "An eye for an eye...and I loved Anais."

She went to push her hand forward, and Ricky Holt's fist swung around. Whether Katia reacted fast or Ricky's aim was sloppy, Katia managed to avoid contact with the fist. She shot her hand out, away from Mickey's throat, running the tiny blade across Holt's face. He yelled, fell back, hands to his face, blood gushing between his fingers.

Katia was off the bed, ramming straight into Darlene, knocking her into Johnny with surprising force. Johnny grabbed for Katia as he tumbled, but she sidestepped him and was out the door.

CHAPTER 102

"DARLENE, STAY HERE. Call 000," Johnny snapped, turned and headed after Katia.

She'd vanished, but there weren't that many places she could run to. Johnny turned a corner and saw Katia yank open the door to Mickey's suite. He heard screams from the room. Johnny ran to the door.

She was like a storm trooper plowing through the party, pushing people aside. Drinks were flying, men and women falling. She turned, saw Johnny no more than ten feet behind her, and pushed even harder against the crowd. A woman fell to the carpet, smashing her head on a chair leg. Katia almost tripped over Hemi, who had been left to sleep on the floor.

For a second she didn't seem to know what to do. Then she reached for a champagne bottle, smashed it on a table, gripping it in her right hand. She spun round. Half the people were so stoned, they moved like zombies. A few looked petrified. Katia grabbed the closest girl to hand, a near-naked waif with cocaine powdered all over her tiny breasts. The kid screamed as she was pulled back and Katia held the jagged spikes of the shattered champagne bottle to her face.

"Get back!" Katia bellowed as Johnny approached.

Someone killed the music, and the place fell silent.

"What are you doing, Katia?" Johnny took a step toward her.

The woman was glowing sweat, her

eyes wild, hair stuck to her exquisite face.

"This isn't you, Katia."

"Get back, I said. NOW!"

"Katia." Johnny stopped and crouched down a few feet in front of her. The young girl in Katia's grip began to sob violently.

"This is a young kid," he said, flicking his eyes toward the terrified girl. "Just like your sister...just like Anais."

"You don't know anything about Anais," she spat.

"I know she suffered. You said so yourself."

Katia screamed suddenly. "Shut up! I don't want to hear it." She went to move her hand to cut the girl's face to shreds. Johnny dived forward, grabbed Katia's hand with his left hand, and smashed his right fist into her gut.

She lost her grip on the bottle. The weeping girl slumped to one side, and Katia groaned but kept on her feet, stumbling backward. Johnny rushed forward, smacked her across the face, hard. She fell backward and slammed into a cabinet of glass shelves. The shelves tumbled down on top of her, shards cascading all around her unconscious body.

CHAPTER 103

THE COPS ARRIVED within minutes, and Mickey Spencer's twenty-seventh birthday party came to a close.

Katia was taken into custody. Mickey and two cops rode with Ricky Holt in the ambulance to St. Vincent's Hospital. Darlene and Johnny, because of their Private connection, managed to talk their way out of going to police headquarters for a grilling. Even so, the interviews at the club took hours.

So it wasn't until 3:30 a.m. that Darlene pulled her VW Beetle away from the club to take Johnny to the station in the CBD. He had five minutes to catch the first morning train home.

"Funny how this goofy little bug of a car can make the world seem like a happy cartoon," Johnny said.

"I'm glad you feel that way. Maybe I'm just used to riding in this goofy little bug. So the evening seemed pretty horrible to me."

"I understand," Johnny, suddenly the wise old man, said. "I understand."

The station was in sight when Darlene spoke next.

"I've just got to say something, Johnny. And I mean it. You did some fantastic work tonight."

"Thanks," he said softly.

They arrived at the station, and Darlene went with him to the ticket machine.

"One more thing I need to tell you," Darlene said as the train was coming into view.

"I'm ready for it," Johnny said.

"We make a great team." And she leaned in and pecked him on the cheek.

He blushed.

"You're doing it again, dude!" Darlene laughed.

He turned and went for the stairs down to the platform, raising his hand as a wave good-bye. His back was to her, and he had a big smile on his face.

Darlene walked to the goofy little bug of a car. She sat behind the wheel, staring silently for a moment at the leafy, deserted avenue running toward the bridge. She hadn't felt this good and fresh and alive in a long time. She wasn't certain what the future was going to bring. In fact, she was only sure of one thing. Sleeping tonight was absolutely out of the question.

CHAPTER 104

EVERYTHING IN THE lab was as Darlene had left it over four hours earlier. On the central counter lay the collection of singed papers and piles of crispy, black remnants from Julie O'Connor's scrapbook. A few feet above it hung the microscope.

Darlene slipped on her lab coat, took her glasses from the right pocket, and stared into the eyepiece, pulling over the last page she'd viewed earlier, the words "WHORE NUMBER 5." The name wiped by fire, the entire page brown.

"Superficial burn, though," Darlene noted to herself. "Which makes it all the more frustrating. If only..."

Her mind was racing. She could try solvents. "No—too dangerous. Might destroy the thing entirely.

"Ultraviolet?" she whispered. "What about ultraviolet?"

She pulled up a chair and sat down. Sighed heavily. The sigh was so dramatic and loud that she smiled at herself.

"No...wouldn't work...wrong sort of disruption of the paper fibers."

All these processes involved "peeling away" the upper layer of flame damage. But if she could do that, she could see what was underneath. But, no, there was no way...

She froze.

"Yes!"

She pushed back her chair and got to her feet. Suddenly she was feeling a little bit giddy.

The giddiness turned to mild elation. She just had to say something out loud.

"Easy, girl...But you *are* a genius, Darlene Cooper."

She smiled as she walked—almost ran—across the room to the storage cupboard.

"You are a bloody genius."

CHAPTER 105

IT WAS CALLED a Saser, and two months earlier, when Darlene was giving Craig a wish list of equipment for the lab, she'd almost deleted it as something they should requisition. It was, she thought, terribly expensive, and she wouldn't be using it very much, and if she really needed it she could . . .

Now she was thanking all that was holy that she had kept the Saser on the list. Even with a price tag of twenty grand, tonight it might prove to be a very solid investment.

For all that money, the machine didn't

look like very much. It wasn't even very large, just a couple of shiny buttons on the front of a six-by-six-inch steel box. It looked something like a small photo-copier.

But talk about appearances being deceptive. The Saser was an amazing invention, and there were maybe only half a dozen in the world.

She found it on the second shelf on the right of the storeroom. It was quite light, easy to lift down. She placed it on the counter below the overhead microscope, plugged it in, and watched a small screen light up.

She pulled over her chair and started programming the device. She remembered the spec. A Saser, she recalled, was, according to the technical review she'd read in *Forensics* magazine, a little like an X-ray machine. But—and this was its USP—it didn't see *right* through things to show the bones of the body or the contents of a suitcase, like an airport scanner. A Saser could be finely adjusted to penetrate beneath the surface to any predetermined depth. In skilled hands, it could reveal layer upon layer of any object. It was exactly what she needed now.

She lifted the lid and picked up the final pages of Julie O'Connor's scrapbook.

The contents appeared on the screen. The pages were covered with scorches; almost all the writing was totally wrecked. Darlene adjusted a few parameters and pushed the Scan button.

The Saser made a hissing sound. It was almost human. Darlene studied the screen. The initial image appeared to be almost identical to the original. But after a few seconds, small patches of scorched paper cleared. She could trace the lines of a few letters that had been invisible.

She altered the penetration depth and upped the resolution. She pushed Scan again. A new image appeared.

"That's better," she said, stunned by the quality. The picture had sharpened dramatically. She could see numbers, letters, a few entire words. She scrolled up. The top of the page was looking better, but still not enough to show what she was after—the damn name.

Darlene adjusted the parameters a third time. Now she was anxious and nervous and had numbers and quotients running through her brain. She had to get the depth right or she would overshoot, go straight through.

She pulled back on the resolution and doubled the depth of penetration to one five-hundredth of a millimeter, pushed the Scan button again.

The waiting actually made her stomach hurt. Her eyes held almost madly to the screen. She could hear her own heart thumping.

As she read two words at the top of the page, Darlene felt a shudder move up her spine and through the nape of her neck.

GRETA...THOROGOOD.

CHAPTER 106

JULIE HAD SET her phone to wake her at 4:30 a.m. It went off on time, but she was already awake. She hadn't slept—too lost in wonderful thoughts, thoughts of blood, rolled-up banknotes, revenge… sweet, sweet revenge. She got up, changed into fresh clothes, threw the wig, mustache, and men's things into a plastic bag.

It was still dark as she tugged open the door onto the parking lot at the rear of SupaMart. Totally deserted, of course. Just two cars left over from the previous night. She tossed the plastic bag into a nearby dumpster.

It was three miles from here to where the silly bitch went jogging every morning. Six a.m. Parsley Beach.

"How typical," Julie said aloud. "Just when she goes out *running*, I'm on the frigging train from Sandsville to wait on stupid bitches like her."

She turned onto Sebastian Road and just kept walking, anger building with each step. She could feel the long knife through the lining of her jacket, and her smile broadened as she contemplated what she would do to Greta Thorogood.

CHAPTER 107

THE PHONE RANG, and I did what any normal person who'd been woken from a deep sleep would do.

I looked at the bedside clock. It said 5:14 a.m.

I said a grumpy "Oh, shit."

Then I said a grumpy "Hello?"

"Craig?" the voice asked.

"Don't you ever go to bed, Darlene?"

"Sorry, Craig. But I think you'll wanna hear this."

I was out the door in ten minutes, cell phone to my ear as I pressed the remote for the car.

Greta's cell just rang and rang and finally went to voice mail. I left a message.

"Greta. It's Craig. If you get this message at home, stay where you are. Got that? Stay put and call me back. I'm heading over to your place right now."

I searched for the Thorogoods' home number as I pulled onto Military Road and headed toward the bridge, found it, punched the preset. No one picked up. I disconnected, tried again. Waited, waited...still nothing.

Even at that early hour, the traffic was beginning to build. I put my foot down. *Bugger the cameras,* I thought—and if I got stopped? Well, then, I got stopped.

I sped left onto Warringah Freeway, with the black colossus of Sydney Harbour Bridge in the distance, the towers of North Sydney to my right, a much-larger collection of skyscrapers directly ahead, over the bridge.

Three minutes later, I was on the Cahill Expressway. I shot down the off-ramp, weaving between slower cars, ignoring the blaring horns, ignoring the speedometer. I tore down onto New South Head Road and just went for it. I saw two traffic surveillance cameras go off. Slowing, I pulled onto Stockton Boulevard, the Thorogoods' house a little way down on the right.

Lights were on. I tried the home number again as I stepped out of the car and ran along the sidewalk. No response. I reached the doorbell, leaned on it. Nothing. Tried again. Banged on the big mahogany door.

The door opened, and I almost fell into the hall. Brett was standing barefoot in a bathrobe, hair wet, bewildered.

"What the—?"

"Where's Greta?"

"What do you—?"

"Where is Greta?" I yelled.

"She's out on her run...why?"

"Your wife's the next victim."

His expression changed to one of horror. I didn't wait for a response.

"Where does she run?" I yelled.

"Parsley Bay, about three miles away," he said, his voice cracking with shock.

"I know it."

"Always the same route—along the beach, up through the reserve, along to the parking lot. Jesus Christ! Look, go . . . Craig! I'll get a team there immediately."

CHAPTER 108

GRETA CLOSED THE door of her BMW and turned to the path down to the beach. It was already seventy degrees plus, and she loved the summer.

She ran down the path, and two minutes later, she was on the sand, the sun casting a fresh morning glow all around. The ocean was so perfect—little whitecaps on the waves, a fat orange ball of sun over the horizon—it looked like it had been Photoshopped.

She found her rhythm and ran close to the water, where the sand was hard. To her right, a line of palms. Running was solitude. *Watch out for the broken seashells. Reapply the sunblock when necessary. Consider the day ahead.*

When she got home she'd have to sort out the kids' breakfast, pack the bags ready for camp, get the children into the car, drive the mile to the drop-off.

Later she would meet friends for lunch at Tony's or Oasis. Then it would be the mad dash to the bus stop—the three-thirty pickup. Back home, dinner for the kids. Later, after the children were in bed, it would be dinner for her and Brett, a glass of wine—thank God! Then, into bed, and Brett probably... Well, she'd see. She laughed. Then she ran on, focusing on her rhythm, her pace, the ocean.

This day was so perfect—and Julie O'Connor so well hidden behind a cluster of three palm trees. Yep, Greta could not possibly notice the sloppy, sneering, smiling crazy lady.

CHAPTER 109

I SCREECHED OFF down Bexham Boulevard and back out onto New South Head Road. I knew Parsley Beach—it was in Vaucluse. Parsley Beach had a panoramic view. It had miles of soft, clean sand. And now it had Julie O'Connor.

Averaging ninety miles per hour, the faithful Mas took a little over two minutes to reach the turn-off. I saw another speed camera flash as I shot past. I swung a hard left off the main highway, onto a smaller road, followed the curves, descended a steep hill, and almost overshot the parking lot. I knew that the path down to the beach lay on the far side of the scrap of sandy ground. This morning, only one car was parked there—Greta's BMW 320i convertible.

I ran across the open space and down the first steps of the path that led through the reserve. I could hear the crashing of waves directly ahead. I stopped for a moment to yank off my shoes.

Then I slowed down. Julie O'Connor could be anywhere. I shoved away the awful thought that I might already be too late.

There was a bend in the path. I gripped the wooden handrail on one side as the descent became steep. Stopped, listened. Nothing but the sound of birds, waves, the breeze rustling the eucalyptus.

I glimpsed sand, a flash of blue water. The beach was less than a hundred feet ahead down the sloping, curving stairway.

A tight bend. I held the rail with both hands, eased down two steps, and there was Greta. She'd just reached the wooden pathway that led to the same wooden stairs I'd just walked down.

I was about to call to her, when I saw movement to her right. Julie O'Connor sprang from a cluster of weeds and high grass and palm trees.

"Greta!" I shouted.

She looked up, saw me, began to smile, and O'Connor was on her.

I felt my stomach flip, and for a second I froze.

Julie grabbed her around the neck, pulling her back. Greta stared at me, eyes wide, and screamed.

I took half a dozen steps toward them.

"*YOU BETTER STOP!*" Julie shouted.

I kept going.

"*STOP!* I've got a very, very big knife

'ere. And the tip of it is just touching this whore's spine."

Greta screamed again.

"Shut the fuck up!" Julie hissed in Greta's ear. Then Julie turned back to me.

"Let her go," I said.

Julie laughed. "Oh, yeah. I'd like to, but her and me have girl business to discuss. Don't we?" She twisted Greta's face round, her fingers digging into her cheeks.

I took another step forward. They were only about ten yards away now.

"STOP! I SAID STOP!"

I walked down two more steps.

"I TOLD YOU TO FUCKING STOP!"

Greta began shaking violently.

"Oh, look, she puked on her lovely little T-shirt," Julie said.

Greta's face had drained. She was panting, her eyes like black dishes.

Julie held up her knife. She smiled a hideous smile and waved the knife in the air. Drops of blood splattered from the knife onto Greta's white T-shirt.

"Just look. A little blood came to visit the puke on the shirt." She stopped talking for a moment. The mad smile on her face vanished.

"That was a little stab in the back. The next one's going in all the way," Julie said.

I stopped. Put my hands up. It was then that I caught a glimpse of movement behind Greta and O'Connor. My cousin Mark was waiting no more than ten steps behind Julie and Greta. Two officers were with him, guns drawn.

"Look, can we talk?" I said.

Julie laughed again, a nasty rasp.

"Why would I wanna talk? I have this bitch under my control. She's *mine*... She's mine. I can do what I want with her. Make her beg, make her squirm. She's a whore... right? She gives it up for Brett the Big Policeman. She gives it up and she gets her Chanel, her Prada, her holidays on Hamilton Island. Two kids, and her husband might screw around on the quiet, but it's a deal...right? What has this stupid bitch ever done for herself?"

"She's a human being, Julie."

CHAPTER 110

"THE KNIFE'S IN a bit further," Julie O'Connor said. She followed this statement with a witchlike cackle.

I could tell from the drowsy eyes and pale skin on Greta that Julie wasn't just bragging. Greta looked bad.

Julie looked down behind Greta's back.

"Oh, yeah...more blood. Lots more," she said. The crazy grin had returned to her face.

She stopped talking. But not for long.

"Oh, yeah, this bitch—"

Then Mark and his men rushed forward.

Greta fell to one side. She groaned softly as she fell. I saw Julie swing to her left, her knife slicing the air, nicking Mark's ear—a small wound, it seemed, but with a good amount of blood. Julie had a hateful look on her face as she ran off the path and back into the wooded area where the palm trees and weeds grew thick.

"You go up into the wooded area," Mark said to me, quietly. "I'll take the path and go in the other way."

Then he turned to his men. "Nichols—go back down to the beach and around. Taylor, stay with Mrs. Thorogood."

I could see that Greta wasn't badly hurt, so I ran up the steps. Mark's plan was a good one. Between us, we'd have the woman. I was sure of it.

Ten seconds later and I was at the top step, the parking lot ahead of me. I ran onto the sandy rectangle, skirted the edge, found the next path down to the beach, and headed onto it. I guessed

Mark would be about thirty yards below on an adjacent path.

I took the steps down two at a time. Turned right, then left, another tight left. Drew up in the sand.

Mark was coming toward me along a sandy path. I caught a movement to my left. Julie charged through the bush and smacked into him, knocking him off balance. He stumbled to his left, pistol flying from his hand.

Julie was on him in a second, her right hand raised, the horrifying blade raised over Mark's face.

I didn't pause to think, just rushed forward and grabbed the woman by the shoulders. I was stunned by how powerful she was. Using all my strength, I managed to yank her away, but I was sent sprawling onto my back. She was incredibly agile and got to her feet ahead of me. I propelled myself upright, watched as she came for me. Mark began to pull up, but he was slow. Julie lunged at me, growling like an animal. I misjudged her thrust and felt a screech of pain rip through my abdomen.

I dropped to my knees and saw the point of the woman's knife coming toward my face. A moment later I heard the crack of a gun going off and felt a heavy weight slamming down on top of me.

CHAPTER 111

I SAW SOME sort of face on a man's body. The features on the face appeared and disappeared and then reappeared as if it were the face of a mannequin wearing and removing a series of masks. It was the face of my dead uncle, which morphed into the face of Brett, which then—I think—became my father and then my high-school tennis coach and then my cousin Mark. And then I waited. I waited for the next face.

But Mark's face did not disappear like the others. I waited. For whom? Who would be next? Jack Morgan? Johnny?

But the change never happened. It was Mark. It stayed Mark.

I could feel the coolness of cotton sheets. The room was filled with natural light, curtains pulled back across the window in the opposite wall.

Everything came to me slowly. But also very surely.

"You were very lucky," Mark said.

"I feel like I've been here before," I managed to say.

Mark smiled an unlikely warm smile.

"What's the damage?" I asked.

"Oh, perforated colon, slight nick to the spleen... The surgeon put you back together again. And that colostomy bag is only temporary. It comes off in two months."

"Great. A good news–bad news joke, only in reverse."

I lifted the sheet and saw a bunch of bloodied bandages across my abdomen.

"I got off light," I said cautiously.

"I got off lighter, Craig... Thanks to you. And you were unarmed."

"So, I'm either the brave one or the dumb one."

Mark grinned and looked down at his feet. Then he spoke.

"I don't think we're ready to be best friends or roommates or...Craig, to be honest...I'm worn out with this constant war. I may have fired the first shot. I may have been the asshole. But, God. It was all a long time ago. We've both lost..." He trailed off and looked into my face.

I sensed the numbing effect of painkillers.

"How long have I been out?"

He glanced at his watch. "Thirty-five, thirty-six hours."

I slowly pulled myself up in the bed.

"Sweet Jesus, this hurts," I said.

From the side of the bed, Mark handed me a long tube with a blue button attached.

"Press that little button. It'll pump you with some morphine."

I pressed. Within thirty seconds the pain diminished.

"Let's not talk ancient history, Mark. Tell me about what just happened. How'd it all end?"

"Julie O'Connor's under armed guard in intensive care."

"And Greta?"

"She's fine. In shock but unharmed. Nasty cut in her back. She was here earlier—but you were still out. You're her hero."

I produced a small laugh. "Ow! Christ. That morphine doesn't work on laughter."

"You know what they say. It only hurts when... Well, you know, Craig."

I lay back on the pillow and spoke.

"So, look... The O'Connor woman... she was just driven by pure jealousy, yeah?"

"More or less. Seems there was a little more to it than that. One of my guys found out something very interesting this morning. Three months ago, Julie O'Connor was the victim of a botched operation—a tuboplasty, they called it.

It was to . . . unblock her fallopian tubes. Guess who the gynecologist was?"

"Cameron Granger . . . Of course."

"Your resident genius, Darlene, managed to get a lot of background stuff from the woman's scrapbook. Julie O'Connor was desperate to have children, and when the op went wrong, it went spectacularly wrong!"

"She became infertile?"

"Totally. Her life fell apart. She was already living on the breadline in Sandsville. Her boyfriend, Bruce Frimmel, left her. She killed him. It gave her the taste for doing it some more, I guess. And it was all made worse because of where she worked. In her scrapbook she refers to the Bellevue Hill women as 'whores.' She thought they were little more than prostitutes—leading lavish lives, thanks to rich husbands."

"Now we understand the ritual of the money inserted in the victims—a symbolic gesture."

"Not just money, Craig. *Fake money* . . .

for what she saw as fake women, fake wives."

"Isn't it amazing, though?" I said. "The killer takes it out on other women. She didn't try to kill the person who caused all the trouble in the first place, Dr. Granger."

"We've both seen it before. O'Connor displaced the blame. That's why I said earlier it was exaggerated by the place she worked in. Deep down, repressed for years, she *was* envious of the women she saw each day in Bellevue Hill. Being messed up by Granger pushed her over the edge."

"We'll probably never know what the original spark could have been."

"Actually, we do. That Darlene really *is* a goddamn genius. She found a file on Julie O'Connor at St. Joseph's Psychiatric Hospital. Julie spent some time there almost ten years ago—she'd been living on the streets, raped, mugged. According to the reports, she claimed her mother had tortured her as a child. The authorities tried to check the story, but the woman, Sheila O'Connor, had moved abroad."

"The other ugly thing is that Cameron Granger got away with it."

"Don't be a downer, Craig. While you've been getting sewn up, the Sydney police have been working hard. Now that the truth is out, the hospital can't cover up for Dr. Granger anymore. We found two nurses and a resident who'll testify that he botched the operation."

I nodded, sighed heavily.

"Yeah, it's good news," Mark said.

I closed my eyes. I was ready to drift off. But first I opened my eyes and spoke.

"Yes, good news," I said. "Today I got a lot of good news."

CHAPTER 112

IT WAS THREE days after I was released from hospital.

I was driving Justine to the airport. I was moving along the same road I'd traveled three years ago when my family and I were headed for a vacation.

Now I was going to the airport again. Once again I had a beautiful woman sitting beside me. But there was no child in the backseat. And I was a very different person.

The airport was packed. Justine checked in, and I walked with her to the departure zone, the scanners and security guys just a few feet away.

"It's been . . . well, it's been 'eventful,' "
I said.

"To say the least, Craig."

"You've been a great help. If you ever
feel like a break from the L.A. office . . ."

"I might seriously consider that."

"Wish Jack my best, yeah?"

"I will."

She kissed me on the cheek, and I
kissed the air next to her cheek. I
watched her walk toward passport control. It was then that I called her name.

She stopped, turned, and walked
back quickly toward me.

"There's one other thing I want you to
tell Jack," I said.

She looked nervous, a little teary.

"What's that, Craig?"

"Tell him that he's the luckiest guy in
the world."

She bit her lower lip. She blinked her
eyes a few times. She shook her head up
and down.

Then she turned and walked away.

CHAPTER 113

NO CAKE. NO balloons. No Welcome Back signs.

I had warned Darlene and Johnny to alert their colleagues at Private HQ that my coming back should be treated professionally and casually. Not that I wasn't happy to be back. It's just that I'm not a cake-and-balloons kind of guy. If you know *me,* then you know *that.*

I walked into the conference room at Private HQ. Well, come on. No balloons, but maybe there could have been more than a few grunts of "Morning, Craig" and "Hey, Craig." It was as if I'd never been away, never been wounded, never been— Then the entire room stood and applauded for fifteen seconds that felt to me like a half hour.

I surveyed the gathering—Johnny, Mary, Darlene, Cookie—the core of my work, the soul of my life.

"Thanks," I said rather weakly. "Not sure why I deserve it..."

"I think we all deserve a pat on the back, actually," Johnny said.

"You do. We all do. It's been one hell of a first week," I replied.

There was laughter and talk, and then I had to be the one to bring it back to business.

"So, how about an update? How's it all wound up?" I asked. "What about Ricky Holt?"

Johnny took that question.

"Seventeen stitches and a very rock 'n' roll scar. Mickey's fine. Spent a night in the hospital after his party, but no permanent harm done. He's rehearsing for a big tour. Out of gratitude, he's helping Holt settle his gambling debts. Katia is being deported."

"But what about the song you told me about? The one describing his own death. Why did he even write that?"

Johnny shrugged.

"I think it was purely an artistic gesture. He felt helpless, controlled by 'the suits.' He's a sensitive bloke, but that sensitivity must have slipped into paranoia. His phenomenal drug intake couldn't have helped!"

"And what about Hemi?"

"He's up and waddling."

We all laughed.

"He's mighty pissed, though—as you would imagine!" Johnny added.

"I heard you bumped into Al Loretto's boy, Jerry."

"Yeah, I did. He was helpful in making me realize what a mess Holt had gotten himself into."

I nodded. "What about the Ho family, Mary?"

"Dai is under observation at a psychiatric retreat in the suburbs. His father has booked him in for plastic surgery. The police here are liaising with the Hong Kong authorities. They're after the big boss, Fong Sum, but I'm pretty sure they're wasting their time."

"Untouchable?"

"For the moment. Ho is just relieved his son was saved. He now wants to have the case of his wife's murder reopened. He's convinced that twelve years ago a bent cop was in the pocket of the Sydney Triads and buried the evidence. He's sure the Triads killed Jiao. He wants to prove it *and* to identify those involved."

"Good luck to him!" I said, leaning back in my chair. Then I said, "Well, sounds like we've had a few positive results. Not bad for our first week!"

"And why shouldn't we have?" Darlene asked.

I smiled and said, "Yeah, you're right. Why shouldn't we have?"

"Oh, there are these, though," Johnny added, and pushed a small pile of papers toward me across the conference table.

I glanced at the top one. The words ROADS AND MARITIME SERVICES were printed in the top right corner.

"They're speeding tickets," Johnny said.

I lifted my eyes and saw that the entire team was grinning at me. Then I spoke.

"Looks like I might have to call in a favor."

Books by James Patterson

FEATURING ALEX CROSS

Cross My Heart
Alex Cross, Run
Merry Christmas, Alex Cross
Kill Alex Cross
Cross Fire
I, Alex Cross
Alex Cross's Trial (with Richard DiLallo)
Cross Country
Double Cross
Cross (also published as *Alex Cross*)
Mary, Mary
London Bridges
The Big Bad Wolf
Four Blind Mice
Violets Are Blue
Roses Are Red
Pop Goes the Weasel
Cat & Mouse
Jack & Jill
Kiss the Girls
Along Came a Spider

THE WOMEN'S MURDER CLUB

Unlucky 13 (with Maxine Paetro)
12th of Never (with Maxine Paetro)
11th Hour (with Maxine Paetro)
10th Anniversary (with Maxine Paetro)
The 9th Judgment (with Maxine Paetro)
The 8th Confession (with Maxine Paetro)
7th Heaven (with Maxine Paetro)
The 6th Target (with Maxine Paetro)
The 5th Horseman (with Maxine Paetro)
4th of July (with Maxine Paetro)
3rd Degree (with Andrew Gross)
2nd Chance (with Andrew Gross)
1st to Die

FEATURING MICHAEL BENNETT

Gone (with Michael Ledwidge)
I, Michael Bennett (with Michael Ledwidge)
Tick Tock (with Michael Ledwidge)
Worst Case (with Michael Ledwidge)
Run for Your Life (with Michael Ledwidge)

Step on a Crack (with Michael
  Ledwidge)

THE PRIVATE NOVELS

Private Down Under (with Michael
  White)
Private L.A. (with Mark Sullivan)
Private Berlin (with Mark Sullivan)
Private London (with Mark Pearson)
Private Games (with Mark Sullivan)
Private: #1 Suspect (with Maxine Paetro)
Private (with Maxine Paetro)

NYPD RED

NYPD Red 2 (with Marshall Karp)
NYPD Red (with Marshall Karp)

STANDALONE BOOKS

Invisible (with David Ellis)
First Love (with Emily Raymond,
  photographs by Sasha Illingworth)
Mistress (with David Ellis)

Second Honeymoon (with Howard Roughan)

Zoo (with Michael Ledwidge)

Guilty Wives (with David Ellis)

The Christmas Wedding (with Richard DiLallo)

Kill Me If You Can (with Marshall Karp)

Now You See Her (with Michael Ledwidge)

Toys (with Neil McMahon)

Don't Blink (with Howard Roughan)

The Postcard Killers (with Liza Marklund)

The Murder of King Tut (with Martin Dugard)

Swimsuit (with Maxine Paetro)

Against Medical Advice (with Hal Friedman)

Sail (with Howard Roughan)

Sundays at Tiffany's (with Gabrielle Charbonnet)

You've Been Warned (with Howard Roughan)

The Quickie (with Michael Ledwidge)

Judge & Jury (with Andrew Gross)

Beach Road (with Peter de Jonge)
Lifeguard (with Andrew Gross)
Honeymoon (with Howard Roughan)
Sam's Letters to Jennifer
The Lake House
The Jester (with Andrew Gross)
The Beach House (with Peter de Jonge)
Suzanne's Diary for Nicholas
Cradle and All
When the Wind Blows
Miracle on the 17th Green (with Peter de Jonge)
Hide & Seek
The Midnight Club
Black Friday (originally published as *Black Market*)
See How They Run (originally published as *The Jericho Commandment*)
Season of the Machete
The Thomas Berryman Number

FOR READERS OF ALL AGES

Homeroom Diaries (with Lisa Papademetriou, illustrated by Keino)

Middle School: Save Rafe! (with Chris
    Tebbetts, illustrated by Laura Park)
Middle School: Ultimate Showdown (with
    Julia Bergen, illustrated by Alec
    Longstretch)
I Even Funnier (with Chris Grabenstein,
    illustrated by Laura Park)
Confessions: The Private School Murders
    (with Maxine Paetro)
Treasure Hunters (with Chris
    Grabenstein and Mark Shulman,
    illustrated by Juliana Neufeld)
*Middle School: How I Survived Bullies,
    Broccoli, and Snake Hill* (with Chris
    Tebbetts, illustrated by Laura Park)
*Middle School: My Brother Is a Big, Fat
    Liar* (with Lisa Papademetriou,
    illustrated by Neil Swaab)
Witch & Wizard: The Kiss (with Jill
    Dembowski)
Maximum Ride: The Manga 6 (with
    NaRae Lee)
I Funny: A Middle School Story (with
    Chris Grabenstein, illustrated by
    Laura Park)

Zoo: The Graphic Novel (with Andy
    MacDonald)
Daniel X: Armageddon (with Chris
    Grabenstein)
Confessions of a Murder Suspect (with
    Maxine Paetro)
Nevermore: A Maximum Ride Novel
Middle School: Get Me out of Here! (with
    Chris Tebbetts, illustrated by Laura
    Park)
Daniel X: The Manga 3 (with SeungHui
    Kye)
Witch & Wizard: The Manga 1 (with
    Svetlana Chmakova)
Witch & Wizard: The Fire (with Jill
    Dembowski)
Daniel X: Game Over (with Ned Rust)
Daniel X: The Manga 2 (with SeungHui
    Kye)
Middle School: The Worst Years of My Life
    (with Chris Tebbetts, illustrated by
    Laura Park)
Maximum Ride: The Manga 4 (with
    NaRae Lee)
Angel: A Maximum Ride Novel

Saving the World and Other Extreme Sports: A Maximum Ride Novel
School's Out—Forever: A Maximum Ride Novel
Maximum Ride: The Angel Experiment
santaKid

For previews of upcoming books and more information about James Patterson, please visit JamesPatterson .com or find him on Facebook or at your app store.

THE WORLD'S #1 BESTSELLING WRITER

JAMES PATTERSON has created more
enduring fictional characters than any
other novelist writing today. He is the
author of the Alex Cross novels, the
most popular detective series of the past
twenty-five years. He also writes the
bestselling Women's Murder Club novels, set in San Francisco, and the top-
selling New York detective series of all
time, featuring Detective Michael Bennett. James Patterson has had more *New
York Times* bestsellers than any other
writer, ever, according to *Guinness World*

Records. Since his first novel won the Edgar Award in 1977, James Patterson's books have sold more than 300 million copies. For previews of his upcoming books and more information about the author, visit www.JamesPatterson.com.